# Life of Fred®

## Linear Algebra
### Expanded Edition

# Life of Fred®
## Linear Algebra
### Expanded Edition

Stanley F. Schmidt, Ph.D.

Polka Dot Publishing

ISBN: 978-1-937032-20-3


Printed and bound in the United States of America




Polka Dot Publishing                                   Reno, Nevada


*To order copies of books in the Life of Fred series,*

*visit our website* PolkaDotPublishing.com




*Questions or comments? Email the author at* lifeoffred@yahoo.com




First Printing


*Life of Fred: Linear Algebra Expanded Edition* was illustrated by the author with additional clip art furnished under license from Nova Development Corporation, which holds the copyright to that art.

*for Goodness' sake*

or as J.S. Bach—who was
never noted for his plain
English—often expressed it:

*Ad Majorem Dei Gloriam*
(to the greater glory of God)

If you happen to spot an error that the author, the publisher, and the printer missed, please let us know with an email to: lifeoffred@yahoo.com

As a reward, we'll email back to you a list of all the corrections that readers have reported.

# A Note to Classroom Students and Autodidacts

In calculus, there was essentially one new idea. It was the idea of the limit of a function. Using that idea, we defined the derivative and the definite integral. And then we played with that idea for two years of calculus.

In linear algebra, we dip back into high school algebra and begin with the idea of solving a system of linear equations like

$$\begin{cases} 3x + 4y = 18 \\ 2x + 5y = 19 \end{cases}$$

and for two or three hours on a Saturday while Fred goes on a picnic, we will play with that idea.

✓ We can change the variables: $\begin{cases} 3x_1 + 4x_2 = 18 \\ 2x_1 + 5x_2 = 19 \end{cases}$

✓ We can consider three equations and three unknowns.

✓ We can look at the coefficient matrix $\begin{pmatrix} 3 & 4 \\ 2 & 5 \end{pmatrix}$

✓ We can consider the case in which the system has exactly one solution (Chapter 1), or when it has many solutions (Chapter 2), or when it has no solution (Chapter 3).

✓ Etc.

Of course, the "Etc." covers all the new stuff such as model functions, orthogonal complements, and vector spaces. But they are all just ideas that you might encounter in the Kansas sunshine as you went on a picnic.

Enjoy!

# A Note to Teachers

Is there some law that math textbooks have to be dreadfully serious and dull?  Is there a law that students must be marched through linear algebra shouting out the cadence count Definition, Theorem and Proof,
Definition, Theorem and Proof,
Definition, Theorem and Proof,
Definition, Theorem and Proof,
as if they were in the army.  If there are such laws, then *Life of Fred: Linear Algebra* is highly illegal.

Besides being illegal, this book is also fattening.  Instead of heading outside and going skateboarding, your students will be tempted to curl up with this textbook and read it.  In 368 pages, they will read how Fred spent three hours on a Saturday picnic with a couple of his friends.  I think that Mary Poppins was right: a spoonful of sugar can make life a little more pleasant.

So your students will be fat and illegally happy.

But what about you, the teacher?  Think of it this way: *If your students are eagerly reading about linear algebra, your work is made easier.  You can spend more time skateboarding!*

This book contains linear algebra—lots of it.  All the standard topics are included.  A good solid course stands admixed with the fun.
At the end of every chapter are six sets of problems giving the students plenty of practice.  Some are easy, and some are like:

> If T, T′ ∈ Hom($\mathcal{V}$, $\mathcal{V}$), and if TT′ is the
> identity homomorphism, then prove that T′T
> is also the identity homomorphism.

*Life of Fred: Linear Algebra* also has a logical structure that will make sense to students.  The best teaching builds on what the student already knows.  In high school algebra they (supposedly) learned how to solve systems of linear equations by several different methods.

The four chapters that form the backbone of this book all deal with systems of linear equations:

> Chapter 1—Systems with Exactly One Solution
> Chapter 2—Systems with Many Solutions
> Chapter 3—Systems with No Solution
> Chapter 4—Systems Evolving over Time

These chapters allow the students to get their mental meat hooks into the less theoretical material.

Then in the interlarded Chapters (1½, 2½, 2¾, 3½) we build on that foundation as we ascend into the more abstract topics of vector spaces, inner product spaces, etc.

Lastly, your students will love you even before they meet you. They will shout for joy in the bookstore when they discover you have adopted a linear algebra textbook that costs only $52.

# Contents

# Chapter One
## Systems of Equations with One Solution
$$Ax = b \; ☺$$

F red had never really been on a lot of picnics in his life. Today was special. Today at noon he was going to meet his two best friends, Betty and Alexander, on the Great Lawn on campus, and they were going to have a picnic.

One good thing about being at KITTENS University* is that just about everything imaginable is either on campus or nearby.

**Wait! Stop! I, your loyal reader, need to interrupt. In your old age, dear author, you're getting kind of foggy-brained.**

What do you mean?

**I'm reading this stuff very carefully, since it's a math book and I have to pay attention to every word. Isn't it obvious that KITTENS would have "just about everything imaginable . . ." since you are doing the imagining?**

Good point. I spoke the truth and plead as John Peter Zenger pleaded.**

**I accept. Please go on with your story.**

Thank you.

Fred knew that food is one important part of a picnic. He picked up the local newspaper and read . . .

---

\* KITTENS University. Kansas Institute for Teaching Technology, Engineering and Natural Sciences.

Background information: Professor Fred Gauss has taught math there for over five years. He is now six years old. Betty and Alexander are students of his. They are both 21.

** In his *Weekly Journal*, Zenger criticized the New York governor. Heavens! The government sent him to jail for libel. He had to wait ten months for his trial. At his trial in 1735 he was accused of promoting "an ill opinion of the government." Zenger's defense was that what he had written was true. The judge said that truth is no defense in a libel case. But the jury ignored the judge and set Zenger free. That marked a milestone in American law. Truth then became a legitimate defense in criminal libel suits in America after that trial. In England that idea did not catch on until the 1920s.

# THE KITTEN Caboodle

The Official Campus Newspaper of KITTENS University

Saturday 11:02 a.m. Grocery Shopping Edition   10¢

# PICNIC MANIA— THE NEW RAGE

KANSAS: A new fad is sweeping the country. Everyone is going on picnics. This was announced last night on television.

News of this great surge in popularity has taken the country by surprise. No one here at the Caboodle news center knew picnicking was popular, much less that it was the newest craze. (continued on p. 31)

"We must picnic," our university president declared in an exclusive Caboodle interview. (continued on p. 24)

Perfect! thought Fred. I'm sure that Butter Bottom's Sack-o-Picnic Food will do the trick. I don't want to disappoint Betty and Alexander.

In a jiffy,* Fred walked to Butter Bottom Foods. And there at the front of the store was a Sack-o-Picnic Food display.

---

\* *In a jiffy* (or *in a jiff*) used to be a common expression meaning "in a very short period of time." Those fun-loving physicists have redefined a *jiffy* as the time it takes for light to travel the radius of an electron.

Wow! Fred thought. They sure make it easy. All I gotta do is choose a sack, and I'm ready to head off to see Alexander and Betty.

Fred was curious. He opened the first sack and looked inside. There were a can, five bottles, and three jars.

He took out the can  . . .  a bottle  . . .  and a jar.

In the first sack, one can, five bottles, and three jars cost $2.65.
$$c + 5b + 3j = 2.65$$

Fred knew That's not enough to tell me what each of the items cost.

He opened the second sack. Two cans, 3 bottles, and 4 jars.
$$2c + 3b + 4j = 2.75$$

The third sack: Five cans, 32 bottles, and 3 jars. Wow. That's a lot of Sluice!

$$5c + 32b + 3j = 10.20$$

With three equations and three unknowns, Fred could use his high school algebra and solve this system of equations:

$$\begin{cases} c + 5b + 3j = 2.65 \\ 2c + 3b + 4j = 2.75 \\ 5c + 32b + 3j = 10.20 \end{cases}$$

In high school algebra, we were often more comfortable using x, y, and z:

$$\begin{cases} x + 5y + 3z = 2.65 \\ 2x + 3y + 4z = 2.75 \\ 5x + 32y + 3z = 10.20 \end{cases}$$

---

*Intermission*

Do your eyes begin to glaze over when you see lines of equations?  In other words: Are you normal?

Unfortunately, much of linear algebra is about solving systems of linear equations like those above.  This book has four main chapters.  The first three chapters deal directly with solving systems of linear equations:
  Chapter 1: Systems with One Solution
  Chapter 2: Systems with Many Solutions
  Chapter 3: Systems with No Solution.

The crux of the matter is that systems of linear equations keep popping up all the time, especially in scientific, business, and engineering situations.

Who knows?  Maybe even in your love life systems of linear equations might be waiting right around the corner as you figure the cost of 6 pizzas, 2 violinists, and 3 buckets of flowers.

---

On the next page is a *Your Turn to Play*.  Even though this is the 27th book in the *Life of Fred* series, the *Your Turn to Play* might be new to some readers.  Let me explain what's coming.

**I, as your reader, would appreciate that.  I hate surprises.**

Psychologists say that the best way to really learn something is to be personally involved in the process.  The *Your Turn to Play* sections give you that opportunity.

The most important point is that you honestly attempt to answer each of the questions before you look at the solutions.  Please, please, please, please, please, please, please, please, please, please, please with sugar on it.

---

*Your Turn to Play*

1. We might as well start off with the eyes-glaze-over stuff.  Pull out your old high school algebra book if you need it.  Solve

$$\begin{cases} c + 5b + 3j = 2.65 \\ 2c + 3b + 4j = 2.75 \\ 5c + 32b + 3j = 10.20 \end{cases}$$

by the "elimination method."  (The other two methods that you may have learned are the substitution method—which works best with two equations and two unknowns—and the graphing method—which, in this case, would involve drawing three planes on the x-y-z axes and trying to determine the point of intersection.)

2. Since this is *linear* algebra, we will be solving *linear* equations.  Which of these equations are linear?

$9x + 3y^2 = 2$

$3x + 2xy = 47$

$7\sin x + 3y = -8$

$5\sqrt{x} = 36$

3. Sometimes linear equations might have four variables.  Then they might be written $3w + 2x + 898y - 5z = 7$.

But what about in the business world?  In your fountain pen factory, there might be 26 different varieties of pens.  Then your linear equation might look like: $2a + 6b + 8c +{}_{9d\,-\,3e\,-\,f\,+\,2g\,+\,14h\,+\,4i\,-\,3j\,+\,2k\,-\,1\text{l}1\,+\,3m\,+\,8n\,+\,20o\,+\,5p\,+\,2q\,+}$ ${}_{3r\,+\,9s\,+\,2t\,+\,99u\,+\,3v} + 8w + 8x - y + 2z = 98723$.  Even then, we might get into a little trouble with the 1l1 (eleven "el") term or the 20o (twenty "oh") term.

If you are in real estate, and there are 40 variables involved in determining the price of a house (e.g., number of bedrooms, size of the lot, age of the house . . . .), you could stick in some of the Greek letters you learned in trig: $2a + 6b + 8c + 9d - 3e - f + 2g + \ldots + 8w + 8x - y + 2z + 66\alpha - 5\beta + 2\gamma + \ldots = \$384{,}280$.

---

If you are running an oil refinery, there might be a hundred equations. Then you might dip into the Hebrew alphabet (א בּ צ ) and the Cyrillic alphabet (Д, Ж, И).

One of the major thrusts of linear algebra is to make your life easier. Certainly, $6x + 5צ – 9 Д + 2ξ = 3$ doesn't look like the way to go.

Can you think of a way out of this mess?

4.  How many solutions does $x + y = 15$ have?

5.  [Primarily for English majors] What's wrong with the definition: "A linear equation is any equation of the form $a_1x_1 + a_2x_2 + \ldots a_nx_n = b$ where the $a_i$ (for $i = 1$ to n) and the b are real numbers and n is a natural number"?

Recall, the natural numbers are $\{1, 2, 3, \ldots\}$ and they are often abbreviated by the symbol $\mathbb{N}$.

### .......COMPLETE SOLUTIONS.......

1.  Now I hope that you hauled out a sheet of paper and attempted this problem before looking here.  I know it's *easier* to just look at my answers than to do it for yourself.

> And it's easier to eat that extra slice of pizza than to diet.
> And it's easier to cheat on your lover than to remain faithful.
> And it's easier to sit around than to do huffy-puffy exercise.
> But *easier* can make you fat, divorced, and flabby.

My solution may be different than yours since there are several ways to attack the problem.  However, our final answers should match.

I'm going to use x, y, and z instead of c, b, and j.  The letter x will stand for the cost of one can of Picnic Rice™, y will stand for the cost of one bottle of Sluice™, and z will stand for the cost of a jar of mustard.

$$\begin{cases} x + 5y + 3z = 2.65 \\ 2x + 3y + 4z = 2.75 \\ 5x + 32y + 3z = 10.20 \end{cases}$$

If I take the first two equations, multiply the first one by –2, and add them together I get

$$-7y - 2z = -2.55$$

If I take the first and third equations, multiply the first one by –5, and add them together I get

$$7y - 12z = -3.05$$

Now I have two equations in two unknowns. If I add them together I get one equation in one unknown

$$-14z = -5.60$$

so z = 0.40 (which means that a jar of mustard costs 40¢).

The last part of the process is to **back-substitute**. Putting z = 0.40 into

$$7y - 12z = -3.05$$

we get           $7y - 12(0.40) = -3.05$

so y = 0.25 (which means that a bottle of Sluice costs 25¢).

Back-substituting z = 0.40 and y = 0.25 into any one of the three original equations, will give x = 0.20 (so a can of Picnic Rice costs 20¢). This may be the last time you ever have to work with all those x's, y's, and z's (unless, of course, you become a high school math teacher). As we progress in linear algebra, the process of solving systems of linear equations will become easier and easier. Otherwise, why in the world would we be studying this stuff?

2.

$9x + 3y^2 = 2$        is not linear because of the $y^2$.

$3x + 2xy = 47$        is not linear because of the 2xy.

$7\sin x + 3y = -8$   is not linear because of the sin x.

$5\sqrt{x} = 36$           is not linear because of the $\sqrt{x}$.

3. The place where we dealt with an arbitrarily large number of variables was in *Life of Fred: Statistics*, but you might not remember the Wilcoxon Signed Ranks Test in which we had a sample $x_1, x_2, x_3, x_4 \ldots$. We used variables with subscripts. Now it doesn't make any difference whether we have three variables or 300.

And you'll never have to face $6\mathfrak{z} - 8\mathfrak{Э} + 2\psi = 98.3$ unless you really want to.

4. We could have x = 3 and y = 12, or x = 9.3 and y = 5.7, etc. There is an infinite number of solutions. That is what we'll study in Chapter 2.

5. Under that definition, y = 3x + 5 wouldn't be a linear equation. English majors would be much happier with "A linear equation is any equation *that can be put in* the form $a_1x_1 + a_2x_2 + \ldots + a_nx_n = b$ where the $a_i$ (for i = 1 to n) and the b are real numbers and n is a natural number."

As promised on the previous page, as we study linear algebra the process of solving

$$\begin{cases} x + 5y + 3z = 2.65 \\ 2x + 3y + 4z = 2.75 \\ 5x + 32y + 3z = 10.20 \end{cases}$$

will become easier and easier. Now is a good time to begin, don't you think?

**I, your reader, couldn't agree more. Let the "easier" begin.**

First of all, let's get rid of all the variables. It's only the coefficients that count. Are you ready for some magic? Poof!

$$\begin{array}{rrrl} 1 & 5 & 3 & = 2.65 \\ 2 & 3 & 4 & = 2.75 \\ 5 & 32 & 3 & = 10.20 \end{array}$$

**I like that. But we really don't need the equal signs either.**

Good point. Poof!

$$\begin{array}{rrrl} 1 & 5 & 3 & 2.65 \\ 2 & 3 & 4 & 2.75 \\ 5 & 32 & 3 & 10.20 \end{array}$$

Now when we start working with these three equations, we don't want loose numbers just splashed on the page. →

So we put some giant parentheses around them.

$$\begin{pmatrix} 1 & 5 & 3 & 2.65 \\ 2 & 3 & 4 & 2.75 \\ 5 & 32 & 3 & 10.20 \end{pmatrix}$$

$$\begin{array}{rrrr} 1 & 5 & 3 & 2.65 \\ 2 & 3 & 4 & 2.75 \\ 5 & 32 & 3 & 10.20 \\ 1 & 5 & 3 & 2.65 \\ 0 & -1 & -2 & -2.55 \\ 0 & 7 & -12 & -3.05 \\ 1 & 5 & 3 & 2.65 \\ 0 & 1 & 2 & 2.55 \\ 0 & 7 & -12 & -3.05 \end{array}$$

An example of the mess we don't want.

20

Just for the record, mathematicians called a rectangular array of numbers a **matrix** long before the movie came out.

Matrices (that's the plural of *matrix*) will be at the heart of linear algebra.  They really don't look very scary.

Here's the coefficient matrix:
$$\begin{pmatrix} 1 & 5 & 3 \\ 2 & 3 & 4 \\ 5 & 32 & 3 \end{pmatrix}$$

By tradition, the **coefficient matrix** is often called  A.

Now let's solve
$$\begin{cases} x + 5y + 3z = 2.65 \\ 2x + 3y + 4z = 2.75 \\ 5x + 32y + 3z = 10.20 \end{cases}$$

using matrices.

Start with the coefficient matrix and then augment it with the constant terms and we get
$$\begin{pmatrix} 1 & 5 & 3 & 2.65 \\ 2 & 3 & 4 & 2.75 \\ 5 & 32 & 3 & 10.20 \end{pmatrix}$$

This matrix is called the **augmented matrix**.
**I, your loyal reader, like that name.  It's very logical.**

This matrix has three **rows**.  The rows are the horizontal strings of numbers.  For example, the second row is 2   3   4   2.75.
**Hey, this stuff is easy-peasy.**

Now we're going to transform this augmented matrix until we get it into the form
$$\begin{pmatrix} 1 & 0 & 0 & 0.20 \\ 0 & 1 & 0 & 0.25 \\ 0 & 0 & 1 & 0.40 \end{pmatrix}$$

and then we can immediately see the answers.

This transformed matrix is the same as
$$\begin{cases} x & = 0.20 \\ y & = 0.25 \\ z & = 0.40 \end{cases}$$

**Whoa! Stop the show! You went from**
$$\begin{pmatrix} 1 & 5 & 3 & 2.65 \\ 2 & 3 & 4 & 2.75 \\ 5 & 32 & 3 & 10.20 \end{pmatrix}$$
**to**
$$\begin{pmatrix} 1 & 0 & 0 & 0.20 \\ 0 & 1 & 0 & 0.25 \\ 0 & 0 & 1 & 0.40 \end{pmatrix}$$

**but you didn't show me how you did that. I hope you're not going to just say** Poof! **again.**

I was just about to show you how I did that transformation when you interrupted. This procedure is called the **Gauss-Jordan elimination** procedure. With the Gauss-Jordan procedure you can get all those pretty little ones and zeros in the coefficient matrix.

Using the Gauss-Jordan elimination you are allowed to do three things:

① Interchange any two rows.
② Multiply any row by a nonzero number.
③ Take any row,
　　multiply it by any number,
　　　and add it to another row.

These (①,②, and ③) are called **elementary row operations**.

Using the elementary row operations, you can go from

$$\begin{pmatrix} 1 & 5 & 3 & 2.65 \\ 2 & 3 & 4 & 2.75 \\ 5 & 32 & 3 & 10.20 \end{pmatrix} \quad \text{to} \quad \begin{pmatrix} 1 & 0 & 0 & 0.20 \\ 0 & 1 & 0 & 0.25 \\ 0 & 0 & 1 & 0.40 \end{pmatrix} \quad \text{in little steps.}$$

Before we take those little steps, it would be nice to notice that those three elementary row operations are legal things that you can do to a system of equations. Using ②, for example, if we take row 1 of our original matrix and multiply it by, say, 7, we go from

$$\begin{pmatrix} 1 & 5 & 3 & 2.65 \\ 2 & 3 & 4 & 2.75 \\ 5 & 32 & 3 & 10.20 \end{pmatrix} \quad \text{to} \quad \begin{pmatrix} 7 & 35 & 21 & 18.55 \\ 2 & 3 & 4 & 2.75 \\ 5 & 32 & 3 & 10.20 \end{pmatrix}$$

That's equivalent to taking our original equations and going from

$$\begin{cases} x + 5y + 3z = 2.65 \\ 2x + 3y + 4z = 2.75 \\ 5x + 32y + 3z = 10.20 \end{cases} \quad \text{to} \quad \begin{cases} 7x + 35y + 21z = 18.55 \\ 2x + 3y + 4z = 2.75 \\ 5x + 32y + 3z = 10.20 \end{cases}$$

**Okay. I'm ready. Let's do the Gauss-Jordan elimination thing.**

We start with

$$\begin{pmatrix} 1 & 5 & 3 & 2.65 \\ 2 & 3 & 4 & 2.75 \\ 5 & 32 & 3 & 10.20 \end{pmatrix}$$

We take –2 times the first row and add it to the second row. (That's elementary row operation ③)

$$\begin{pmatrix} 1 & 5 & 3 & 2.65 \\ 0 & -7 & -2 & -2.55 \\ 5 & 32 & 3 & 10.20 \end{pmatrix}$$

We take –5 times the first row and add it to the third row. (③: $-5r_1 + r_3$)

$$\begin{pmatrix} 1 & 5 & 3 & 2.65 \\ 0 & -7 & -2 & -2.55 \\ 0 & 7 & -12 & -3.05 \end{pmatrix}$$

**Hey! You used the "1" in the first row like a vacuum cleaner and swept out all the junk in the first . . . what do you call that vertical string of numbers?**

The vertical line of numbers in a matrix is called a **column**.

**Thank you. Now, as I was saying, you used that "killer one" to wipe out all the enemies in the first column.**

Dear Reader, it sounds like you're turning my linear algebra book into a war movie.

**I know what you're doing. You're now going to use the –7 in the second row, second column as the new "killer." You'll use it to destroy the other entries in the second column. Do you permit your readers to offer their art?**

$$\begin{pmatrix} 1 & ☎ & 3 & 2.65 \\ 0 & -7 & 2 & -2.55 \\ 0 & ☎ & -12 & -3.05 \end{pmatrix}$$

Art by the Reader

Geep! Are those mushroom clouds you drew? Did you just go nuclear?

23

**Yup. I bought your book, and I can do whatever I want with it.**

In the meantime, let me take the matrix from the previous page

$$\begin{pmatrix} 1 & 5 & 3 & 2.65 \\ 0 & -7 & -2 & -2.55 \\ 0 & 7 & -12 & -3.05 \end{pmatrix}$$

and <u>nicely</u> eliminate the 7 in the third row.

Multiply the second row by 1 and add it to the third row.

$$\begin{pmatrix} 1 & 5 & 3 & 2.65 \\ 0 & -7 & -2 & -2.55 \\ 0 & 0 & -14 & -5.60 \end{pmatrix}$$

**Don't look now, but I see trouble. If we try to wipe out the 5 in the first row, we'll have to add 5/7 of row 2 to row 1. $(5/7)r_2 + r_1$**

**That will give us horrible arithmetic.**
$$\begin{pmatrix} 1 & 0 & 1.5714286 & 0.8285715 \\ 0 & -7 & -2 & -2.55 \\ 0 & 0 & -14 & -5.60 \end{pmatrix}$$

I agree. Let's not go that way. Instead, remember what our goal is in using Gauss-Jordan elimination: We want the coefficient matrix to be

$$\begin{pmatrix} 1 & 0 & 0 \\ 0 & 1 & 0 \\ 0 & 0 & 1 \end{pmatrix}$$

having 1s on the **diagonal**, and 0s elsewhere.*

---

✱ The only rule of thumb is that once you create a 1 on the main diagonal or a 0 off the main diagonal, don't mess that up.

Main diagonals go from the upper left to the lower right just like most coats of arms:

Avoiding the bad arithmetic, we'll create a 1 in the third row, third column.  Starting with

$$\begin{pmatrix} 1 & 5 & 3 & 2.65 \\ 0 & -7 & -2 & -2.55 \\ 0 & 0 & -14 & -5.60 \end{pmatrix}$$

we multiply the third row by $(-1/14)$
(②: $(-1/14)r_3$)

$$\begin{pmatrix} 1 & 5 & 3 & 2.65 \\ 0 & -7 & -2 & -2.55 \\ 0 & 0 & 1 & 0.40 \end{pmatrix}$$

The next easiest thing to do is to use the 1 in the third row to eliminate the other entries in the third column.  We can do both rows at the same time.  That eliminates extra writing.

$-3r_3 + r_1$    and   $2r_3 + r_2$

$$\begin{pmatrix} 1 & 5 & 0 & 1.45 \\ 0 & -7 & 0 & -1.75 \\ 0 & 0 & 1 & 0.40 \end{pmatrix}$$

Now is a good time to create a 1 in the second row on the diagonal.
$(-1/7)r_2$

$$\begin{pmatrix} 1 & 5 & 0 & 1.45 \\ 0 & 1 & 0 & 0.25 \\ 0 & 0 & 1 & 0.40 \end{pmatrix}$$

$-5r_2 + r_1$ finishes the job.

$$\begin{pmatrix} 1 & 0 & 0 & 0.20 \\ 0 & 1 & 0 & 0.25 \\ 0 & 0 & 1 & 0.40 \end{pmatrix}$$

  costs 20¢

 costs 25¢

 costs 40¢

*Your Turn to Play*

1. [For English majors]  The matrix $\begin{pmatrix} 1 & 0 & 0 & 0.20 \\ 0 & 1 & 0 & 0.25 \\ 0 & 0 & 1 & 0.40 \end{pmatrix}$

is in **reduced row-echelon form**.  How would you describe that in English?

2. When it came to food, Betty was a little more health conscious than Fred.  At Jack LaRoad's Healthy Foods, their sign read "Is sugar worth dying for?"ˢᴹ

The shopping carts at LaRoad's were the latest thing. When you put your food into the cart and pressed a button, it would tell you the total amount of your purchase.

Betty put 3 whole-wheat bagels, 5 slices of turkey, and 2 peaches into her cart and pushed the button.  The cart read $3.20.

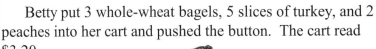

She tried 5 whole-wheat bagels, 3 slices of turkey, and 1 peach, and the cart read $2.85.

Finally, 1 bagel, 4 slices of turkey, and 3 peaches, and the cart read $2.35.

Use Gauss-Jordan elimination to find the individual prices.  (Please attempt this on your own on a sheet of paper until you get to the answers of 30¢, 40¢, and 15¢.  The important part is learning the skill of doing elementary row operations, not finding the answer.)

3. How many elementary row operations would it take to put this matrix into reduced row-echelon form?

$$\begin{pmatrix} 1 & 0 & 0 & 0 & 6 \\ 0 & 0 & 1 & 0 & 9 \\ 0 & 1 & 0 & 0 & 0 \\ 0 & 0 & 0 & 1 & 8 \end{pmatrix}$$

**. . . . . . . COMPLETE SOLUTIONS . . . . . . .**

1. For this chapter (in which every system of equations has exactly one solution), we would say that a reduced row-echelon matrix is a matrix in which there are 1s on the diagonal of the coefficient matrix and 0s above and below those 1s.

2. The system of linear equations is
$$\begin{cases} 3x_1 + 5x_2 + 2x_3 = 3.20 \\ 5x_1 + 3x_2 + x_3 = 2.85 \\ x_1 + 4x_2 + 3x_3 = 2.35 \end{cases}$$

The augmented matrix is
$$\begin{pmatrix} 3 & 5 & 2 & 3.20 \\ 5 & 3 & 1 & 2.85 \\ 1 & 4 & 3 & 2.35 \end{pmatrix}$$

If we use the 3 in the first row to wipe out the 5 and the 1, we'll have to deal with fractions. Instead, let's interchange rows 1 and 3:

$$\begin{pmatrix} 1 & 4 & 3 & 2.35 \\ 5 & 3 & 1 & 2.85 \\ 3 & 5 & 2 & 3.20 \end{pmatrix}$$

$$-5r_1 + r_2$$
$$-3r_1 + r_3$$

$$\begin{pmatrix} 1 & 4 & 3 & 2.35 \\ 0 & -17 & -14 & -8.9 \\ 0 & -7 & -7 & -3.85 \end{pmatrix}$$

$$(-1/7)r_3$$

$$\begin{pmatrix} 1 & 4 & 3 & 2.35 \\ 0 & -17 & -14 & -8.9 \\ 0 & 1 & 1 & 0.55 \end{pmatrix}$$

Interchange rows 2 and 3
(which I abbreviate I $r_2 r_3$)

$$\begin{pmatrix} 1 & 4 & 3 & 2.35 \\ 0 & 1 & 1 & 0.55 \\ 0 & -17 & -14 & -8.9 \end{pmatrix}$$

Copying the matrix from the previous page, so that you don't have to have flip back and forth between the pages to see "the action."

$$\begin{pmatrix} 1 & 4 & 3 & 2.35 \\ 0 & 1 & 1 & 0.55 \\ 0 & -17 & -14 & -8.9 \end{pmatrix}$$

$$-4r_2 + r_1$$
$$17r_2 + r_3$$

$$\begin{pmatrix} 1 & 0 & -1 & 0.15 \\ 0 & 1 & 1 & 0.55 \\ 0 & 0 & 3 & 0.45 \end{pmatrix}$$

$$(1/3)r_3$$

$$\begin{pmatrix} 1 & 0 & -1 & 0.15 \\ 0 & 1 & 1 & 0.55 \\ 0 & 0 & 1 & 0.15 \end{pmatrix}$$

$$r_3 + r_1$$
$$-r_3 + r_2$$

$$\begin{pmatrix} 1 & 0 & 0 & 0.30 \\ 0 & 1 & 0 & 0.40 \\ 0 & 0 & 1 & 0.15 \end{pmatrix}$$

So the whole-wheat bagel costs 30¢.
A slice of turkey costs 40¢.
A peach costs 15¢.

3.  Interchange rows 2 and 3.  It takes one step.  I $r_2$ $r_3$

Alexander knew that Betty and Fred would take care of getting the food for the picnic.  His responsibility was supplying the plastic flatware.[*]

At Butter Bottom Foods ("It's always a better buy at Butter Bottom!"SM), there were three assortments:

Sandy's Salad 'Sortment: 1    4    2         for 46¢.

Bobby's BBQ: 4    4    and no        for 60¢.

Full Family: 8    8    8         for 160¢.

_____

[*] I was going to write "plastic silverware," but that didn't seem quite right.

**Hey! I, your reader, recognize this. We've already done this kind of problem. You just set up**
$$\begin{pmatrix} 1 & 4 & 2 & 46 \\ 4 & 4 & 0 & 60 \\ 8 & 8 & 8 & 160 \end{pmatrix}$$
**and then use elementary row operations. When you get it in reduced row-echelon form, you can read off the answers. Doing the work on scratch paper, I got**
$$\begin{pmatrix} 1 & 0 & 0 & 8 \\ 0 & 1 & 0 & 7 \\ 0 & 0 & 1 & 5 \end{pmatrix}$$
**so a ⎸ is 8¢, a ⎸ is 7¢ and a ⎸ is 5¢.**

You didn't let me finish the story. Alexander went to the other grocery store—Alfredo's Foods ("Never be afraid at Alfredo's!"SM *). Alfredo's had the same three packages of flatware, but with different prices:

Sandy's Salad 'Sortment: 1 ⎸ 4 ⎸ 2 ⎸ for 43¢.

Bobby's BBQ: 4 ⎸ 4 ⎸ and no ⎸ for 60¢.

Full Family: 8 ⎸ 8 ⎸ 8 ⎸ for 160¢.

**So now you're telling me I gotta do it all over again. This time with**
$$\begin{pmatrix} 1 & 4 & 2 & 43 \\ 4 & 4 & 0 & 60 \\ 8 & 8 & 8 & 160 \end{pmatrix}$$
**and I'm going to have to do those same elementary row operations—the very same ones—in order to get the coefficient matrix in the form**
$$\begin{pmatrix} 1 & 0 & 0 \\ 0 & 1 & 0 \\ 0 & 0 & 1 \end{pmatrix}$$
**again. And then you'll have Alexander go to a third grocery store like Jonda Joods ("Joods you couldn't be fonder of"SM) and do it again.**

---

✱ SM is an abbreviation for service mark. It's like a trademark.

Perhaps, the best known service mark belongs to Sluice. As everyone knows, Sluice is a clear lemon-lime soda not to be confused with Slice® / Sprite® / Storm® / Seven-Up® / Sierra Mist® / Squirt® which all start with *S* and which some people tend to confuse with each other. No straws are ever served with Sluice because of its high viscosity. It's a bit thicker than motor oil. That is because of the amount of $C_{12}H_{22}O_{11}$ (sugar) it contains. Its service mark is "the World's Sweetest Soft drink."

There's no such place as Fonda Foods. I invented that name and service mark. As everyone knows, the *Life of Fred* series attempts to keep the amount of fiction down to a bare minimum.

Now you were worried that you would have to solve both

$$\begin{cases} x_1 + 4x_2 + 2x_3 = 46 \\ 4x_1 + 4x_2 \quad\quad = 60 \\ 8x_1 + 8x_2 + 8x_3 = 160 \end{cases} \quad \text{and} \quad \begin{cases} x_1 + 4x_2 + 2x_3 = 43 \\ 4x_1 + 4x_2 \quad\quad = 60 \\ 8x_1 + 8x_2 + 8x_3 = 160 \end{cases}$$

Or, in matrices, you were worried that you would have to solve both

$$\begin{pmatrix} 1 & 4 & 2 & 46 \\ 4 & 4 & 0 & 60 \\ 8 & 8 & 8 & 160 \end{pmatrix} \quad \text{and} \quad \begin{pmatrix} 1 & 4 & 2 & 43 \\ 4 & 4 & 0 & 60 \\ 8 & 8 & 8 & 160 \end{pmatrix}$$

You have forgotten a theme of this book ("easier and easier") that I mentioned on page 20. By giving you more knowledge of linear algebra, we will make whatever work needs to be done easier and easier.

You remember in elementary school when the teacher told you that each crab has 8 legs and asked you how many legs were on 5 crabs?

The hard way was to count all the legs: 1, 2, 3. . . . It was easier to add: $8 + 8 + 8 + 8 + 8$. And the teacher showed you an even easier way: $8 \times 5 = 40$.

In that spirit we take the coefficient matrix and augment it twice:

$$\begin{pmatrix} 1 & 4 & 2 & 46 & 43 \\ 4 & 4 & 0 & 60 & 60 \\ 8 & 8 & 8 & 160 & 160 \end{pmatrix}$$ and now both answers will pop out together.

We begin with . . .    $-4r_1 + r_2$

$-8r_1 + r_3$

$$\begin{pmatrix} 1 & 4 & 2 & 46 & 43 \\ 0 & -12 & -8 & -124 & -112 \\ 0 & -24 & -8 & -208 & -184 \end{pmatrix}$$ If I were now to turn the $-12$ in the second row into a 1, by multiplying the second row by $(-1/12)$, I would get fractions.

Instead, how about $-2r_2 + r_3$?

1. If one goal is for you to learn how to do elementary row operations on a doubly-augmented matrix, this might be a good time for you to take over.

$$\begin{pmatrix} 1 & 4 & 2 & 46 & 43 \\ 0 & -12 & -8 & -124 & -112 \\ 0 & -24 & -8 & -208 & -184 \end{pmatrix}$$  and continue using $-2r_2 + r_3$

and don't stop until you're done. Again, I ask you, take out a piece of paper and copy the matrix and do the work. Just reading the solution won't give you the facility in doing the elementary row ops.

. . . . . . . **COMPLETE SOLUTIONS** . . . . . . .

1.
$$\begin{pmatrix} 1 & 4 & 2 & 46 & 43 \\ 0 & -12 & -8 & -124 & -112 \\ 0 & 0 & 8 & 40 & 40 \end{pmatrix}$$

To me, the next easiest step would be $(1/8)r_3$

$$\begin{pmatrix} 1 & 4 & 2 & 46 & 43 \\ 0 & -12 & -8 & -124 & -112 \\ 0 & 0 & 1 & 5 & 5 \end{pmatrix}$$

Using the 1 in row 3 as the annihilator:
$-2r_3 + r_1$
$8r_3 + r_2$

$$\begin{pmatrix} 1 & 4 & 0 & 36 & 33 \\ 0 & -12 & 0 & -84 & -72 \\ 0 & 0 & 1 & 5 & 5 \end{pmatrix}$$

Now we can turn the $-12$ in the second row into a 1 without getting fractions. $(-1/12)r_2$

$$\begin{pmatrix} 1 & 4 & 0 & 36 & 33 \\ 0 & 1 & 0 & 7 & 6 \\ 0 & 0 & 1 & 5 & 5 \end{pmatrix}$$

$$-4r_2 + r_1$$

$$\begin{pmatrix} 1 & 0 & 0 & 8 & 9 \\ 0 & 1 & 0 & 7 & 6 \\ 0 & 0 & 1 & 5 & 5 \end{pmatrix}$$

and now we can read off both solutions.

The vector $\begin{pmatrix} 8 \\ 7 \\ 5 \end{pmatrix}$

tells us that at Butter Bottom Foods, a ⎸ is 8¢, a ⎸ is 7¢ and a ⎸ is 5¢.

The vector $\begin{pmatrix} 9 \\ 6 \\ 5 \end{pmatrix}$

tells us that at Alfredo's Foods, a ⎸ is 9¢, a ⎸ is 6¢ and a ⎸ is 5¢.

One definition of a **vector** is a matrix that has only one row or only one column. The definition you may have seen in physics is that a vector is an arrow that has length and direction.

As a preview of coming attractions, in Chapter 2½ we will give a third definition of vector. (A vector v is any element of a vector space **V**.)

Fred wanted to do the right thing for his two best friends. He bought the big Butter Bottom Sack-o-Picnic Food: 5 cans of cooked white rice, 32 bottles of Sluice, and 3 jars of mustard. He thought to himself *Betty will probably bring soy sauce for the rice and hotdogs for the mustard. Alexander is six feet tall, and he'll want plenty of Sluice to drink.*

At the age of 6, Fred had not yet figured out exactly how the world worked. It's true he was good at math. You don't get to be a professor of math at such an early age otherwise. But there was no reason to expect Betty to bring soy sauce, and Alexander didn't get to be six feet tall by drinking lots of sugar water.

You might say that Fred was a shrimp. He was only three feet tall. He had spent the last five years living off the vending machines down the hall from his math office. Sluice and candy bars were his main staples. Occasionally, he had had pizza with Alexander and Betty.

So there Fred stood, outside Butter Bottom's Foods.

Butter Bottom's
Sack-o-Picnic Food

$10.20

He knew how much each of the three sacks cost. He knew what was in each sack. He wanted to find out how much a can of rice, a bottle of Sluice, and a jar of mustard cost.

But he didn't use high school algebra.

He didn't use Gauss-Jordan elimination.

There's no use working my head to the bone he thought. I know a shorter, easier way.

Fred didn't need paper. In his head he started doing elementary row operations on the augmented matrix—just as if he were doing the Gauss-Jordan elimination—but he stopped in the middle.

He stopped when he got to:
$$\begin{pmatrix} 1 & 5 & 3 & 2.65 \\ 0 & -7 & -2 & -2.55 \\ 0 & 0 & 1 & 0.40 \end{pmatrix}$$

If you'll take a peek back on page 25, you'll see this matrix. It's the second from the top.

Now everyone knows that in order to do Gauss-Jordan, you have to transform the matrix into reduced row-echelon form where the coefficient matrix is
$$\begin{pmatrix} 1 & 0 & 0 \\ 0 & 1 & 0 \\ 0 & 0 & 1 \end{pmatrix}$$

He stopped when he had created zeros *below* the diagonal. That was enough. He didn't need to get rid of . . .

**I, your reader, want to do some more illustrating.**

You're not going to draw mushroom clouds, are you?

**No, I already did that. What I'm going to draw is the entries that Fred didn't have to eliminate. He was at the point**

$$\begin{pmatrix} 1 & 5 & 3 & 2.65 \\ 0 & -7 & -2 & -2.55 \\ 0 & 0 & 1 & 0.40 \end{pmatrix}$$ **and he didn't have to eliminate**

Art by the Reader

Yikes! What's that 🕷?

**You said you didn't like mushroom clouds. Don't you think he's cute?** 🕷 **is my drawing of an Ebola virus. It's a fresh new virus and highly lethal.**[*]

When I use the word *eliminate*, that doesn't mean *exterminate*.

When Fred got this far (creating zeros below the diagonal), he figured he was virtually done.

$$\begin{pmatrix} 1 & 5 & 3 & 2.65 \\ 0 & -7 & -2 & -2.55 \\ 0 & 0 & 1 & 0.40 \end{pmatrix}$$ corresponds to the equations $$\begin{cases} x_1 + 5x_2 + 3x_3 = 2.65 \\ \quad\quad -7x_2 - 2x_3 = -2.55 \\ \quad\quad\quad\quad\quad x_3 = 0.40 \end{cases}$$

which can be solved by **back-substitution**.

---

**Back-Substitution**
(from high school algebra)

We know $x_3$ is 0.40.
Put that value in the previous equation
$$-7x_2 - 2(0.40) = -2.55$$
and solve. Then $x_2$ is 0.25.
Put these two values in the first equation
$$x_1 + 5(0.25) + 3(0.40) = 2.65$$
and solve. Then $x_1$ is 0.20.

It's called *back*-substitution because we keep going backwards starting from the last equation.

---

[*] Let's put this stuff in a footnote so that we can get on with Fred's shorter, easier technique for solving systems of linear equations.

For the record, the Ebola virus was first discovered in 1976. The Zaire Ebola form of the virus is especially deadly. Eighty-eight percent died of those stricken in an outbreak in 1976, 100% in 1977, 59% in 1994, 81% in 1995, 73% in 1996, 80% in 2001-2002 and 90% in 2003.

Some notes:

♪#1: Fred's new shorter method (of only creating zeros below the diagonal and then back-substituting) can't, of course, be called Gauss-Jordan elimination. It's just called **Gaussian elimination**.

Here's Gauss.              Here's Jordan.

We just chop off the Jordan.

♪#2: *Gauss* rhymes with *house.*       Here's Fred.

♪#3: I prefer the expression *Gaussian elimination* to *Gauss elimination.* Although both terms are used by mathematicians, I have a very personal reason why I don't use *Gauss elimination.* You see, Fred's full name happens to be Fred Gauss. And I don't want him eliminated.

♪#4: For "small stuff" such as three equations with three unknowns, it really doesn't matter much whether you use Gauss-Jordan which turns the matrix into reduced row-echelon form with the coefficient matrix $\begin{pmatrix} 1 & 0 & 0 \\ 0 & 1 & 0 \\ 0 & 0 & 1 \end{pmatrix}$

or you use the Gaussian elimination, which creates zeros below the diagonal.

But when it comes to the real-life situations in which you may have ten equations with ten unknowns, chop off the Jordan. Gauss-Jordan takes more arithmetic than Gaussian elimination.

The details: If you solve by Gaussian elimination n equations and n unknowns and you sit down and count the number of multiplications and divisions you use, it works out to $\frac{n^3}{3}$ plus or minus some terms of lower power to get the matrix in reduced row-echelon form and back-substituting.

For Gauss-Jordan, it will cost you $\frac{n^3}{2}$ plus or minus some terms of lower power. (For large n, only the term with the largest power is important.)

Since $\frac{n^3}{2}$ is 50% larger than $\frac{n^3}{3}$, Gauss-Jordan takes about 50% more time.

Do you want to retire at age 50 or at age 75?

*Your Turn to Play*

1. How was Fred going to get that Sack-o-Picnic food to the Great Lawn picnic area?

   If he put the sack on top of his head (as he had seen in National Geographic), he would probably shrink from three feet tall to one foot.

 ☜ place load here

   I know thought Fred. If I need to move something, I should call a moving company.

   The first moving company that Fred contacted told him it would be $26.80 to do the job. They explained to Fred that their estimate depended on the number of minutes (t) that it would take, on the weight (w) in pounds, and on the distance (d) in miles. They used the formula

$$0.5t + 0.2w + 0.4d = \$26.80$$

   That's a lot of money! Fred got an estimate from a second company. Their formula was     $0.3t + 0.3w + 0.3d = \$21.60$

   A third company thought that distance was the most important factor. Their formula was     $0.1t + 0.01w + 10d = \$24.30$

   You know what we're going to ask:

   How many minutes does it take to move that sack to the Great Lawn? (That's t, which we are going to rename $x_1$.)
   How much does that sack weigh? (That's w, which we rename $x_2$.)
   How far is it to the Great Lawn? (That's d, which we rename $x_3$.)

   You also can guess that you are to find $x_1$, $x_2$, and $x_3$ by Gaussian elimination.

   Get out a sheet of paper and work until you get 40 minutes, 30 pounds, and 2 miles. Please don't just gander at the answer.

······· **COMPLETE SOLUTIONS** ·······

1. The augmented matrix is

$$\begin{pmatrix} 0.5 & 0.2 & 0.4 & 26.80 \\ 0.3 & 0.3 & 0.3 & 21.60 \\ 0.1 & 0.01 & 10 & 24.30 \end{pmatrix}$$

There are many choices of which elementary operations you could use. My first preference is to get rid of the decimals and simplify row 2.

$10r_1$
$10r_2$
$100r_3$
$(1/3)r_2$

$$\begin{pmatrix} 5 & 2 & 4 & 268 \\ 1 & 1 & 1 & 72 \\ 10 & 1 & 1000 & 2430 \end{pmatrix}$$

If I were to use the 5 in the first row as the annihilator of the 1 below the 5, I would have to use the elementary row operation $(-1/5)r_1 + r_2$. That would give me fractions. Instead I'll interchange row 1 and row 2.

$I\ r_1\ r_2$

$$\begin{pmatrix} 1 & 1 & 1 & 72 \\ 5 & 2 & 4 & 268 \\ 10 & 1 & 1000 & 2430 \end{pmatrix}$$

$-5r_1 + r_2$
$-10\ r_1 + r_3$

$$\begin{pmatrix} 1 & 1 & 1 & 72 \\ 0 & -3 & -1 & -92 \\ 0 & -9 & 990 & 1710 \end{pmatrix}$$

And one last elementary row operation to make the matrix **upper triangular** (zeros below the diagonal). (And $-r_2$ to get rid of minus signs.)

$-3r_2 + r_3$
$-r_2$

$$\begin{pmatrix} 1 & 1 & 1 & 72 \\ 0 & 3 & 1 & 92 \\ 0 & 0 & 993 & 1986 \end{pmatrix}$$

Back-substituting:     $993x_3 = 1986$ so $x_3 = 2$ miles.
$3x_2 + 2 = 92$  so $x_2 = 30$ lbs.
$x_1 + 30 + 2 = 72$  so $x_1 = 40$ minutes

Just before the *Your Turn to Play* I wrote, ". . . real-life situations in which you may have ten equations and ten unknowns. . . ." We may need to modify that a bit.

Fred was concerned about how cold it would be at the Great Lawn picnic area on the KITTENS campus at noon for the picnic. His bare head was vulnerable to extremes of temperature. He wondered whether he would need a safari hat                    or a woolen cap.

Across the country there are thousands of weather stations which report their temperatures. I have placed a dot on one of them and labeled it $x_1$. If you have a moment, add 999 more dots to this map and label them $x_2, x_3, \ldots, x_{1000}$.

If Fred wanted to know what the temperature would be an hour from now at the Great Lawn picnic area, he wouldn't care what the temperature reading was at •$x_1$ in Reno, Nevada. Instead, he would want to know the reading at •$x_{347}$, which is seven miles west of the picnic area, and at •$x_{348}$, which is three miles east of the picnic area.

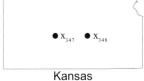

Kansas

**I've got a question. Why not make things simple and use the closest weather station which is three miles east?**

It might be that the weather systems in that part of Kansas tend to move from west to east. In that case, •$x_{347}$, even though it is farther away, might have more predictive power.

**Well, how do you tell?**

One way is to let the past be a guide to the future.[*]

Yesterday at this time it was 58° at •$x_{347}$ and 52° at •$x_{348}$, and at noon on that day the temperature at the Great Lawn was 57°.

**So wouldn't the •$x_{347}$ be the best predictor?**

---

[*] This sentence was an epigraph in Professor Eldwood's *History and What It's Good For*, 1846.

An epigraph is a quotation that an author sticks at the beginning of a chapter.

But two days ago, at this time it was 51° at •$x_{347}$ and 58° at •$x_{348}$, and at noon the temperature at the Great Lawn was 56°.

One possibility is to set up a system of linear equations:

$$58x_{347} + 52x_{348} = 57$$
$$51x_{347} + 58x_{348} = 56$$

---

*Your Turn to Play*

1. [Harder question] If you solved this system of linear equations, what would $x_{347}$ and $x_{348}$ represent?

2. Solve the system. Use Gaussian elimination. (I'm going to round off the entries to about four significant digits. If I were being paid big bucks as a weather forecaster, I'd use all the digits my calculator would give me.)

3. If the readings at $x_{347}$ and $x_{348}$ right now were 57° and 54°, what would you predict the noontime temperature would be at the Great Lawn?

**. . . . . . . C O M P L E T E   S O L U T I O N S . . . . . . .**

1. The $x_{347}$ and $x_{348}$ represent the relative weights we should assign the readings at the two weather stations for predicting the noontime temperature at the Great Lawn.

2. We start with
$$\begin{pmatrix} 58 & 52 & 57 \\ 51 & 58 & 56 \end{pmatrix}$$

$(1/58)r_1$
$$\begin{pmatrix} 1 & 0.8966 & 0.9828 \\ 51 & 58 & 56 \end{pmatrix}$$

$-51r_1 + r_2$
$$\begin{pmatrix} 1 & 0.8966 & 0.9828 \\ 0 & 12.27 & 5.8772 \end{pmatrix}$$

$(1/12.27)r_2$
$$\begin{pmatrix} 1 & 0.8966 & 0.9828 \\ 0 & 1 & 0.4790 \end{pmatrix}$$

$$x_{348} = 0.4790$$
$$x_{347} + (0.8966)(0.4790) = 0.9828 \quad \text{so } x_{347} = 0.5533$$

3. $57(0.5533) + 54(0.4790) = 57.4041 \doteq 57°$
$\doteq$ means "rounded off"

---

What weather stations would you have to take into account if you wanted to predict the temperature at the Great Lawn a week from now?  It could be mild and sunny right now at the Great Lawn, but a giant ice storm could be blowing through Reno at •$x_1$ which you should take into account. You wouldn't want to just limit yourself to the weather in Kansas if you are predicting the weather a week from now.

So a real-life situation could involve a thousand equations with a thousand unknowns.

Don't limit yourself.

It's traveling time.  We will have the opportunity to visit six cities in the next several pages.  Each city is an overview of what we have seen in this chapter.

The city names are from around the United States.

(If you haven't worked through the *Your Turn to Play* on your own, these questions may be tougher than they need to be.)

---

**Ely**

---

Pat, the choir director, went shopping for music.

On Monday she bought one copy of the Russian song, four copies of the German song, and two copies of the American. The bill was $20.

On Tuesday, 3 copies of the Russian song, 2 copies of the German song, and 5 copies of the American song cost $45.

On Wednesday, 7 copies of the Russian, 2 of the German, and 6 of the American cost $74.

$$\begin{pmatrix} 1 & 4 & 2 & 20 \\ 3 & 2 & 5 & 45 \\ 7 & 2 & 6 & 74 \end{pmatrix}$$

1. Find out how much each song cost using high school algebra.
2. Solve the system using Gaussian elimination.
3. Solve the system using Gauss-Jordan.

*answers*

The Russian song cost $6, the German cost $1, and the American cost $5.

---

**Paar**

---

1. You are a door-to-door salesman. The number of sales that you make on a particular day depends, in part, on the number of doors you knock on $(x_1)$, the price of the thing you are selling $(x_2)$, and the number of minutes you devote to your sales pitch $(x_3)$.

For example, one day you might knock on 100 doors ($x_1 = 100$), set your price at $6 ($x_2 = 6$), and use your 4-minute presentation ($x_3 = 4$). Explain why this is *not* part of linear algebra. Please do some thinking about this before you take a look at the answer.
2. On the other hand, the number of calories you burn going from door to

door might be a linear function of the number of miles you walk, the number of pounds of merchandise you carry, and the number of dogs that you have to run away from.

If you walked one mile, carried 5 pounds of merchandise and eluded 2 dogs, you would burn 1030 calories. (You are carrying a newly invented Kalorie Kounter.)

If you walked 2 miles, carried 3 pounds and ran away from 1 dog, you would burn 1590 calories.

Eight miles, 2 pounds, and 5 dogs, and you would burn 5900 calories.

$$\begin{pmatrix} 1 & 5 & 2 & 1030 \\ 2 & 3 & 1 & 1590 \\ 8 & 2 & 5 & 5900 \end{pmatrix}$$

Solve this system using Gauss-Jordan elimination.

### answers

1.  The equations in linear algebra are linear. (The exceptions are rare.) They are equations of the form $a_1x_1 + a_2x_2 + \ldots a_nx_n = b$ where the $a_i$ (for $i = 1$ to n) and the b are real numbers and n is a natural number.

Suppose there were a linear equation, $0.02x_1 - 5x_2 + 1.7x_3 = b$, relating the number of doors knocked on ($x_1$), the price of the item ($x_2$), and the length of your presentation ($x_3$) to the number of sales you make (b). (The coefficient of $x_2$ is negative, because lowering the price will increase sales.) Then doubling $x_1$, $x_2$, and $x_3$ should double b.

Is this true? Suppose that instead of knocking on 100 doors, you knocked on 200. Or that you changed your price from $6 to $12, or that you gave 8-minute presentations instead of 4-minute ones. Would that double your sales?

Lots of things could mess that up. Knocking on 200 doors/day might mean that you had to work until 10 p.m. Many people would not be as receptive to you at that time of night. Doubling your price might kill your sales completely. If you had been selling rolls of twenty 35¢ postage stamps (= $7) for only $6, you might have had some buyers, but pricing them at $12 would be a different story.

And giving an 8-minute presentation might not be any more effective than a 4-minute one.

2.  Walking a mile burns 700 calories, carrying one pound of merchandise burns 50 calories, and running away from a dog burns 40 calories.

---

**Vassar**

---

Lee loves to shop at automotive stores.  She bought 3 hub caps, 2 tires, and 5 cans of polish.  The bill was $68.

One hub cap, 6 tires, and a can of polish totaled $144.

Five hub caps, 2 tires, and 4 cans of polish cost $74.

She purchased lots of hub caps, since they kept falling off her car.

$$\begin{pmatrix} 3 & 2 & 5 & 68 \\ 1 & 6 & 1 & 144 \\ 5 & 2 & 4 & 74 \end{pmatrix}$$

1.  Find out how much each item cost using Gaussian elimination.
2.  Why is it incorrect to say that one of the elementary row operations is: "Multiply any row by any number"?
3.  Suppose Lee purchased 3 steering wheels, 4 brake lights, and 2 auto horns and the bill came to $38.  Is there enough information to tell how much a steering wheel cost?

*answers*

1.  $4, $23, $2
2.  If you multiply a row by zero, you will wipe out that row.  If you started with $\begin{cases} 6x + 4y = 15 \\ 8x - 5y = 12 \end{cases}$   and multiplied the second equation by zero,

you would get $\begin{cases} 6x + 4y = 15 \\ 0x + 0y = \ 0 \end{cases}$

This second system of equations has many more solutions than the first system.

3.  $3x_1 + 4x_2 + 2x_3 = 38$ is called an **underdetermined linear system**. There are many solutions.  This is the topic of Chapter 2.

## Zumbro Falls

The number of weeks that Robin held a job depended linearly on the number of times per week that she wasn't late for work, the number of coffee breaks she took each day, and the number of items she stole from the office.

When Robin was working at *Wally's Window Washing*, she arrived on time one day/week, took 4 coffee breaks/day, and stole one item. She was fired after one week.

At *Willy's Walnuts*, she arrived on time 4 days/week, took 2 coffee breaks, and stole nothing. She lasted 44 weeks at Willy's.

At *Wanda's Wallpapering*, she bought an alarm clock and showed up on time 5 days/week, took 5 coffee breaks/day, and stole 5 items. It was 35 weeks before Wanda told Robin she needed to find other work.

When Robin was hired at *William's Welding*, she arrived on time 3 days/week, took no coffee breaks, and stole 11 items.

1. Estimate how long Robin would last at William's Welding.
2. True or False: "When you solve a system of linear equations by Gaussian elimination, the final matrix is always upper triangular."

Robin as a welder

3. What does it mean for a matrix to be **lower triangular**?
4. Complete this sentence: When you use Gauss-Jordan elimination, the matrix will be both upper triangular and _____.

### answers

1. Solving $\begin{pmatrix} 1 & 4 & 1 & 1 \\ 4 & 2 & 0 & 44 \\ 5 & 5 & 5 & 35 \end{pmatrix}$ by Gauss-Jordan gives $\begin{pmatrix} 1 & 0 & 0 & 12 \\ 0 & 1 & 0 & -2 \\ 0 & 0 & 1 & -3 \end{pmatrix}$

so $x_1 = 12$, $x_2 = -2$, and $x_3 = -3$.

At William's Welding, $3(12) + 0(-2) + 11(-3) = 3$ weeks she lasted.

2. True. After solving by Gaussian elimination, a matrix is always upper triangular.

3. Here's a lower triangular matrix:
A lower triangular matrix has zeros above the diagonal.

$\begin{pmatrix} 3 & 0 & 0 & 0 \\ 2 & 1 & 0 & 0 \\ 0 & 8 & 2 & 0 \\ 9 & \pi & \frac{1}{2} & -4 \end{pmatrix}$

4. When you use Gauss-Jordan elimination, the matrix will be both upper triangular and lower triangular.

---

**Inwood**

---

Madison holds the den meetings for her Cub Scouts at her apartment. For snack time she gives them Sluice to drink, which makes the boys slightly nuts. They run around more, they cry more, and they break things.

She kept a diary of the three recent den meetings.

*March 3 meeting.  I gave each boy one oz. of Sluice.  There were 5 boys, and our meeting lasted 40 minutes.  Thirteen items were broken.*

*March 10 meeting.  Each of the 7 boys got 3 oz. of Sluice.  It was the usual 40-minute meeting.  I found 19 broken items in my apt.*

*March 17 meeting.  4 oz. Sluice.  4 boys.  60 minutes (since none of the other mothers showed up to claim their boys on time).  Fifteen broken items.*

$$\begin{pmatrix} 1 & 5 & 40 & 13 \\ 3 & 7 & 40 & 19 \\ 4 & 4 & 60 & 15 \end{pmatrix}$$

1. Solve the system by Gauss-Jordan.
2. Using the values found in the previous question, estimate how many items would be broken in Madison's apartment if she gave the boys 2 ounces of Sluice, there were 6 boys, and the meeting lasted 40 minutes.

***answers***

1. Number of items broken as related to ounces of Sluice = 1.
Number of items broken as related to boys in attendance = 2.
Number of items broken as related to minutes of meeting = 0.05.
2. $2(1) + 6(2) + 0.05(40) = 16$ items broken.

---

**Johnston**

---

Jackie was new on the police force.  Jackie had three choices of which assortment of equipment to carry:

There was The Every Day assortment which consisted of 1 pair of handcuffs, 1 can of pepper spray, and 1 pistol.  It weighed 8 pounds.

There was The Mass Arrest assortment which consisted of 5 pairs of handcuffs, 2 cans of pepper spray, and 2 pistols.  It weighed 22 pounds.

There was The Make-'em-Cry assortment which consisted of 1 pair of handcuffs, 10 cans of pepper spray and 1 pistol.  It weighed 17 pounds.

1.  Being new on the force, Jackie wanted to impress everyone.  "I'll take the Big Daddy assortment: 6 pairs of handcuffs, 1 can of pepper spray, and 6 pistols."

The quartermaster exclaimed, "Six pistols!  That's a lot of iron.  May I ask why you will be packing that much heat?"

Jackie smiled and said, "I think it will look cool.  And besides, you can never be too careful."

How much does the Big Daddy assortment weigh?

2.  Paperwork is the bane* of many police officers.  Jackie wanted to be on the street catching crooks, not sitting at a desk filling out arrest forms.

Jackie's main beat as a beginning officer was A.D.A., which stands for auto theft, drunkenness, and assault.

In the first week of work, Jackie made 0 auto theft arrests, 2 arrests for public drunkenness, and 5 arrests for assault, and spent 66 minutes doing the paperwork.

In the second week, 3 auto, 1 drunk, and 2 assault ➟ 61 minutes.

In the third week, 1 auto, 7 drunk, and 6 assault ➟ 127 minutes.

$$\begin{pmatrix} 0 & 2 & 5 & 66 \\ 3 & 1 & 2 & 61 \\ 1 & 7 & 6 & 127 \end{pmatrix}$$

---

✴ *Bane* is a word that is more than a thousand years old.  Bane is anything that causes misery, death, or ruin.  In Old English *bana* meant slayer.

Solve this by Gaussian elimination.

The zero in the row 1, column 1 position presents a problem: How do you use it to eliminate the 3 and the 1 directly below it? You can't. What is the first elementary row operation you should use?

3. How many minutes would it take to process the paperwork for 1 auto theft, 4 public drunkenness, and 3 assault? (Hint: the sum of the digits of your answer will equal ten.)

*answers*

1. A pair of handcuffs weighs 2 lbs. A can of pepper spray is 1 lb. A pistol is 5 lbs.
So the Big Daddy assortment weighs $6(2) + 1(1) + 6(5) = 43$ lbs.
2. To get the zero out of the row 1, column 1 position, use the elementary row operation "① Interchange any two rows." Interchanging rows 1 and 3 seem like a good bet.

$$\begin{pmatrix} 0 & 2 & 5 & 66 \\ 3 & 1 & 2 & 61 \\ 1 & 7 & 6 & 127 \end{pmatrix} \xrightarrow{\text{I } r_1 r_3} \begin{pmatrix} 1 & 7 & 6 & 127 \\ 3 & 1 & 2 & 61 \\ 0 & 2 & 5 & 66 \end{pmatrix} \xrightarrow{-3r_1 + r_2} \begin{pmatrix} 1 & 7 & 6 & 127 \\ 0 & -20 & -16 & -320 \\ 0 & 2 & 5 & 66 \end{pmatrix}$$

In contrast, if we interchange rows 1 and 2, we get fractions:

$$\begin{pmatrix} 0 & 2 & 5 & 66 \\ 3 & 1 & 2 & 61 \\ 1 & 7 & 6 & 127 \end{pmatrix} \xrightarrow{\text{I } r_1 r_2} \begin{pmatrix} 3 & 1 & 2 & 61 \\ 0 & 2 & 5 & 66 \\ 1 & 7 & 6 & 127 \end{pmatrix} \xrightarrow{(-1/3)r_1 + r_3} \begin{pmatrix} 3 & 1 & 2 & 61 \\ 0 & 2 & 5 & 66 \\ 0 & 6\frac{2}{3} & 5\frac{1}{3} & 106\frac{2}{3} \end{pmatrix}$$

Finishing the Gaussian elimination, we get:
    auto = 11 minutes,
    drunkenness = 8 minutes,
    assault = 10 minutes.
3. 1 auto + 4 drunk + 3 assault = $1(11) + 4(8) + 3(10) = 73$ minutes

# *Chapter One and a Half*
## *Matrices*

The moving company carried Fred's 30-pound Sack-o-Pic Food two miles to the Great Lawn picnic grounds. . . .

**Wait a minute. I, your reader, have a big question. Where's Chapter 2? We just finished Chapter One. After Chapter 1 is supposed to come Chapter 2. What's this Chapter 1½?**

Do you remember back on page 24, when you told me, "I bought your book, and I can do whatever I want with it"? If you would like to scratch out the *Chapter 1½* and make it *Chapter 2,* I would have great difficulty stopping you. The only one you'll have to explain anything to is your kid sister who will be reading this book after you're done with it.

**But why do you have this crazy numbering?**

This *Life of Fred: Linear Algebra* book has four chapters:

Chapter 1     Systems of Linear Equations with One Solution
Chapter 2     Systems of Linear Equations with Many Solutions
Chapter 3     Systems of Linear Equations with No Solution
Chapter 4     Long-term Behavior of Systems of Linear Equations.

**Let me guess. Linear algebra is about systems of linear equations.**

In some sense, we could say that you are half right. This Chapter 1½ is the other half of linear algebra. In this chapter we look behind the scenes at what is going on when we work with linear systems of equations. It's a little like the movie *The Wizard of Oz* in which Dorothy got a chance to peek behind the curtain and discover the wizard.

When we are solving linear systems of equations, we are down in the forest. In this chapter, we head to the mountaintop and get an overview of the whole landscape.

Instead of being at the ground level, hacking our way through the arithmetic forest of Gauss-Jordan elimination, we will be looking down on it all.

**Okay. Take me to the heights.**

Down on the ground in Chapter 1, a **matrix** was a rectangular array of numbers.  Actually, the elements of a matrix don't have to be numbers.  They could be various mathematic objects such as variables or functions.  Here's a $2 \times 3$ ("two by three") matrix: $\begin{pmatrix} \theta & \sin \theta & \ln 5.4 \\ x^2 & \pi & \sinh y \end{pmatrix}$

with two rows and three columns.  The rows are always mentioned before the columns.

Matrices are often named by capital letters such as A or B or C.

If a matrix has only one row or only one column, it is called a **vector**, and is often named by lower case letters such as b or x or v.

Fred stood there on the sidewalk outside of Butter Bottom's Foods. What else do I need to get before I head to the picnic grounds? he thought to himself.  I don't want to show up and have forgotten something really important.

He pulled a book out of his back pocket.  (Fred always carried several books with him for emergencies like these.)  He looked at the cover and checked off the various picnic items: people ✓ food ✓ plates ✓ and then he realized: I don't have a picnic blanket! You can't just sit on the bare Kansas ground.

Fred now had a quest: find the perfect picnic blanket.  He ran to that famous blanket store, Blanche's Better Blankets and asked Blanche, "What have you got?"

If Fred had been a doctor, she might have listed her maladies.  Instead, given the context, she said she had 6 very nice red cotton blankets, 1 very nice pink cotton blanket, 2 very nice orange cotton blankets, 3 very nice red wool blankets, 3 very nice pink wool blankets and 5 very nice orange wool blankets. Fred's head started to spin with all the information.  In his head, he organized it all into a matrix and named it B in honor of Blanche.

$$B = \begin{array}{cc} & \begin{array}{ccc} \text{red} & \text{pink} & \text{orange} \end{array} \\ \begin{array}{c} \text{cotton} \\ \text{wool} \end{array} & \begin{pmatrix} 6 & 1 & 2 \\ 3 & 3 & 5 \end{pmatrix} \end{array}$$

Blanche's main competitor in blanket sales was Carrie's Comfy Covers.  Fortunately, Carrie's store was right next to Blanche's.  Fred headed to Carrie's and asked, "What have you got?"  Carrie told him that she had 2 very nice red cotton blankets, 4 very nice pink cotton blanket, 7 very nice orange cotton blankets, 4 very nice red wool blankets, 1 very nice pink wool blanket and 3 very nice orange wool blankets. Fred knew what to do. He popped that data into a matrix and named it in honor of Carrie.

$$C = \begin{array}{c} \\ \text{cotton} \\ \text{wool} \end{array} \begin{array}{ccc} \text{red} & \text{pink} & \text{orange} \\ \left( \begin{array}{ccc} 2 & 4 & 7 \\ 4 & 1 & 3 \end{array} \right) \end{array}$$

One advantage of using matrices is that you can think of C as just a single object and not as six pieces of data.

Carrie said, "If I don't have the perfect blanket for you, then you might try Blanche's."

Fred was surprised. "That's your competitor. Why would you suggest I see her?"

Carrie smiled. "She's my sister. We thought it would be very nice if we had stores right next to each other."

Fred winced when he heard the phrase *very nice*. He thought to himself They should combine their stores and rename it as the Very Nice Blankets store. Mentally, Fred combined their inventory:

$$B + C = \begin{pmatrix} 6 & 1 & 2 \\ 3 & 3 & 5 \end{pmatrix} + \begin{pmatrix} 2 & 4 & 7 \\ 4 & 1 & 3 \end{pmatrix} = \begin{pmatrix} 8 & 5 & 9 \\ 7 & 4 & 8 \end{pmatrix}$$

If you look at this for a moment or two, you will know the secret of **matrix addition**. Other linear algebra books which are paragons of prolixity* will define matrix addition in mind-numbing length: *In order to add two matrices, they must have the same dimensions—the same number of rows and the same number of columns. Then matrix addition is performed by arithmetically adding corresponding entries in the two addends. Thus, for example, two 2 × 3 matrices will add to create a 2 × 3 matrix.* But you knew that just looking at the example. This book attempts to avoid such wordiness.

Fred picked up an orange wool blanket. It felt a little scratchy, and the orange dye colored his hand. He put it back on the table.

A pink cotton blanket felt smooth, and the pink color didn't come off in his hand.

---

✱ Phrases such as paragons of prolixity, Blanche's Better Blankets, or Carrie's Comfy Covers, which have the same initial sounds, are examples of alliteration.

"How much are your very nice blankets?" Fred asked. He was starting to sound like the sisters. Carrie pointed to a price chart on the wall (which happened to look very much like a matrix).

$$P = \begin{array}{cc} & \begin{array}{ccc} \text{red} & \text{pink} & \text{orange} \end{array} \\ \begin{array}{c} \text{cotton} \\ \text{wool} \end{array} & \begin{pmatrix} \$2 & \$5 & \$4 \\ \$5 & \$7 & \$8 \end{pmatrix} \end{array}$$

"And what's the current sales tax?" he asked. He knew that there was a 7% state sales tax, a 1% county tax, a 0.4% stadium tax, and a 0.3% mosquito abatement tax, but he had heard that there was a new county commissioner's pension improvement tax.

Carrie said, "Today the rate is 11.65%. In order to figure the tax, you multiply the purchase price by 0.1165. Then you add that to the purchase price to get the price you pay."

Just because he was six years old, Fred thought that Carrie was "talking down" to him. Besides he thought Carrie is doing it the long way. All you got to do is multiply the purchase price by 1.1165 to get the price I pay.

Fred mentally multiplied 1.1165 times the matrix P.

$$1.1165\,P = 1.1165 \begin{pmatrix} 2 & 5 & 4 \\ 5 & 7 & 8 \end{pmatrix} = \begin{pmatrix} 2.2330 & 5.5825 & 4.466 \\ 5.5825 & 7.8155 & 8.932 \end{pmatrix}$$

---

*Your Turn to Play*

1. Have some fun . . . If you were writing one of those other linear algebra books that are sometimes called "volumes of verbosity," how might you define the multiplication of a matrix by a number?

2. If A and B have the same dimensions, is it always true that
$$A + B \stackrel{?}{=} B + A$$

3. When you multiply a number times a matrix, that is called **scalar multiplication**. That's because in matrix arithmetic, numbers are called **scalars**. I don't know why. It's just a tradition.

Give six examples of the weirdest scalars you can think of.

---

## .......COMPLETE SOLUTIONS.......

1. You might write something like: *In order to multiply a number times a matrix, you multiply each entry in the matrix by that number. The dimensions of the matrix are unaffected by this operation.*

2. If you try it out with a couple of matrices, you can see that it's always true. For example,

$$\begin{pmatrix} 3 & 5 & 1 \\ 8 & 2 & 5 \end{pmatrix} + \begin{pmatrix} 2 & 4 & 7 \\ 1 & 3 & 3 \end{pmatrix} = \begin{pmatrix} 3+2 & 5+4 & 1+7 \\ 8+1 & 2+3 & 5+3 \end{pmatrix} = \begin{pmatrix} 2 & 4 & 7 \\ 1 & 3 & 3 \end{pmatrix} + \begin{pmatrix} 3 & 5 & 1 \\ 8 & 2 & 5 \end{pmatrix}$$

In a nutshell, matrix addition is commutative. If you have a highlighter and this is your book, the previous sentence has four words that you may wish to highlight.

3. Your list might be different than mine. Here is a list of my six favorite weird scalars: $\pi$, 93793492392364290642396.07, csc 23°, $\sqrt{-1}$ (which is the complex number i), $\log_3 7$, and 0.

Virtually all of the scalars in this book will be real numbers. As you go further in linear algebra, we may toss in complex numbers like $\sqrt{-1}$.

We have added two matrices.    A + B

We have multiplied a scalar times a matrix.    aA

What's left? To multiply two matrices.    AB

Matrix multiplication isn't what you would expect. If we just defined it as multiplying together corresponding entries—similar to matrix addition—the definition would be easy. And it would be virtually worthless.

The real definition of AB is more complicated, but it turns out to be much more useful.

The easiest case to start with is multiplying together two skinny matrices—a row vector times a column vector. We'll take Carrie's cotton inventory $\begin{pmatrix} \overset{red}{2} & \overset{pink}{4} & \overset{orange}{7} \end{pmatrix}$ times their corresponding prices $\begin{pmatrix} \$2 \\ \$5 \\ \$4 \end{pmatrix}\begin{smallmatrix} red \\ pink \\ orange \end{smallmatrix}$ expressed as a column vector.

$$(2 \quad 4 \quad 7) \begin{pmatrix} \$2 \\ \$5 \\ \$4 \end{pmatrix} \quad \text{equals} \quad (\, 2\times2 \,+\, 4\times5 \,+\, 7\times4 \,) = (52)$$

Some notes:

♪#1: Optometrists really hate this. We're asking you to move your left eye this way ↔ and your right eye this way ↕.

♪#2: You can't multiply $(3 \quad 9 \quad 1 \quad 8) \begin{pmatrix} 2 \\ 1 \\ 7 \end{pmatrix}$

because the column vector doesn't have enough rows.

♪#3: The answer at the top of this page is (52) which is a 1×1 matrix. When you multiply two matrices, you get a matrix. You don't get the scalar 52 as an answer.

♪#4: We multiplied Carrie's cotton inventory by the corresponding prices: the 2 red blankets times the cost of a red blanket

plus

the 4 pink blankets times the cost of a pink blanket

plus

the 7 orange blankets times the cost of an orange blanket.

The result was the total cost of her inventory of cotton blankets.

What matrix multiplication does is <sub>squish down</sub> a lot of data into a more compact form.

If Carrie's cotton blankets were destroyed by a flood, the insurance agent would do the computation $(2 \quad 4 \quad 7) \begin{pmatrix} \$2 \\ \$5 \\ \$4 \end{pmatrix} = (52)$

and hand Carrie a check for $52.

---

*Your Turn to Play*

1. When you multiply a 1 × 7 vector times a 7 × 1 vector, what are the dimensions of the answer?

2. $(3 \quad -4 \quad 20 \quad \pi) \begin{pmatrix} 8 \\ 3 \\ 1 \\ 5 \end{pmatrix} = ?$

---

3.  Here's Carrie's inventory red    pink    orange

$$C = \begin{array}{c}\text{cotton}\\\text{wool}\end{array}\begin{pmatrix}2 & 4 & 7\\4 & 1 & 3\end{pmatrix}$$

and her prices red    pink    orange

$$P = \begin{array}{c}\text{cotton}\\\text{wool}\end{array}\begin{pmatrix}\$2 & \$5 & \$4\\\$5 & \$7 & \$8\end{pmatrix}$$

Suppose her wool inventory were destroyed.* Set up the row vector and the column vector and compute her loss.

## . . . . . . . COMPLETE SOLUTIONS . . . . . . .

1.  You get a $1 \times 1$ matrix.
2.  $(32 + 5\pi)$  This is a $1 \times 1$ matrix.
3.  $(4\ \ 1\ \ 3)\begin{pmatrix}\$5\\\$7\\\$8\end{pmatrix} = (\ 4{\times}5\ +\ 1{\times}7\ +\ 3{\times}8\ ) = (51)$. A loss of \$51.

Now it's time for **matrix multiplication**.

Let's just look at cotton blankets. Here are Blanche's and Carrie's cotton inventories:

red    pink    orange

$$\begin{array}{c}\text{Blanche's cotton}\\\text{Carrie's cotton}\end{array}\begin{pmatrix}6 & 1 & 2\\2 & 4 & 7\end{pmatrix}$$

The usual prices for the cotton blankets was $\begin{array}{c}\text{red}\\\text{pink}\\\text{orange}\end{array}\begin{pmatrix}\$2\\\$5\\\$4\end{pmatrix}$

but Congress enacted a tariff against foreign cotton. This tax on imported cotton does what every import tax does: (1) it gives consumers fewer choices, and (2) it drives up prices. Blanche didn't benefit from the tariff. She paid more to buy the blankets and sold them at a higher price. The consumer was hurt by the higher prices. Who benefitted? The local cotton growers.

---

∗ I was trying to think of something that would destroy wool. All I could come up with was a wooly mammoth.

Every industry would love to see a 1000% tariff against foreign competition. It's one reason they send lobbyists to the offices of Congressmen and Congresswomen.

Suppose the price matrix is

$$\begin{array}{c} \\ \text{red} \\ \text{pink} \\ \text{orange} \end{array} \begin{pmatrix} \overset{\text{before}}{\underset{\text{tariff}}{}} & \overset{\text{after}}{\underset{\text{tariff}}{}} \\ \$2 & \$3 \\ \$5 & \$7 \\ \$4 & \$6 \end{pmatrix}$$

Now we multiply

$$\begin{array}{c} \\ \text{Blanche's cotton} \\ \text{Carrie's cotton} \end{array} \begin{pmatrix} \overset{\text{red}}{} & \overset{\text{pink}}{} & \overset{\text{orange}}{} \\ 6 & 1 & 2 \\ 2 & 4 & 7 \end{pmatrix} \begin{array}{c} \\ \text{red} \\ \text{pink} \\ \text{orange} \end{array} \begin{pmatrix} \overset{\text{before}}{\underset{\text{tariff}}{}} & \overset{\text{after}}{\underset{\text{tariff}}{}} \\ \$2 & \$3 \\ \$5 & \$7 \\ \$4 & \$6 \end{pmatrix}$$

We take each row vector of the first matrix times each column vector of the second matrix.

For example, the first row of the first matrix times the first column of the second matrix:

$$\begin{pmatrix} 6 & 1 & 2 \\ 2 & 4 & 7 \end{pmatrix} \begin{pmatrix} 2 & 3 \\ 5 & 7 \\ 4 & 6 \end{pmatrix} = \begin{pmatrix} 6\times2 + 1\times5 + 2\times4 & \bullet \\ \bullet & \bullet \end{pmatrix} = \begin{pmatrix} 25 & \bullet \\ \bullet & \bullet \end{pmatrix}$$

The first row times the first column gives an answer in the row 1, column 1 position.

Now we do the other three.

$$\begin{pmatrix} 6 & 1 & 2 \\ 2 & 4 & 7 \end{pmatrix} \begin{pmatrix} 2 & 3 \\ 5 & 7 \\ 4 & 6 \end{pmatrix} = \begin{pmatrix} 6\times2 + 1\times5 + 2\times4 & 6\times3 + 1\times7 + 2\times6 \\ 2\times2 + 4\times5 + 7\times4 & 2\times3 + 4\times7 + 7\times6 \end{pmatrix} = \begin{pmatrix} 25 & 37 \\ 52 & 76 \end{pmatrix}$$

Let's label the answer

$$\begin{array}{c} \\ \text{Blanche's cotton} \\ \text{Carrie's cotton} \end{array} \begin{pmatrix} \overset{\text{before}}{\underset{\text{tariff}}{}} & \overset{\text{after}}{\underset{\text{tariff}}{}} \\ 25 & 37 \\ 52 & 76 \end{pmatrix}$$

The matrix multiplication eliminated the reds, pinks, and oranges and gave us the total value of the two cotton inventories before and after the tariff.

Matrix multiplication gives us a bigger picture. We do that all the time in mathematics. Instead of saying all the little pieces: 3 + 5 = 5 + 3   1 + 7 = 7 + 1   6 + 10 = 10 + 6   5 + 4 = 4 + 5   7 + 2 = 2 + 7   44 + 53 = 53 + 44   π + 9 = 9 + π   6.2 + 9 = 9 + 6.2   ½ + ⅔ = ⅔ + ½   ln 2 + 8935 = 8935 + ln 2   we say that addition of numbers is commutative.

Instead of listing the sizes of all the classes that Fred has taught at KITTENS University over the years 38, 52, 33, 71, 46, 45, 35, 82, . . . we say that his average class size was 48.3 students.

*Your Turn to Play*

1. Here's one practice problem so that you can get used to having your left eye go ↔, and your right eye go ↕.

$$\begin{pmatrix} 2 & 7 & 0 & -3 \\ -3 & 1 & 4 & 10 \\ 0 & 2 & -3 & 8 \end{pmatrix} \begin{pmatrix} 1 & 4 \\ 9 & -2 \\ 0 & 6 \\ -4 & 5 \end{pmatrix}$$

> In matrix multiplication, the i[th] row times the j[th] column gives an answer in the i-j[th] position.

2. Suppose we are multiplying together two matrices.

$$\begin{pmatrix} \phantom{xxxx} \end{pmatrix} \begin{pmatrix} \phantom{xx} \end{pmatrix}$$

Fill in the blanks:

*The number of* ___?___ *in the first*
[rows or columns]
*matrix must match the number of*
___?___ *in the second matrix.*
[rows or columns]

3. When we say that multiplication of real numbers is commutative, we mean that for *every* pair of real numbers $r_1$ and $r_2$, it is true that $r_1 r_2 = r_2 r_1$.

Here comes a surprise.

Multiply $\begin{pmatrix} 1 & 3 \\ 5 & 4 \end{pmatrix} \begin{pmatrix} 2 & 3 \\ 9 & 0 \end{pmatrix}$ and then do them in reverse order $\begin{pmatrix} 2 & 3 \\ 9 & 0 \end{pmatrix} \begin{pmatrix} 1 & 3 \\ 5 & 4 \end{pmatrix}$

4. With the real numbers, we know that if $r_1 r_2 = 0$, then either $r_1$ or $r_2$ must be zero. Get ready for **ANOTHER SHOCK**.

Multiply $\begin{pmatrix} 1 & 1 \\ 2 & 2 \end{pmatrix} \begin{pmatrix} -1 & 1 \\ 1 & -1 \end{pmatrix}$

5. If we multiply an m×n matrix times an n×p matrix, what are the dimensions of the answer?

56

........**COMPLETE SOLUTIONS**........

1. $\begin{pmatrix} 77 & -21 \\ -34 & 60 \\ -14 & 18 \end{pmatrix}$

2. The length of those dashed lines ⸻ must be equal. That means that the number of columns in the first matrix must equal the number of rows in the second matrix.

3. In the first case you get $\begin{pmatrix} 29 & 3 \\ 46 & 15 \end{pmatrix}$ and in the second case $\begin{pmatrix} 17 & 18 \\ 9 & 27 \end{pmatrix}$

So matrix multiplication is not commutative. But, of course, there are lots of things in life that are not commutative. For example, putting on your shoes and socks gives you a different result than putting on your socks and then your shoes.

4. You get the **zero matrix** $\begin{pmatrix} 0 & 0 \\ 0 & 0 \end{pmatrix}$

(A zero matrix is a matrix in which all the entries are zero.)

5. $m \times n$ times $n \times p$ yields an $m \times p$. It's as if the $n$'s canceled.

Back in Chapter 1 (starting on page 28), Alexander was trying to find the prices of the plastic flatware at Butter Bottom Foods. Two pages later he had the system of equations $\begin{cases} x_1 + 4x_2 + 2x_3 = 46 \\ 4x_1 + 4x_2 \quad\;\; = 60 \\ 8x_1 + 8x_2 + 8x_3 = 160 \end{cases}$

Here's where we can use matrix multiplication.

Let A be the coefficient matrix $\begin{pmatrix} 1 & 4 & 2 \\ 4 & 4 & 0 \\ 8 & 8 & 8 \end{pmatrix}$ and let x be the column vector $\begin{pmatrix} x_1 \\ x_2 \\ x_3 \end{pmatrix}$

Then the whole left side of the system of equations can be written very neatly as Ax. As I promised at the beginning of this chapter, instead of hacking our way through a forest, we can get a clear-eyed view from above. Ax is neat and clean.

To finish the cleanup process, let $b = \begin{pmatrix} 46 \\ 60 \\ 160 \end{pmatrix}$

Then the whole system $\begin{array}{l} x_1 + 4x_2 + 2x_3 = 46 \\ 4x_1 + 4x_2 \phantom{+ 2x_3} = 60 \\ 8x_1 + 8x_2 + 8x_3 = 160 \end{array}$ shrinks to $Ax = b$.

**Finally I, your reader, feel relieved. I always wanted to know what that Ax = b at the top of each page of Chapter 1 meant. Thank you.**

You're welcome. The little happy face ☺ that I put there meant that we were dealing with systems that had exactly one solution. In Chapter 2 the header will look like $Ax = b$ ☺☺☺. . . since we will be working with systems that have many solutions.

To continue our overview, suppose we had some m×n matrix $A$ that looked like

$$\begin{pmatrix} a_{11} & a_{12} & a_{13} & \cdots & a_{1n} \\ a_{21} & a_{22} & a_{23} & \cdots & a_{2n} \\ \vdots & & & & \\ a_{m1} & a_{m2} & a_{m3} & \cdots & a_{mn} \end{pmatrix}$$

$a_{23}$ is a scalar. $a_{23}$ is the entry in $A$ that sits in row 2, column 3.

If we want to talk about the entry in $A$ that is in the i[th] row and j[th] column, we write $(A)_{ij}$. So in the above matrix we have $(A)_{23} = a_{23}$.

**Double subscripts! You gotta be kidding.**

Yes, but it's so much easier writing $(A)_{23}$ than "the entry in $A$ that sits in row 2, column 3.

**I guess I should polish up my subscriptsmanship. I just made up that word. It's my contribution to linear algebra.**

I like that. That's how new words enter our language. When Shakespeare wrote his plays, he invented a lot of new words. Here are some of the words that he made up: *amazement, apostrophe, assassination, bump, countless, courtship, critic, critical, exposure, frugal, generous, gloomy, hurry, impartial, lapse, laughable, lonely, majestic, misplaced, obscene, pious, radiance, reliance, road, submerge, suspicious.*

Suppose we have   A = $\begin{pmatrix} a_{11} & a_{12} & a_{13} & \ldots & a_{1n} \\ a_{21} & a_{22} & a_{23} & \ldots & a_{2n} \\ & \vdots & & & \\ a_{m1} & a_{m2} & a_{m3} & \ldots & a_{mn} \end{pmatrix}$

Then A is sometimes written as $(a_{ij})$. This is the matrix whose i-j[th] entry is $a_{ij}$. Writing $(a_{ij})$ is a lot easier than writing

$$\begin{pmatrix} a_{11} & a_{12} & a_{13} & \ldots & a_{1n} \\ a_{21} & a_{22} & a_{23} & \ldots & a_{2n} \\ & \vdots & & & \\ a_{m1} & a_{m2} & a_{m3} & \ldots & a_{mn} \end{pmatrix}$$

*Your Turn to Play*

1. If C = $(c_{ij})$ where $c_{ij} = i + 2j$ and C is a 2×3 matrix, fill in all the entries in $\begin{pmatrix} \bullet & \bullet & \bullet \\ \bullet & 6 & \bullet \end{pmatrix}$

2. If D = $(d_{ij})$ and D = $\begin{pmatrix} 1 & 2 & 3 \\ 4 & 8 & 12 \\ 9 & 18 & 27 \end{pmatrix}$   what would $d_{ij}$ equal in terms of i and j? Please try to figure this out before you look at the answer.

3. [Harder question] If A = $(a_{ij})$, B = $(b_{ij})$, C = $(c_{ij})$, and AB = C, what would $(C)_{ij}$ equal (in terms of $a_{ij}$ and $b_{ij}$). Suppose that A is m×n and that B is n×p.

(To do this problem, you may need to write out A as I did at the top of this page. Then write out B in a similar way, making it an n×p matrix. The take the i[th] row of A times the j[th] column of B.)

4. It may have been several weeks (years?) since you used sigma notation. Here are some examples to refresh your memory.

$$\sum_{i=1}^{4} x^i = x + x^2 + x^3 + x^4 \qquad \sum_{k=2}^{3} \log_k(ky) = \log_2(2y) + \log_3(3y)$$

Do the previous problem, expressing $c_{ij}$ in terms of $a_{ij}$ and $b_{ij}$, using sigma notation.

---

.......**COMPLETE SOLUTIONS**.......

1. $\begin{pmatrix} 3 & 5 & 7 \\ 4 & 6 & 8 \end{pmatrix}$    When using some computer programs to do linear algebra, you may need to input all the entries in some big matrix—some matrix that may be 20×30. Rather than type in all 600 entries, other programs often allow you to enter just the arbitrary $c_{ij}^{th}$ entry, and the program will fill in all the entries.

2. $d_{ij} = i^2 j$

3. $c_{ij} = a_{i1}b_{1j} + a_{i2}b_{2j} + \ldots + a_{in}b_{nj}$

4. $c_{ij} = \sum\limits_{k=1}^{n} a_{ik}b_{kj}$

---

"All your blankets are very nice," Fred told Carrie, "but this pink cotton one is my favorite."

"You have made a nice choice, young man," Carrie said as she took the blanket out of Fred's hands and spread it out on the counter. "Now what shall we put on it? I can embroider nice little jerboas,[*] or nice little froggies, or nice little walking ducks. Jerboas take the longest to embroider. You have got to get their tan fur just right."

Fred interjected, "I'll need the blanket by noon, because that's when the picnic will be."

"Oh," said Carrie. She turned to the pages in her sewing diary.

*One jerboa, three froggies, and two ducks took 12 minutes to do.*
*Three jerboas, four froggies and one duck took 21 minutes to do.*
*Two jerboas and eight ducks took 16 minutes to do.*

---

[*] You have to go out at night to see a jerboa. You also have to be in the right spot, which is the deserts of North Africa, Mongolia, and China. Jerboas are really weird: up to six inches long, ears that are 35% longer than their heads, kangaroo legs, and can leap three feet into the air or ten feet horizontally. They are not just mice with big feet and ears but a very distinct mammal. Dr. Jonathon Baille of the Zoological Society of London said, "There's no other animal of its type."

60

**Wait! Stop! I, your loyal reader, know where you're going. Let me cut things short. Here's the system of linear equations.**

$$x_1 + 3x_2 + 2x_3 = 12$$
$$3x_1 + 4x_2 + x_3 = 21$$
$$2x_1 \qquad + 8x_3 = 16$$

**I have only one question. Why are we looking at this again? We already have three ways to solve that system. Let me count them for you: (1) high school algebra, (2) Gauss-Jordan elimination, and (3) Gaussian elimination.**

You forgot one.

**I did not! Which one did I forget?**

Well, actually, you didn't forget one. I just haven't presented it yet.

**What are you waiting for?**

Here goes. We have a system $Ax = b$. In this case of embroidery,

$$A = \begin{pmatrix} 1 & 3 & 2 \\ 3 & 4 & 1 \\ 2 & 0 & 8 \end{pmatrix} \text{ and } x = \begin{pmatrix} x_1 \\ x_2 \\ x_3 \end{pmatrix} \text{ and } b = \begin{pmatrix} 12 \\ 21 \\ 16 \end{pmatrix}$$

You know in algebra that if we have $5w = 20$, we can solve it by multiplying both sides by $\frac{1}{5}$
$$5^{-1}(5w) = 5^{-1}(20)$$
$$w = 4$$

We do the same thing in linear algebra.

| | |
|---|---|
| We start with | $Ax = b$ |
| We multiply both sides by the inverse matrix $A^{-1}$ | $A^{-1}(Ax) = A^{-1}b$ |
| And we're done | $x = A^{-1}b$ |

*Your Turn to Play*

1. Using the above values for A, x, and b, find $A^{-1}$ and solve $Ax = b$.

**Wait! Hold your horses Mr. Author. Cancel this** *Your Turn to Play.* **You haven't told me how to find $A^{-1}$ yet.**

You're right.  I do need to explain how to get $A^{-1}$.

When you multiply 5 by its inverse $5^{-1}$ you get 1.
When you multiply $A$ by its inverse $A^{-1}$ you get the **identity matrix** I.

$$I = \begin{pmatrix} 1 & 0 & 0 \\ 0 & 1 & 0 \\ 0 & 0 & 1 \end{pmatrix}$$

Identity matrices are square matrices that come in all sizes.  The $1\times1$ identity matrix is (1).  The $500\times500$ identity matrix has 1s on the diagonal and 0s elsewhere.  Using the notation introduced on page 58:

$$(I)_{ij} = \begin{cases} 1 & \text{if } i = j \\ 0 & \text{if } i \neq j \end{cases}$$

To find $A^{-1}$   Start with $A$ and augment the matrix with I.  Then using elementary row operations, work until the left half of the matrix becomes I.  The right half will be $A^{-1}$.  We'll do it with the embroidery

matrix $A = \begin{pmatrix} 1 & 3 & 2 \\ 3 & 4 & 1 \\ 2 & 0 & 8 \end{pmatrix}$.  Augment $A$ with I and get $\begin{pmatrix} 1 & 3 & 2 & 1 & 0 & 0 \\ 3 & 4 & 1 & 0 & 1 & 0 \\ 2 & 0 & 8 & 0 & 0 & 1 \end{pmatrix}$

Now the arithmetic that only a computer could love.

$$\begin{pmatrix} 1 & 3 & 2 & 1 & 0 & 0 \\ 3 & 4 & 1 & 0 & 1 & 0 \\ 2 & 0 & 8 & 0 & 0 & 1 \end{pmatrix} \Rightarrow \begin{pmatrix} 1 & 3 & 2 & 1 & 0 & 0 \\ 0 & -5 & -5 & -3 & 1 & 0 \\ 0 & -6 & 4 & -2 & 0 & 1 \end{pmatrix} \Rightarrow \begin{pmatrix} 1 & 3 & 2 & 1 & 0 & 0 \\ 0 & -5 & -5 & -3 & 1 & 0 \\ 0 & 30 & -20 & 10 & 0 & -5 \end{pmatrix}$$

$$\Rightarrow \begin{pmatrix} 1 & 3 & 2 & 1 & 0 & 0 \\ 0 & -5 & -5 & -3 & 1 & 0 \\ 0 & 0 & -50 & -8 & 6 & -5 \end{pmatrix} \Rightarrow \begin{pmatrix} 1 & 3 & 2 & 1 & 0 & 0 \\ 0 & 1 & 1 & 0.6 & -0.2 & 0 \\ 0 & 0 & 1 & 0.16 & -0.12 & 0.1 \end{pmatrix} \Rightarrow$$

$$\begin{pmatrix} 1 & 3 & 0 & 0.68 & 0.24 & -0.2 \\ 0 & 1 & 0 & 0.44 & -0.08 & -0.1 \\ 0 & 0 & 1 & 0.16 & -0.12 & 0.1 \end{pmatrix} \Rightarrow \begin{pmatrix} 1 & 0 & 0 & -0.64 & 0.48 & 0.1 \\ 0 & 1 & 0 & 0.44 & -0.08 & -0.1 \\ 0 & 0 & 1 & 0.16 & -0.12 & 0.1 \end{pmatrix}$$

1. I used an ⇒ between each of the two matrices on the previous page. That is because the matrices are not equal. Two **matrices are equal** if they have the same dimension, and corresponding entries are equal.

Complete the definition: $A = B$ if for every i and every j, $(A)_{ij}$ =. . . .

2. Above the first ⇒ I should have written the elementary row operation(s) that I used. Looking at the first two matrices, it only takes a moment to see that I multiplied the first row by $-3$ and added it to the second row ($-3r_1 + r_2$), and also did $-2r_1 + r_3$.

Looking at the second and third matrices, guess why I didn't eliminate the $-6$ in the third row using $(-6/5)r_2 + r_3$.

3. That leaves four ⇒'s for you to figure out which of the three elementary row operations I used.

4. After all the elementary row operations we transformed $\left( \begin{matrix} A & I \end{matrix} \right)$ into $\left( \begin{matrix} I & A^{-1} \end{matrix} \right)$

I won't ask you to multiply $A^{-1}$, which is $\begin{pmatrix} -0.64 & 0.48 & 0.1 \\ 0.44 & -0.08 & -0.1 \\ 0.16 & -0.12 & 0.1 \end{pmatrix}$

times A, which is $\begin{pmatrix} 1 & 3 & 2 \\ 3 & 4 & 1 \\ 2 & 0 & 8 \end{pmatrix}$

in order to show that it equals I which is $\begin{pmatrix} 1 & 0 & 0 \\ 0 & 1 & 0 \\ 0 & 0 & 1 \end{pmatrix}$.

Two pages back in that pretty ‖ double-bordered box ‖ we started

with $Ax = b$ and finished with $x = A^{-1}b$.

To finish this problem, we need to multiply $A^{-1}b$. So you don't have

to turn back two pages, here is b: $\begin{pmatrix} 12 \\ 21 \\ 16 \end{pmatrix}$    Do the multiplication.

```
. . . . . . .COMPLETE SOLUTIONS . . . . . . .
```

1. A = B if for every i and every j, $(A)_{ij} = (B)_{ij}$.

2. I am lazy. Plain lazy. If I did $(-6/5)r_2 + r_3$, I would be facing either fractions or decimals, and I want to avoid them as long as possible. Being lazy is one reason I chose to be a math major in college. English majors have to read zillions of books. My eyes would wear out doing that. Math majors read three pages for every hundred that English majors read.

History majors have to write long papers. They type so much that they get short fingers.

In biology and French, you have to memorize tons of words. Do you remember the difference between *mitosis* and *meiosis*? [Mitosis is normal cell division. Meiosis is like mitosis except it's more like a divorce. Each of the new cells gets half of the "goodies." Did your biology teacher explain it that way?]

Culinary arts is even worse than biology. In both fields you cut up dead things, but in culinary arts you have to eat them afterward.

So in going from the second to the third matrix, I multiplied the third row by –5, which I abbreviate as $-5r_3$.

3. Third arrow: $6r_2 + r_3$
Fourth arrow: $(-1/5)r_2$   and   $(-1/50)r_3$
Fifth arrow: $-2r_3 + r_1$   and   $-r_3 + r_2$
Sixth arrow: $-3r_2 + r_1$

4. Did you take out a piece of paper and actually do the multiplication? I hope so. Some books have a million matrix multiplication problems so that you can get used to having your left eye go ↔, and your right eye go ↕. In the math textbook writing field, it's called drill & kill.

I'm saying *please*.

$$\begin{pmatrix} -0.64 & 0.48 & 0.1 \\ 0.44 & -0.08 & -0.1 \\ 0.16 & -0.12 & 0.1 \end{pmatrix} \begin{pmatrix} 12 \\ 21 \\ 16 \end{pmatrix}$$

Having done so, we find that it takes 4 minutes to embroider a jerboa,

2 minutes for a froggie, and 1 minute for a duck.

We need a word to describe two matrices that are connected by an arrow ➠ where one matrix was obtained from the other by one or more elementary row operations. We can't call them equal. Instead, they are called **row-equivalent**.

We now have four ways of solving a system of linear equations: (1) high school algebra, (2) Gauss-Jordan elimination, (3) Gaussian elimination, and (4) starting with $Ax = b$, find $A^{-1}$ and then compute $A^{-1}b$.

Let me spend a couple more seconds on the stuff in the pretty

| double-bordered box | from page 61.

We started with $Ax = b$.

Then we multiplied both sides by $A^{-1}$ and got $A^{-1}(Ax) = A^{-1}b$.

At this point, I had skipped a step. The next line should have read $(A^{-1}A)x = A^{-1}b$.

The left side of the equation has three matrices multiplied together. I then moved the parentheses. In algebra, we know that a(bc) = (ab)c. It is called the associative law of multiplication. But is it also true with matrix multiplication? In contrast, we know that the commutative law of multiplication doesn't hold. $AB$ doesn't in general equal $BA$.*

The answer is . . . yes. Matrix multiplication is associative.

That means, for any three matrices $A$, $B$, and $C$, $(AB)C = A(BC)$. This gives us permission to drop the parentheses if we want to and just write $ABC$.

Here's the double-bordered box in all its glory.

| | |
|---|---|
| We start with | $Ax = b$ |
| We multiply both sides by the inverse matrix $A^{-1}$ | $A^{-1}(Ax) = A^{-1}b$ |
| By the associative law of multiplication for matrices | $(A^{-1}A)x = A^{-1}b$ |
| Using the definition of $A^{-1}$ | $Ix = A^{-1}b$ |
| Using the property of the identity matrix (see next page) | $x = A^{-1}b$ |

---

✱ If you graduate from college with a Bachelor of Arts degree, it is often abbreviated B.A. (This can also stand for batting average or Buenos Aires.) But it is also abbreviated as A.B. (from *Artium Baccalaureus*). Either way is correct. So in contrast to matrix multiplication, A.B. does equal B.A.

1. The identity element in a particular context depends on which operation you are using.

   If we are using multiplication in the real numbers, the identity is the number 1. So, for example, $5 \cdot 1 = 1 \cdot 5 = 5$.

   What is the identity for multiplication in the complex numbers? [The complex numbers are any numbers that can be written as a + b$i$ where a and b are real numbers and $i = \sqrt{-1}$.]

2. In general, the **identity element** for some operation, call it $\odot$, and some set, call it A, is some element i contained in A, written i $\in$ A, such that for every a $\in$ A, i$\odot$a = a$\odot$i = a.

   Using even more symbols, we could make the definition: i is the identity element for some set A with operation $\odot$ iff a$\in$A $\Rightarrow$ i$\odot$a = a$\odot$i = a. [Translation: "iff" is mathematical shorthand for "if and only if." The word *iff* can be found in most good dictionaries. $\Rightarrow$ is the symbol for "implies."]

   What's the identity element for addition in the integers? [The integers, which are often abbreviated $\mathbb{Z}$, are . . . –2, –1, 0, 1, 2, 3, 4. . . .]

3. Is there an identity for subtraction in $\mathbb{Z}$?

4. We have asserted that $\begin{pmatrix} 1 & 0 & 0 \\ 0 & 1 & 0 \\ 0 & 0 & 1 \end{pmatrix}$ is the identity for matrix multiplication for 3×3 matrices. That means that for any 3×3 matrix **A** where $(\mathbf{A})_{ij} = a_{ij}$, we have

$$\begin{pmatrix} 1 & 0 & 0 \\ 0 & 1 & 0 \\ 0 & 0 & 1 \end{pmatrix}\begin{pmatrix} a_{11} & a_{12} & a_{13} \\ a_{21} & a_{22} & a_{23} \\ a_{31} & a_{32} & a_{33} \end{pmatrix} = \begin{pmatrix} a_{11} & a_{12} & a_{13} \\ a_{21} & a_{22} & a_{23} \\ a_{31} & a_{32} & a_{33} \end{pmatrix}\begin{pmatrix} 1 & 0 & 0 \\ 0 & 1 & 0 \\ 0 & 0 & 1 \end{pmatrix} = \begin{pmatrix} a_{11} & a_{12} & a_{13} \\ a_{21} & a_{22} & a_{23} \\ a_{31} & a_{32} & a_{33} \end{pmatrix}$$

   Look over these multiplications mentally for the next 23 seconds and satisfy yourself that $\begin{pmatrix} 1 & 0 & 0 \\ 0 & 1 & 0 \\ 0 & 0 & 1 \end{pmatrix}$ really is the identity.

5. (Genius-level question) Show matrix multiplication is associative. (AB)C = A(BC)    Let **A** be m×n, **B** be n×p, and **C** be p×q.

Hints: From what we did on page 59 in problem 4, we have that

$$(AB)_{ij} = \sum_{\alpha=1}^{n} a_{i\alpha} b_{\alpha j}$$    I'm using $\alpha$ as the indexing variable instead of k,

otherwise, I start to run out of letters.  The other item you will need is the distributive law.  In algebra, it was expressed as $a(b+c) = ab + ac$.

The extended distributive law is

$a(x_1 + x_2 + \ldots + x_n) = ax_1 + ax_2 + \ldots + ax_n$ which can be written as

$a\sum_{k=1}^{n} x_k = \sum_{k=1}^{n} ax_k$.  If your IQ is less than 150, you are excused from

attempting this problem.

### . . . . . . . **COMPLETE SOLUTIONS** . . . . . . .

1.  The identity for multiplication in the complex numbers is also the number 1.  For example, $1(3+4i) = (3+4i)1 = 3+4i$.

2.  The identity for addition in the integers is the number 0.  For example, $0 + (-7) = (-7) + 0 = -7$.

3.  No there isn't.  If there were some integer, say i, that was the identity for subtraction, then, for example, $i - 5$ would have to equal $5 - i$.

4.  Mentally, check the appropriate box:

☐ I looked and am satisfied that it's the identity.

☐ I didn't really spend the 23 seconds looking.  I trust you, Mr. Author.

☐ Look at what?

5.  To show $(AB)C = A(BC)$   where A is m×n, B is n×p, and C is p×q.

$(AB)_{ij} = \sum_{\alpha=1}^{n} a_{i\alpha} b_{\alpha j}$.  For the moment let $AB = D$.  So $(AB)_{ij} = (D)_{ij} = d_{ij}$.

In other words, $d_{ij} = \sum_{\alpha=1}^{n} a_{i\alpha} b_{\alpha j}$.

We will now compare an arbitrary i-k[th] entry in $(AB)C$ with the i-k[th] entry in $A(BC)$.  If we show that they are equal, then we have shown that the two matrices, $(AB)C$ and $A(BC)$, are equal.  (Back on page 63, we defined two matrices to be equal if their corresponding entries are equal.)

$$((AB)C)_{ik} = ((D)C)_{ik} = \sum_{j=1}^{p} d_{ij} c_{jk} = \sum_{j=1}^{p} \left( \sum_{\alpha=1}^{n} a_{i\alpha} b_{\alpha j} \right) c_{jk}$$

using the extended distributive law $$= \sum_{j=1}^{p} \left( \sum_{\alpha=1}^{n} a_{i\alpha} b_{\alpha j} c_{jk} \right)$$

Now we start with $(A(BC))_{ik}$

For any $\alpha$ and k, by the definition of matrix multiplication,

$$(BC)_{\alpha k} = \sum_{j=1}^{p} b_{\alpha j} c_{jk}$$

$$(A(BC))_{ik} = \sum_{\alpha=1}^{n} a_{i\alpha} \left( \sum_{j=1}^{p} b_{\alpha j} c_{jk} \right)$$

using the extended distributive law $$= \sum_{\alpha=1}^{n} \left( \sum_{j=1}^{p} a_{i\alpha} b_{\alpha j} c_{jk} \right)$$

Comparing this with the double summation at the top of this page, the only difference is that the order of the summations has been switched. All that means is that the order in which the $a_{i\alpha} b_{\alpha j} c_{jk}$ terms are added up has been changed. That's no big deal since addition is commutative.

We have shown that the i-k[th] entry in $(AB)C$ and the i-k[th] entry in $A(BC)$ are equal. Matrix multiplication is associative.

Q.E.D.   (which does not stand for quacky duck)

Fred asked Carrie to embroider three ducks on the pink cotton blanket.

"I'll be back in three minutes," Carrie announced as she headed into the back room to do the work.

Fred noticed a sampler on the wall of the store.

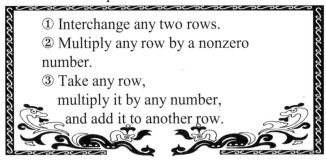

① Interchange any two rows.
② Multiply any row by a nonzero number.
③ Take any row,
    multiply it by any number,
    and add it to another row.

Wow Fred thought to himself. I thought that the only samplers that people embroidered were the ones that read "Home Sweet Home." I wonder if it's for sale. I think it would be neat to show to my linear algebra class on Monday. Good old elementary row ops.

68

Elementary row operations were around long before computers. In the old days, you could give a class a matrix A and ask them to do an elementary row operation. They would respond:

Hey! That's easy!

But if A were, say, $\begin{pmatrix} 1 & 2 & 3 \\ 4 & 5 & 6 \\ 7 & 8 & 9 \end{pmatrix}$ and you asked a computer to

interchange rows 2 and 3, you would get a different response:

Does not compute. I do not speak English.* Please place your request in a language I speak.

But computers can multiply matrices. That language they do speak. The question is can we express the three elementary row operations in terms of multiplication by some special matrices?

**I, your reader, am going to guess, "Yes." Otherwise, you would have never brought this up.**

Good guess.

So what matrix do I multiply $\begin{pmatrix} 1 & 2 & 3 \\ 4 & 5 & 6 \\ 7 & 8 & 9 \end{pmatrix}$ by in order to interchange

rows 2 and 3? Let's call that mystery matrix M.

We want MA to equal $\begin{pmatrix} 1 & 2 & 3 \\ 7 & 8 & 9 \\ 4 & 5 & 6 \end{pmatrix}$

The hard way to figure out what M is, would be to just experiment around, trying various 3×3 matrices until we found one that worked.

---

✱ Computers are getting to be more and more like humans. They have their "bad days" where nothing you do can please them. They also have their illogical moments. How else could you explain a computer saying in English, "I do not speak English"?

Here's an easier approach that will work for all three elementary row operations.

We want MA to be the result of interchanging rows 2 and 3 of A.

First, replace A by IA where I is the identity matrix. We know that A = IA.

We now have MIA. I don't have to use parentheses, because we have shown that matrix multiplication is associative.

What's MI? Since M is the mystery matrix that interchanges rows 2 and 3,

$$\text{MI will be} \begin{pmatrix} 1 & 0 & 0 \\ 0 & 0 & 1 \\ 0 & 1 & 0 \end{pmatrix}$$

Putting that all together: $\text{MA} = \text{MIA} = \begin{pmatrix} 1 & 0 & 0 \\ 0 & 0 & 1 \\ 0 & 1 & 0 \end{pmatrix} \text{A}$

In English, for all my human readers, if you want the matrix that will perform an elementary row operation, just do that elementary row operation on I.

The matrix that will multiply the second row of A by 17 and add it to the first row is $\begin{pmatrix} 1 & 17 & 0 \\ 0 & 1 & 0 \\ 0 & 0 & 1 \end{pmatrix}$

These matrices are called **elementary matrices**.

Now we can talk to computers in a language they can understand.

## 𝕬 𝕿𝖍𝖊𝖔𝖗𝖊𝖒

On page 62, we started with some matrix A and augmented it with the identity matrix $\begin{pmatrix} \text{A} & \text{I} \end{pmatrix}$ and then we did elementary row operations on that augmented matrix until the left half of the matrix became I. Then I wrote, "The right half will be $\text{A}^{-1}$." Those were my very words. And you believed me. $\begin{pmatrix} \text{I} & \text{A}^{-1} \end{pmatrix}$

Now, using elementary matrices, I can prove that this is true.

# Proof

In order to turn A into I, we did a bunch of elementary row operations. Each elementary row operation had the same result as multiplying on the left by an elementary matrix. If we call those elementary matrices $E_1$, $E_2$, . . . , $E_n$, then $E_nE_{n-1} \ldots E_2E_1A$ equals I. But $E_nE_{n-1} \ldots E_2E_1A = I$ means that $E_nE_{n-1} \ldots E_2E_1$ has to equal $A^{-1}$, because $E_nE_{n-1} \ldots E_2E_1$ times A gives you I.

Now what will those same elementary row operations do to the identity matrix?

Or, in terms of matrix multiplication, what does $E_nE_{n-1} \ldots E_2E_1I$ equal? It equals $A^{-1}I$, which equals $A^{-1}$.

$$w^5 \quad (= \text{which was what was wanted})^*$$

Carrie came back into the store humming a tune about three nice duckies. She placed the blanket on a table. "I wonder where the little guy went."

Fred was nowhere to be seen.

Of course, you wouldn't be visible either if you were covered with a blanket.

---

*Your Turn to Play*

1. What is the elementary $3 \times 3$ matrix that corresponds to the elementary row operation *multiply row 3 by 26*?

2. Find the matrix that is the inverse to the answer you gave for the previous question.

3. What is the elementary $4 \times 4$ matrix that corresponds to *add 5 times row 4 to row 2*?

4. Find the matrix that is the inverse to the answer you gave for the previous question. As usual, please don't peek at the answer until you have worked it out for yourself on a piece of paper.

---

\* There are several ways to announce the end of a proof so that your reader will know that you are done talking. Q.E.D. is the most traditional. Some mathematicians have never heard of $w^5$. Many textbooks use the symbol ∎ to mark the end of a proof. On the blackboard, teachers often write ⊠. That saves the effort of having to color in the whole square.

.......**COMPLETE SOLUTIONS**.......

1. $\begin{pmatrix} 1 & 0 & 0 \\ 0 & 1 & 0 \\ 0 & 0 & 26 \end{pmatrix}$

2. There are two ways you might go about solving this problem. The first way is to set up some mystery matrix M so that M times your answer for question 1 would equal I. In other words, $\begin{pmatrix} ? & ? & ? \\ ? & ? & ? \\ ? & ? & ? \end{pmatrix}\begin{pmatrix} 1 & 0 & 0 \\ 0 & 1 & 0 \\ 0 & 0 & 26 \end{pmatrix} = I$

and then try to figure out the values of the nine ?'s in M.

A second way would be to try and figure out what the inverse operation to *multiply row 3 by 26* would be.

In either case, the answer will be $\begin{pmatrix} 1 & 0 & 0 \\ 0 & 1 & 0 \\ 0 & 0 & \frac{1}{26} \end{pmatrix}$

3. $\begin{pmatrix} 1 & 0 & 0 & 0 \\ 0 & 1 & 0 & 5 \\ 0 & 0 & 1 & 0 \\ 0 & 0 & 0 & 1 \end{pmatrix}$

4. $\begin{pmatrix} 1 & 0 & 0 & 0 \\ 0 & 1 & 0 & -5 \\ 0 & 0 & 1 & 0 \\ 0 & 0 & 0 & 1 \end{pmatrix}$

Fred didn't know what was the polite thing to do. Having a head that looked a little like a table top and being accidentally covered by a blanket with three ducks on it wasn't exactly something you could look up in the index of some book on manners.

But Fred tried. Out of his back pocket, he had pulled his copy of Prof. Eldwood's *Authoritative Guide to the Polite Way to Act in Difficult Situations*, 1845.

Using a flashlight, he looked at the index to the book:

Nothing in the index seemed to fit exactly. Fred just pretended nothing had happened. He said to Carrie, "Excuse me. How much is that lovely sampler you have on the wall, the one with the elementary row operations embroidered on it?"

"Goodness gracious young man! What are you doing playing underneath the blanket?" She pulled the blanket off of him, ripping it slightly on his sharp nose. "Oh that sampler. My daughter did that sampler when she was taking a linear algebra class at KITTENS University. She told me about a funny little man named Professor Gauss who was teaching the course. Some day when you get old enough, you might take a class from him.

"Oh. You asked about the price," she continued. "I don't know what that ①, ② and ③ thing is all about. I can let you have it for two bits.* "

Fred nodded, and Carrie put the blanket and the sampler in a large bag.

Fred stood outside Carrie's Comfy Covers in a very familiar situation.

Just purchased from Carrie's

Fred called the three moving companies again. They were still using the same old formulas for estimating the cost of moving a sack.

t = minutes it would take
w = pounds
d = miles.

---

* Old people like Carrie, who is in her 40s, sometimes use old-fashioned language. Two bits = 25¢.

The first company still used $0.5t + 0.2w + 0.4d$ and gave him an estimate of $17.40.

The second company still used $0.3t + 0.3w + 0.3d$ and gave him an estimate of $11.70.

The third company still used $0.1t + 0.01w + 10d$ and gave him an estimate of $33.06.

We rename the t, w, and d as $x_1$, $x_2$, and $x_3$. When the moving companies made their estimates of moving the sack of Butter Bottom's Sack-o-Picnic Food back on page 36, we had

$$\text{A}x = \begin{pmatrix} \$26.80 \\ \$21.60 \\ \$24.30 \end{pmatrix} \quad \text{but now we have } \text{A}x = \begin{pmatrix} \$17.40 \\ \$11.70 \\ \$33.06 \end{pmatrix}$$

**I, your reader, would like to guess which method you'll use.**

**It won't be the old high school algebra approach in which you have to write out all the variables.**

**It won't be the Gauss-Jordan elimination method in which you create zeros below and above the diagonal. That takes too long. As you pointed out, we should chop off the Jordan to get Gaussian elimination where you only get zeros below the diagonal.**

**If we had to solve the two systems that you list above, then you would make the double-augmented matrix** $\begin{pmatrix} 0.5 & 0.2 & 0.4 & 26.80 & 17.40 \\ 0.3 & 0.3 & 0.3 & 21.60 & 11.70 \\ 0.1 & 0.01 & 10 & 24.30 & 33.06 \end{pmatrix}$

**but I have the sneaking suspicion that we may see the moving companies many times. Back on page 36, we had** $\text{A}x = b_1$ **and now we have** $\text{A}x = b_2$ **and tomorrow we'll have** $\text{A}x = b_3$, **and the day after that we'll have** $\text{A}x = b_4$.

**I know! We'll use our most recent method where we started with** $\text{A}x = b$ **and found** $\text{A}^{-1}$. **We noted that** $\text{A}x = b$ **was the same as** $x = \text{A}^{-1}b$. **So now having found** $\text{A}^{-1}$, **we could use it over and over again to find** $\text{A}^{-1}b_1$ **and** $\text{A}^{-1}b_2$ **and** $\text{A}^{-1}b_3$ **and** $\text{A}^{-1}b_4$ **for as many different sacks as you want to have hauled. One simple matrix multiplication by** $\text{A}^{-1}$ **and we solve each new linear system of equations for every new b in** $\text{A}x = b$.

**Am I brilliant or am I brilliant?**

I would have to say that you know the four methods of solving systems of linear equations. But I know a fifth method that's even easier than finding $A^{-1}$ and multiplying $A^{-1}b$.

**Easier? I'm all ears.**

The fifth method is known as the lower-upper matrix decomposition method or **LU-decomposition** for short. Or even shorter, **LU**. (Some of Fred's students started calling it the "Love You" method.)

Whenever you have $Ax = b_i$ for more than one i, the *LU*-decomposition is the shortest approach.

*LU* is really just the Gaussian elimination method with a little extra bookkeeping.

We are going to solve the linear system for the moving companies' bids to move the Carrie's Comfy Covers sack to the picnic ground.

$$Ax = \begin{pmatrix} \$17.40 \\ \$11.70 \\ \$33.06 \end{pmatrix} \quad \text{where A is} \begin{pmatrix} 0.5 & 0.2 & 0.4 \\ 0.3 & 0.3 & 0.3 \\ 0.1 & 0.01 & 10 \end{pmatrix}$$

To decompose A by *LU*-decomposition, we perform Gaussian elimination with <u>three tiny changes</u> in the procedure:

This → → → →
is how we are
going to break A
into two matrices
L and U.

L will be
lower-triangular.

> (i) Never use ① Interchange any two rows. *You can only use ② Multiply any row by a nonzero number or ③ Take any row, multiply it by any number, and add it to another row.*
>
> (ii) Work on it column-by-column. Get a 1 on the diagonal and zeros below. *Do the first column, then do the second column, then, the third, etc.*
>
> (iii) Just before you work on a column, make a copy of the elements of that column that are at or below the diagonal. *Some students use their digital cameras, but that seems a little extreme.*

U will be **unit-upper-triangular**, which means it will have 1s on the diagonal.

It will be a lot clearer when you see it done on the next page.

Here's the **A** that we are going to decompose $\begin{pmatrix} 0.5 & 0.2 & 0.4 \\ 0.3 & 0.3 & 0.3 \\ 0.1 & 0.01 & 10 \end{pmatrix}$

Using (iii) in the box on the previous page, we take a picture of the first column at or below the diagonal.

$\begin{pmatrix} 0.5 & 0.2 & 0.4 \\ 0.3 & 0.3 & 0.3 \\ 0.1 & 0.01 & 10 \end{pmatrix}$

**Hey, Mr. Author. Not all of us are rich and can afford a camera that can take pictures of matrices. That's one reason why we bought your Life of Fred: Linear Algebra—it was cheap.**

The less expensive approach is just to copy those entries on a piece of scratch paper.

**I like that. Cheap and easy. Here goes:**

0.5
0.3
0.1

Now we start with the first column and put a 1 on the diagonal. The elementary row operation is $2r_1$.

$\begin{pmatrix} 1 & 0.4 & 0.8 \\ 0.3 & 0.3 & 0.3 \\ 0.1 & 0.01 & 10 \end{pmatrix}$

Clean out the rest of column one.
$-0.3r_1 + r_2$
$-0.1r_1 + r_3$

$\begin{pmatrix} 1 & 0.4 & 0.8 \\ 0 & 0.18 & 0.06 \\ 0 & -0.03 & 9.92 \end{pmatrix}$

It's photo time before we start working on column two. We copy only entries at or below the diagonal.

$\begin{pmatrix} 1 & 0.4 & 0.8 \\ 0 & 0.18 & 0.06 \\ 0 & -0.03 & 9.92 \end{pmatrix}$

Put a 1 on the diagonal of column two.   $(1/0.18)r_2$

$\begin{pmatrix} 1 & 0.4 & 0.8 \\ 0 & 1 & ⅓ \\ 0 & -0.03 & 9.92 \end{pmatrix}$

**Hey! You are mixing fractions and decimals! Is that legal?**

I know that computers don't do that. They print out everything in decimals. I'm trying to make it easy on myself. I would not like to have to write: $\begin{pmatrix} 1 & 0.4 & 0.8 \\ 0 & 1 & 0.333333333333333333333333333333333333333333333333333333333333333333 \\ 0. & -0.03 & 9.92 \end{pmatrix}$

0.03$r_2$ + $r_3$ and it's photo time again.
We take the picture just before we start
working on a new column.  We copy only
the entries at or below the diagonal.

$$\begin{pmatrix} 1 & 0.4 & 0.8 \\ 0 & 1 & \frac{1}{3} \\ 0 & 0 & 9.93 \end{pmatrix}$$

Put a 1 on the diagonal
of column three.   (1/9.93)$r_3$
And this is the unit-upper-triangular matrix U.

$$\begin{pmatrix} 1 & 0.4 & 0.8 \\ 0 & 1 & \frac{1}{3} \\ 0 & 0 & 1 \end{pmatrix}$$

And L?  The lower-triangular matrix is

$$\begin{pmatrix} 0.5 & 0 & 0 \\ 0.3 & 0.18 & 0 \\ 0.1 & -0.03 & 9.93 \end{pmatrix}$$

---

*Your Turn to Play*

1.  We have now decomposed A into LU.  I.e., A = LU.
Let's check that that is true.

Multiply $\begin{pmatrix} 0.5 & 0 & 0 \\ 0.3 & 0.18 & 0 \\ 0.1 & -0.03 & 9.93 \end{pmatrix}\begin{pmatrix} 1 & 0.4 & 0.8 \\ 0 & 1 & \frac{1}{3} \\ 0 & 0 & 1 \end{pmatrix}$

and see if you really do get A.

. . . . . . . **COMPLETE SOLUTION** . . . not

1.  Many linear algebra books require their readers to do a million of these
problems in order to win their M.M.M. degree (Master of Matrix
Multiplication).  It does take some actual practice to learn to make your
left eye go ↔, and your right eye go ↕.
 **But I know how to do matrix multiplication!**
    And a 16-year-old who has been behind the wheel twice knows how to
drive.

---

We have now gone from Ax = b to LUx = b.

We can throw **A** into the garbage can.  Solving **LUx = b** is so easy that your kid brother who just started beginning algebra can do it.  The steps are very mechanical and always the same.

   **LUx = b** becomes **Ly = b**        . . . temporarily replacing **Ux** by **y**

so **Ly = b**  is
$$\begin{pmatrix} 0.5 & 0 & 0 \\ 0.3 & 0.18 & 0 \\ 0.1 & -0.03 & 9.93 \end{pmatrix} \begin{pmatrix} y_1 \\ y_2 \\ y_3 \end{pmatrix} = \begin{pmatrix} 17.40 \\ 11.70 \\ 33.06 \end{pmatrix}$$

which in the language almost anyone can understand is
$$\begin{cases} 0.5y_1 & = 17.40 \\ 0.3y_1 + 0.18y_2 & = 11.70 \\ 0.1y_1 - 0.03y_2 + 9.93y_3 & = 33.06 \end{cases}$$

---

from high school algebra
### Forward-Substitution

We know $y_1$ is $\dfrac{17.40}{0.5}$ which is 34.8.

Put that value in the next equation
$$0.3(34.8) + 0.18y_2 = 11.70$$
and solve.  Then $y_2$ is 7.

Put these two values in the last equation
$$0.1(34.8) - 0.03(7) + 9.93y_3 = 33.06$$
and solve.  Then $y_3$ is 3.

   It's called *forward*-substitution because we keep going forward starting from the first equation.

---

And since **Ux = y** we get
$$\begin{pmatrix} 1 & 0.4 & 0.8 \\ 0 & 1 & \frac{1}{3} \\ 0 & 0 & 1 \end{pmatrix} \begin{pmatrix} x_1 \\ x_2 \\ x_3 \end{pmatrix} = \begin{pmatrix} 34.8 \\ 7 \\ 3 \end{pmatrix}$$

which in the language your kid brother can understand is
$$\begin{cases} x_1 + 0.4x_2 + 0.8x_3 & = 34.8 \\ x_2 + (\frac{1}{3})x_3 & = 7 \\ x_3 & = 3 \end{cases}$$

which by back-substitution gives $x_3 = 3$ miles to haul the sack from Carrie's to the picnic ground.  $x_2 = 6$-pound sack.  $x_1 = 30$ minutes to haul.

*Your Turn to Play*

1. Solve $\begin{cases} 2x_1 + x_2 + 4x_3 = 20 \\ 3x_1 - 6x_2 + 8x_3 = 10 \\ 2x_1 + 7x_2 - 5x_3 = -1 \end{cases}$

using *LU*-decomposition.

. . . . . . . **COMPLETE SOLUTIONS** . . . . . . .

1. First, we decompose A into LU.

Photo time
$$\begin{pmatrix} 2 & 1 & 4 \\ 3 & -6 & 8 \\ 2 & 7 & -5 \end{pmatrix}$$

Put a 1 on the diagonal.
$(1/2)r_1$
$$\begin{pmatrix} 1 & 0.5 & 2 \\ 3 & -6 & 8 \\ 2 & 7 & -5 \end{pmatrix}$$

Clean out the rest of column one.
$-3r_1 + r_2$
$-2r_1 + r_3$
$$\begin{pmatrix} 1 & 0.5 & 2 \\ 0 & -7.5 & 2 \\ 0 & 6 & -9 \end{pmatrix}$$

Photo time
$$\begin{pmatrix} 1 & 0.5 & 2 \\ 0 & -7.5 & 2 \\ 0 & 6 & -9 \end{pmatrix}$$

Put a 1 on the diagonal.
$r_2 /(-7.5)$
which could have been written $(1/(-7.5))r_2$
$$\begin{pmatrix} 1 & 0.5 & 2 \\ 0 & 1 & -0.2667 \\ 0 & 6 & -9 \end{pmatrix}$$

It looks like we got stuck with some big decimals in row 2. I have rounded it off to four decimal places. How much you round off depends on several things.  If they are paying you big bucks to do linear algebra, then give them $-0.26666666666666666666666666667$.

Clean out the rest of column two.

$-6r_2 + r_3$

$$\begin{pmatrix} 1 & 0.5 & 2 \\ 0 & 1 & -0.2667 \\ 0 & 0 & -7.4 \end{pmatrix}$$

Photo time

$$\begin{pmatrix} 1 & 0.5 & 2 \\ 0 & 1 & -0.2667 \\ 0 & 0 & -7.4 \end{pmatrix}$$

Put a 1 on the diagonal.

$r_3 / (-7.4)$

$$\begin{pmatrix} 1 & 0.5 & 2 \\ 0 & 1 & -0.2667 \\ 0 & 0 & 1 \end{pmatrix}$$

So A = LU = $\begin{pmatrix} 2 & 0 & 0 \\ 3 & -7.5 & 0 \\ 2 & 6 & -7.4 \end{pmatrix} \begin{pmatrix} 1 & 0.5 & 2 \\ 0 & 1 & -0.2667 \\ 0 & 0 & 1 \end{pmatrix}$

LUx = b   becomes Ly = b    ... temporarily replacing Ux by y.

Using forward-substitution, we solve Ly = b

$$\begin{pmatrix} 2 & 0 & 0 \\ 3 & -7.5 & 0 \\ 2 & 6 & -7.4 \end{pmatrix} \begin{pmatrix} y_1 \\ y_2 \\ y_3 \end{pmatrix} = \begin{pmatrix} 20 \\ 10 \\ -1 \end{pmatrix}$$

and we obtain $y_1 = 10$, $y_2 = 2.6667$, $y_3 = 5$.

Using back-substitution, we solve Ux = y

$$\begin{pmatrix} 1 & 0.5 & 2 \\ 0 & 1 & -0.2667 \\ 0 & 0 & 1 \end{pmatrix} \begin{pmatrix} x_1 \\ x_2 \\ x_3 \end{pmatrix} = \begin{pmatrix} 10 \\ 2.667 \\ 5 \end{pmatrix}$$

and we obtain $x_3 = 5$, $x_2 = 4$, $x_1 = -2$.

Looking at the solution to the above *Your Turn to Play*, you might notice that about 75% of the solution was devoted to decomposing A into LU, and 25% to doing the forward- and back-substitutions.

Some people who like to do such things have sat down and figured out how many multiplications and how many additions are needed to convert A into LU.  It depends, of course, on how big A is.  If A is an n × n matrix, then the arithmetic varies as the *cube* of n.

Those same people have counted the number of multiplications and additions that are needed to finish the problem once the LU is found.  The arithmetic varies as the *square* of n.

Putting together the previous two paragraphs, we can say that for industrial-sized systems of linear equations (with, say, a hundred equations and a hundred unknowns) 99% of the time will be spent in decomposing A into LU.

**I, your reader, have one last question, before we head off to the Cities at the end of this chapter. Six pages ago you said** (i) Never use ① Interchange any two rows. **Those were your very words when we changed A into LU.  (I'm not going to use the word "decompose.")  Why can't I interchange rows?**

We are solving Ax = b, but *LU* only messes with the A and not the b.  Imagine if I started with $\begin{cases} 2x_1 + 3x_2 = 8 \\ 4x_1 + 5x_2 = 9 \end{cases}$ and interchanged the rows of A only and got $\begin{cases} 4x_1 + 5x_2 = 8 \\ 2x_1 + 3x_2 = 9 \end{cases}$

**Okay.  I have a second question.  Suppose I'm working on the A and I get to** $\begin{pmatrix} 1 & 8 & 7 \\ 0 & 0 & 5 \\ 0 & 3 & 4 \end{pmatrix}$ **How in the world do I get a 1 in the row 2, column 2 position?  Normally, I would interchange row 2 with row 3, but you tell me I'm not allowed to do that when I'm using *LU*-decomposition.**

If you are halfway through the decomposition of A, and you hit a zero on the diagonal, then just back up ( back up ) and rearrange the original equations.  This will move the entries in b as well as those in Ax.

None of your work on A will be lost.  Not a single addition or multiplication.  And now, magically, there won't be a zero on the diagonal.

**What if there are zeros below the zero, like** $\begin{pmatrix} 1 & 8 & 7 \\ 0 & 0 & 5 \\ 0 & 0 & 4 \end{pmatrix}$ **????**

That's what Chapters 2 and 3 are all about.

1. Fred wanted to join Pat's choir. Pat had her choir roster all filled out as a matrix: $\begin{pmatrix} 8 & 9 & 2 & 0 \\ 0 & 0 & 8 & 8 \end{pmatrix}$ where the first row was *female* and where the columns were *soprano, alto, tenor,* and *bass.* At the age of 6, Fred's voice hadn't changed yet. He sang in the soprano range. Show the matrix addition that Pat performed as she added Fred to the choir. As usual, please don't look at the answer until you have tried it on your own.

2. Fred was frightened when he went to his first choir practice. He was afraid that Pat would have him audition in front of everybody. He thought his voice was kind of squeaky, and he didn't want to do any solo performances. Pat announced, "We have a new choir member tonight. His name is Fred." Everyone clapped. Pat motioned to Fred that he should join the sopranos in the second row.

　　　　The seating chart looked like: B B B B - T T - -

　　　　　　　　　　　　　　　　　　S S S S S A A A -

where "-" is a blank seat. Fred looked and gasped. There are no empty seats in the soprano section! he thought to himself. One of the women lifted him up and set Fred on her lap. The population of each chair could be represented by $\begin{pmatrix} 1 & 1 & 1 & 1 & 0 & 1 & 1 & 0 & 0 \\ 1 & 2 & 1 & 1 & 1 & 1 & 1 & 1 & 0 \end{pmatrix}$ which we will call A.

　　　　The big basses in the front row were blocking the view, so Pat had the basses and tenors sit behind the women (and Fred). The population of each chair now looked like $\begin{pmatrix} 1 & 2 & 1 & 1 & 1 & 1 & 1 & 1 & 0 \\ 1 & 1 & 1 & 1 & 0 & 1 & 1 & 0 & 0 \end{pmatrix}$ which we'll call B.

　　　　We need a **permutation matrix** P, so that PA = B. Find P.

3. Pat liked Russian, German, and American songs. In her choir notebook she kept track of how many copies she had of each type of song. She wrote (300  250  400). Fill in the two blanks: This matrix has only one _____, and matrices that have only one row or only one column are called_____. (This was defined on page 49.)

4. With Pat's permission, a bass borrowed 10 copies of the Russian songs, 6 of the German songs, and 20 of the American songs. Pat wrote in her notebook (300  250  400) – (10  6  20) = (290  244  380). Have we ever defined **subtraction of matrices**?

5. Okay. Define subtraction of matrices. I'll start the definition, and you complete it: A − B = ?

***answers***

1.
$$\begin{pmatrix} 8 & 9 & 2 & 0 \\ 0 & 0 & 8 & 8 \end{pmatrix} + \begin{pmatrix} 0 & 0 & 0 & 0 \\ 1 & 0 & 0 & 0 \end{pmatrix} = \begin{pmatrix} 8 & 9 & 2 & 0 \\ 1 & 0 & 8 & 8 \end{pmatrix}$$

2. Back on page 70, was the clue: . . .*if you want the matrix that will perform an elementary row operation, just do that elementary row operation on* I. The permutation matrix that interchanges rows is an elementary row operation. PA = B becomes PIA = B. In this case, the identity matrix I is a 2×2 matrix. PI interchanges the rows of $\begin{pmatrix} 1 & 0 \\ 0 & 1 \end{pmatrix}$ to become $\begin{pmatrix} 0 & 1 \\ 1 & 0 \end{pmatrix}$ = P.

3. Pat's Russian-German-American matrix is a 1×3 matrix. It has one <u>row</u>. If a matrix has only one row or only one column, it is called a <u>vector</u>, and is often named by lower case letters such as b or x or v.

4. We have defined addition of matrices, multiplication of matrices, and multiplication of a scalar times a matrix, but we haven't defined subtraction of matrices.

5. A − B = A + (−1)B

So when Pat wrote (300  250  400) − (10  6  20), her next line would be (300  250  400) + (−1)(10  6  20) and then (300  250  400) + (−10  −6  −20) which finally gives (290  244  380).

But in actual practice, no one writes out those intermediate lines. They just do as Pat did in her notebook.

## Cuba

You are a door-to-door salesman. This week you have a new product to sell: **Fred's Fountain Pens**. There are two warehouses from which you can order your supplies. The original **Fred's Fountain Pens** warehouse is located in Lampasas, Texas. In stock are 1 fine point pen, 4 medium point pens, and 2 broad point pens.

A second **Fred's Fountain Pens** warehouse opened up in Cuba, Kansas. (There is such a place!) On its shelves are 3 fine point pens, 2 medium point pens, and 3 broad point pens. (These aren't large warehouses.)

1. Write up this inventory as a 2×3 matrix in which Lampasas and Cuba are the rows and fine, medium, and broad are the columns.

2. You buy the pens from those warehouses at the wholesale prices of $6, $5, and $7 for the fine, medium, and broad point pens respectively. Show, using matrix multiplication, how much it would cost to buy out the complete inventory of each warehouse.

3. You have bought out the complete inventory from both warehouses. Using matrix addition, show how you would compute your inventory.

4. Door to door you sell the three kinds of **Fred's Fountain Pens** for $40, $30, and $50 respectively. With your sample case filled with the 4 fine point, 6 medium point, and 5 broad point pens, you knock on the first door. You can't believe your ears when the person who answers the door says, "Oh, I've been dying to buy **Fred's Fountain Pens!** I've heard so many good things about them! I'll take them all!"

How much was her total payment?

**answers**

1. $\begin{pmatrix} 1 & 4 & 2 \\ 3 & 2 & 3 \end{pmatrix}$

2. First, you need to set up the prices as a column vector in order to do the matrix multiplication.

$\begin{pmatrix} 1 & 4 & 2 \\ 3 & 2 & 3 \end{pmatrix} \begin{pmatrix} \$6 \\ \$5 \\ \$7 \end{pmatrix} = \begin{pmatrix} \$40 \\ \$49 \end{pmatrix}$     Notice how much neater the matrices are than all the English in the question.

3. $(1 \quad 4 \quad 2) + (3 \quad 2 \quad 3) = (4 \quad 6 \quad 5)$  This example is simple, but it shows how you would do it if you were combining two giant department store inventories.

4. $(4 \quad 6 \quad 5) \begin{pmatrix} 40 \\ 30 \\ 50 \end{pmatrix} = (160 + 180 + 250) = \$590$

## Teller

1. Lee was trying to figure out if her spending was getting to be excessive. In January she made up this matrix:

|  | auto store | jewelry |
|---|---|---|
| cash spent | $436 | $807 |
| credit card | $1180 | $66 |

In February, she discovered sales at Mack's Fishing Store and started shopping there also. Here are her February purchases:

|  | auto store | jewelry | Mack's |
|---|---|---|---|
| cash spent | $224 | $117 | $82 |
| credit card | $498 | $77 | $2445 |

Two-part question: Why couldn't she add these two matrices together to get her spending for the first two months of the year? How might she change the matrices so that she could add them?

2. Suppose we name the augmented matrix for January (as given in answer 1) J, and we name the February matrix F. What would J + F represent?

3. What might 6(J + F) represent?

4. If F represents Lee's purchases in February, and it includes a sales tax of 7%, then rF (where r is some scalar) represents the 2×3 matrix of the sales tax she paid. Find the value of r. ("Scalar" is defined on page 51 in problem 3.)  Hint: r is neither 0.07, (1/0.07), nor (1/1.07).

*answers*

1. As stated on page 50, "*In order to add two matrices, they must have the same dimensions. . . .*" One easy way to fix the situation would be to augment the January matrix with a column for Mack's. Viz.,

|  | auto store | jewelry | Mack's |
|---|---|---|---|
| cash spent | $436 | $807 | $0 |
| credit card | $1180 | $66 | $0 |

[*Viz.* is the standard abbreviation for *videlicet*, which is pronounced wi-DAY-leh-kit or vi-DEL-eh-sit.] **Wait! And what in the world does *videlicet* mean?**

Viz. is one of four standard abbreviations (from the Latin). Here's the list:

etc. = and so on

i.e. = that is to say

e.g. = for example

viz. = namely

2. J + F would represent the Lee's spending for January and February.

3. 6(J + F) might represent our best estimate of how much Lee will spend for the entire year.

4. She spent $224 in cash at the auto store. This includes the retail price plus 7% sales tax. If we let x equal the retail price, then x + 0.07x is the total purchase. I.e., x(1 + 0.07) = 224. Then $x = \dfrac{224}{1.07}$ . Then 0.07x is the sales tax paid. Multiplying both sides of the previous equation by 0.07, we get $0.07x = \dfrac{0.07(224)}{1.07}$ . The r that we are looking for is $\dfrac{0.07}{1.07}$ .

Percent problems can be tougher than linear algebra.

## Idabel

1. Robin was looking for another job. She saw an ad in the help wanted section of the newspaper: *Harry's Hamburgers: "Nothing beats a Harry Hamburger!"*[SM] *Burger flipper wanted.* Robin sat by the phone debating whether she should call them. She doodled her picture of matrix multiplication on a piece of scratch paper.

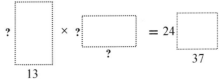

The "13" was the number of columns in the first matrix. (Her drawing was not to scale.) Find the values of the three question marks.

2. She called *Harry's* and went down for the interview. Mr. Harry handed her an employment test. The first question was, "Find the inverse to the 3×3 identity matrix." Robin knew that meant the same as, "Find $I^{-1}$." Do it for Robin.

3. The second question on the employment test had a "little more meat on it" (as Harry expressed it).    If $A = \begin{pmatrix} 0 & -1 & -1 \\ 1 & 0 & 2 \\ 4 & 1 & 8 \end{pmatrix}$    find $A^{-1}$.

Do it so Robin can be a burger flipper at *Harry's*.

### answers

1. The number of rows in the first matrix must be 24 since there are 24 rows in the answer.  The number of rows in the second matrix must match the 13 columns in the first matrix.  Otherwise, you couldn't do the multiplication.  The number of columns in the second matrix must be equal to the number of columns (37) in the answer.

2. This is the same as asking what do you multiply $\begin{pmatrix} 1 & 0 & 0 \\ 0 & 1 & 0 \\ 0 & 0 & 1 \end{pmatrix}$

in order to get $\begin{pmatrix} 1 & 0 & 0 \\ 0 & 1 & 0 \\ 1 & 0 & 1 \end{pmatrix}$

Since $I \times I = I$, we see that $I^{-1} = I$.

3. First, augment A with I. $\begin{pmatrix} 0 & -1 & -1 & 1 & 0 & 0 \\ 1 & 0 & 2 & 0 & 1 & 0 \\ 4 & 1 & 8 & 0 & 0 & 1 \end{pmatrix}$

$I\, r_1\, r_2$ $\begin{pmatrix} 1 & 0 & 2 & 0 & 1 & 0 \\ 0 & -1 & -1 & 1 & 0 & 0 \\ 4 & 1 & 8 & 0 & 0 & 1 \end{pmatrix}$

$-4r_1 + r_3$ $\begin{pmatrix} 1 & 0 & 2 & 0 & 1 & 0 \\ 0 & -1 & -1 & 1 & 0 & 0 \\ 0 & 1 & 0 & 0 & -4 & 1 \end{pmatrix}$

Then, after several more elementary row operations we arrive at

$\begin{pmatrix} 1 & 0 & 0 & 2 & -7 & 2 \\ 0 & 1 & 0 & 0 & -4 & 1 \\ 0 & 0 & 1 & -1 & 4 & -1 \end{pmatrix}$    so $A^{-1} = \begin{pmatrix} 2 & -7 & 2 \\ 0 & -4 & 1 \\ -1 & 4 & -1 \end{pmatrix}$

---

**Kedron**

---

1. Sometime in your math career you should meet the **Kronecker delta** $\delta_{ij}$ which is defined by $\delta_{ij} = \begin{cases} 1 & \text{if } i = j \\ 0 & \text{if } i \neq j \end{cases}$   ("Sometime" = today.)

Suppose we have a square matrix $C$ where $(C)_{ij} = \delta_{ij}$. What is the usual name we give to $C$?

2. A square matrix $D$ is called a **diagonal matrix** if $(D)_{ij} = 0$ wherever $i \neq j$.

   True or False? *A square matrix is diagonal if it is both upper triangular and lower triangular.*

3. Madison is trying to finish up her linear algebra homework before the Cub Scouts arrive for their den meeting. She has 20 matrices to decompose using *LU*. The 3×3 matrices have taken her an average of 5 minutes each to decompose. How long should she expect to spend decomposing a 4×4 matrix? (See the last paragraph on page 80. If she had to decompose a 5×5, it would take her a little more than 23 minutes.)

4. Can you think of an 18×18 matrix that would be easy to *LU*-decompose?

*answers*

1. The identity matrix I.
2. It is true.
3. T = kn³ where T is the time it takes for an n×n matrix. Putting in the given values of T = 5 when n = 3, we find that k (known as the constant of proportionality) is 5/27. So T = (5/27)n³. With n = 4, we have T = (5/27)4³ which is approximately 11.85 minutes.
4. The identity matrix is both lower- and upper-triangular, so I = (I)(I) is an *LU*-decomposition. The zero matrix, in which all the entries are zero, is also easy. In fact, any diagonal matrix is easy.

Lantry

1. If A, B, and C have dimensions so that ABC makes sense, and if A is 5×7, and C is 22×13, what are the dimensions of B?

2. An important aspect of Jackie's work on the police force is stopping off at Waddle's Doughnuts and selecting the right doughnuts. On the wall at Waddle's is the nutrition information sign.

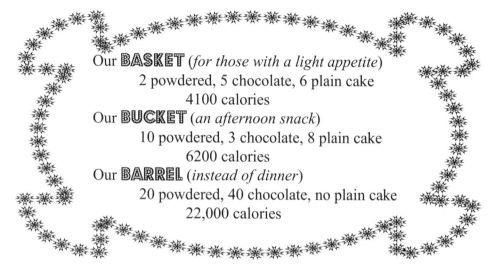

Our **BASKET** (*for those with a light appetite*)
2 powdered, 5 chocolate, 6 plain cake
4100 calories
Our **BUCKET** (*an afternoon snack*)
10 powdered, 3 chocolate, 8 plain cake
6200 calories
Our **BARREL** (*instead of dinner*)
20 powdered, 40 chocolate, no plain cake
22,000 calories

Using *LU*-decomposition, find out how many calories are in each type of doughnut.

3. Next week, Waddle's is changing its recipe. It is adding bacon grease and beef lard ("for extra flavor").

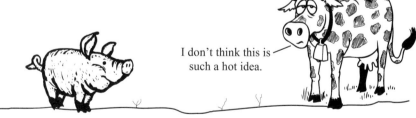

I don't think this is such a hot idea.

The new nutrition sign will be the same as the old one except for calories. Now a basket will deliver 6300 calories, a bucket will have 9500 calories, and a barrel, 38,000 calories.

Since you have already done the *LU*-decomposition in the previous problem, finding the new calorie content of each doughnut will be a cinch.

(Note: You can always check your answers to these kinds of problems by putting them back in the original problem.  E.g., using the answers you found in the basket case with the new recipe, 2 powdered plus 5 chocolate plus 6 plain cake should give you 6300 calories.)

### answers

1. **B** must be 7×22.

2.

| | | |
|---|---|---|
| 2 | 5 | 6 |
| 10 | 3 | 8 |
| 20 | 40 | 0 |

| | | |
|---|---|---|
| 1 | 2.5 | 3 |
| 10 | 3 | 8 |
| 20 | 40 | 0 |

| | | |
|---|---|---|
| 1 | 2.5 | 3 |
| 0 | −22 | −22 |
| 0 | −10 | −60 |

| | | |
|---|---|---|
| 1 | 2.5 | 3 |
| 0 | 1 | 1 |
| 0 | −10 | −60 |

| | | |
|---|---|---|
| 1 | 2.5 | 3 |
| 0 | 1 | 1 |
| 0 | 0 | −50 |

| | | |
|---|---|---|
| 1 | 2.5 | 3 |
| 0 | 1 | 1 |
| 0 | 0 | 1 |

And so **L** and **U** are

$$\begin{pmatrix} 2 & 0 & 0 \\ 10 & -22 & 0 \\ 20 & -10 & -50 \end{pmatrix} \quad \text{and} \quad \begin{pmatrix} 1 & 2.5 & 3 \\ 0 & 1 & 1 \\ 0 & 0 & 1 \end{pmatrix}$$

$\textbf{LU}x = b$

$\textbf{L}y = b \qquad \text{where } \textbf{U}x = y$

$$\begin{cases} 2y_1 & = 4100 \\ 10y_1 - 22y_2 & = 6200 \\ 20y_1 - 10y_2 - 50y_3 & = 22{,}000 \end{cases}$$

Using forward-substitution,
$y_1 = 2050$, $y_2 = 650$, $y_3 = 250$.

$$\begin{cases} x_1 + 2.5x_2 + 3x_3 & = 2050 \quad \text{This is } \textbf{U}x = y \\ x_2 + x_3 & = 650 \\ x_3 & = 250 \end{cases}$$

Using back-substitution,
$x_3 = 250$, $x_2 = 400$, $x_1 = 300$

3.  L and U we found in the previous problem:

$$\begin{pmatrix} 2 & 0 & 0 \\ 10 & -22 & 0 \\ 20 & -10 & -50 \end{pmatrix} \quad \text{and} \quad \begin{pmatrix} 1 & 2.5 & 3 \\ 0 & 1 & 1 \\ 0 & 0 & 1 \end{pmatrix}$$

LUx = b

Ly = b    where Ux = y

$$\begin{cases} 2y_1 & = 6300 \\ 10y_1 - 22y_2 & = 9500 \\ 20y_1 - 10y_2 - 50y_3 & = 38,000 \end{cases}$$

Using forward-substitution,
$y_1 = 3150$, $y_2 = 1000$, $y_3 = 300$.

$$\begin{cases} x_1 + 2.5x_2 + 3x_3 & = 3150 \quad \text{This is } Ux = y \\ x_2 + x_3 & = 1000 \\ x_3 & = 300 \end{cases}$$

Using back-substitution,
$x_3 = 300$, $x_2 = 700$, $x_1 = 500$

# Chapter Two

## Systems of Equations with Many Solutions

$$Ax = b \ \text{☺☺☺}\ldots$$

A s the moving company hauled off the sack from Carrie's that contained the pink cotton blanket and the sampler, Fred had finished the computation in his head. He knew that the picnic ground was three miles away and that the moving company would deliver the sack in 30 minutes.

That's six miles per hour. I'll race them Fred thought. Although Fred's diet was atrocious, his little body was well-exercised. Most mornings, he loved to go jogging. He pointed his nose toward the picnic grounds on the Great Lawn on the KITTENS campus and dashed.

Fred beat the moving company by four minutes. He found the picnic table with the sign: Reserved for Betty, Alexander, and Prof. Gauss.

He climbed up onto the bench and wondered Where's the sack from Butter Bottom's Foods?

What a picnic table
looks like to Fred who
is three feet tall

What that same table looks
like to Alexander who is six
feet tall

Some people tend to panic if you mess with their food. But eating has never been a big part of Fred's life. He has often missed meals and not given it a second thought. This may be one reason why our six-year-old is only three feet tall and weighs 37 pounds.

But it was very important that he not disappoint Betty and Alexander. So Fred began to think.

What if the moving company forgot to deliver the Butter Bottom's Food sack? He realized that wasn't the case since the empty sack was under the table.

What if some homeless hungry person stole the food? Fred looked around and saw a bunch of bunnies on the lawn. They were munching away on . . . rice!

92

What if some vandals had just taken the picnic food and scattered it around? And poured the sticky Sluice all over the picnic table and bench? Fred saw half-empty cans of picnic rice scattered on the ground, yellow splotches of mustard and glass where someone had thrown jars of mustard against the tree trunks, and worst of all, the bottom of his pants felt very sticky. Fred was right.

Fred pried himself off the bench and headed to the edge of the Great Lake near the Great Lawn. He waded into the water. When he was waist-deep, he danced the Twist because that looked a lot like a washing machine. Several nearby fish floated to the surface. (Sluice's sugar content is deadly to most living things.)

**Hey. I, your reader, have my nightmares too. When I was studying algebra, I dreamed that the square root of two was chasing me down the hall.**

You don't have fears about linear algebra, do you?

**Glad you brought that up. Sometimes when I'm doing regular old Gaussian elimination, I get the "What ifs" just like Fred experienced.**

**I know that all I have to do is make the augmented matrix into an upper-triangular matrix using the three elementary row operations.**

Yes, just take some linear system of equations such as

$$\begin{cases} 2x_1 + 3x_2 - x_3 = 22 \\ 4x_1 + 6x_2 + 5x_3 = 65 \\ 8x_1 + 8x_2 + 2x_3 = 78 \end{cases}$$

which we write as Ax = b.

Then we create the augmented matrix $\begin{pmatrix} 2 & 3 & -1 & 22 \\ 4 & 6 & 5 & 65 \\ 8 & 8 & 2 & 78 \end{pmatrix}$

which we make into an upper-triangular matrix using the three elementary row operations:

    ① Interchange any two rows.

    ② Multiply any row by a nonzero number.

    ③ Take any row,

        multiply it by any number,

           and add it to another row.

**Hey. That's a perfect example of a "What if." After you eliminate the 4 and the 8 in the first column (using $-2r_1 + r_2$ and $-4r_1 + r_3$), you get**

**my first nightmare. Look! In the second column there is a stupid zero sitting there on the diagonal.**
$$\begin{pmatrix} 2 & 3 & -1 & 22 \\ 0 & 0 & 7 & 21 \\ 0 & -4 & 6 & -10 \end{pmatrix}$$

**I can't use that zero to get rid of the −4 below it.**

Have no fear. There's an easy solution. If there is a zero on the diagonal, just look for a nonzero entry below it and interchange the two rows. The matrix will become
$$\begin{pmatrix} 2 & 3 & -1 & 22 \\ 0 & -4 & 6 & -10 \\ 0 & 0 & 7 & 21 \end{pmatrix}$$

Nightmare #1: There is a zero on the diagonal.

Banish it by interchanging that row with a row below it that doesn't have a zero in that column. Look south ☞.

**But what if there are zeros all the way down? For example,**

$$\begin{pmatrix} 7 & 8 & -5 & 1 & 62 \\ 0 & 0 & -2 & 7 & 89 \\ 0 & 0 & 4 & 8 & 55 \\ 0 & 0 & 1 & 6 & -3 \end{pmatrix}$$ **Then your interchange trick won't work.**

Then you declare the variable for that column (in this case, $x_2$) to be a **free variable**, and you move on to row 2, column 3. For the time being, you mentally erase that column 2. Now the −2 can be considered to be on the diagonal, and you can eliminate the 4 and the 1 below it.

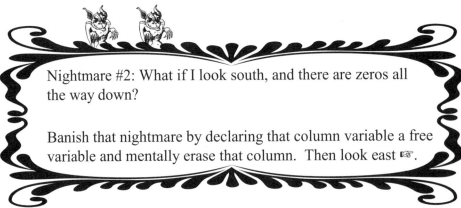

Nightmare #2: What if I look south, and there are zeros all the way down?

Banish that nightmare by declaring that column variable a free variable and mentally erase that column. Then look east ☞.

**I know you are eventually going to tell me what to do with the free variable, but I have another nightmare. What if I look east, as you call it, and all I see are zeros.**

$$\begin{pmatrix} 7 & 8 & -5 & 1 & 62 \\ 0 & 0 & 0 & 0 & 0 \\ 0 & 0 & 4 & 8 & 55 \\ 0 & 0 & 1 & 6 & -3 \end{pmatrix}$$  **Zeros as far as the eye can see.**

What does a row of zeros tell us?

It is the same as the equation $0x_1 + 0x_2 + 0x_3 + 0x_4 = 0$.

Can you name some values of $x_1$, $x_2$, $x_3$, and $x_4$ that make that equation true? How about $x_1 = 32$, $x_2 = \pi$, $x_3 = \log_3 17$, and $x_4 = \sqrt{3.5}$ ? Anything you name will work. Imagine you are on a quiz program and you are asked, "For the grand prize of $593,946, name a solution to the equation $0x_1 + 0x_2 + 0x_3 + 0x_4 = 0$." Some days everything seems to come up roses.

On the other hand, $0x_1 + 0x_2 + 0x_3 + 0x_4 = 0$ is absolutely worthless. It tells you nothing.

Take any row of zeros that you encounter and stick it at the bottom of the matrix—just to get it out of the way. Or just eliminate that worthless row.

Nightmare #3: What if I look east and all I see is zeros?

Banish that nightmare by using the row interchange operation. Stick all the all-zero rows at the bottom of the matrix, or just discard that row.

**I've saved the worst nightmare for last. This is one that would scare a rock.**

Scare a rock?

**Yeah, and rocks are tough to scare.**

Okay. What's your last nightmare?

**Suppose you're working along with Gaussian elimination and one of the rows looks like**

$$\begin{pmatrix} 0 & 0 & 0 & 0 & 7 \end{pmatrix}$$

**That corresponds to the equation** $0x_1 + 0x_2 + 0x_3 + 0x_4 = 7$. **I'll give you a million dollars if you can solve that equation.**

Those kinds of systems of linear equations pop up all the time when you are solving real-life problems. Such systems are called **overdetermined**. That is the topic of Chapter 3 $\text{A}x = \text{b}$ ☹. The sad face means, as you might guess, that they can't be solved.

**Wait! Stop! You are talking about real-life problems that can't be solved. That deserves a lot of sad faces. ☹☹☹☹☹☹☹☹☹☹☹☹☹☹**

Not at all. In one little step in Chapter 3, we will turn $\text{A}x = \text{b}$ ☹ into $\text{A}^\text{T}\text{A}x = \text{A}^\text{T}\text{b}$ which gives us the best solution in an impossible situation.

Nightmare #4: A row with all zeros except for the last column.

Banish that nightmare using the $\text{A}^\text{T}$ trick (called "A transpose") described in Chapter 3.

**I know your tricks. I bet that the $\text{A}^\text{T}$ thing takes four weeks to compute. You are delaying telling me about A transpose because it's a killer.**

Hardly. If A were $\begin{pmatrix} 4 & 5 & 8 & 9 \\ 7 & 2 & 2 & 3 \end{pmatrix}$ then $\text{A}^\text{T}$ would be $\begin{pmatrix} 4 & 7 \\ 5 & 2 \\ 8 & 2 \\ 9 & 3 \end{pmatrix}$

Transposing a matrix just flips rows into columns.

*Your Turn to Play*

1. While the idea of transpose is fresh in your mind, complete the following definition of $\mathbf{A}$ transpose: If $(\mathbf{A})_{ij} = a_{ij}$, then $(\mathbf{A}^T)_{ij} = ?$

2. If $\mathbf{A}$ is a 3×5 matrix, what are the dimensions of $\mathbf{A}^T$?

3. If you are given a system of linear equations like
$$\begin{cases} 5x_1 + 8x_2 + 2x_3 = 7 \\ 10x_1 + 16x_2 + 5x_3 = 2 \\ 15x_1 + 27x_2 + 9x_3 = 11 \end{cases}$$
you can't immediately tell what kind of nightmare you will encounter when you solve it by Gaussian elimination. The first step is to put that system into an augmented matrix
$$\begin{pmatrix} 5 & 8 & 2 & 7 \\ 10 & 16 & 5 & 2 \\ 15 & 27 & 9 & 11 \end{pmatrix}$$
Start solving that matrix until you hit a nightmare and then state what you would do to banish that nightmare. You don't have to complete the whole solution.

4. Repeat problem 3 using this matrix
$$\begin{pmatrix} 5 & 8 & 2 & 7 \\ 10 & 16 & 4 & 14 \\ 15 & 27 & 9 & 11 \end{pmatrix}$$

5. Repeat problem 3 using
$$\begin{pmatrix} 5 & 8 & 3 & 2 & 7 \\ 10 & 16 & 7 & 8 & 24 \\ 15 & 24 & 13 & 9 & 35 \end{pmatrix}$$

. . . . . . . **COMPLETE SOLUTIONS** . . . . . . .

1. If $(\mathbf{A})_{ij} = a_{ij}$, then $(\mathbf{A}^T)_{ij} = a_{ji}$.

2. The dimensions of $\mathbf{A}^T$ would be 5×3.

3. $\begin{pmatrix} 5 & 8 & 2 & 7 \\ 10 & 16 & 5 & 2 \\ 15 & 27 & 9 & 11 \end{pmatrix} \xrightarrow[-3r_1 + r_3]{-2r_1 + r_2} \begin{pmatrix} 5 & 8 & 2 & 7 \\ 0 & 0 & 1 & -12 \\ 0 & 3 & 3 & -10 \end{pmatrix}$

This is the first kind of nightmare—a zero on the diagonal. Interchange rows 2 and 3 to get
$\begin{pmatrix} 5 & 8 & 2 & 7 \\ 0 & 3 & 3 & -10 \\ 0 & 0 & 1 & -12 \end{pmatrix}$

4. $\begin{pmatrix} 5 & 8 & 2 & 7 \\ 10 & 16 & 4 & 14 \\ 15 & 27 & 9 & 11 \end{pmatrix} \xrightarrow[{-3r_1 + r_3}]{-2r_1 + r_2} \begin{pmatrix} 5 & 8 & 2 & 7 \\ 0 & 0 & 0 & 0 \\ 0 & 3 & 3 & -10 \end{pmatrix}$

This is nightmare #3—a row of zeros. Interchange rows to put that row of zeros at the bottom of the matrix and get $\begin{pmatrix} 5 & 8 & 2 & 7 \\ 0 & 3 & 3 & -10 \\ 0 & 0 & 0 & 0 \end{pmatrix}$

5. $\begin{pmatrix} 5 & 8 & 3 & 2 & 7 \\ 10 & 16 & 7 & 8 & 24 \\ 15 & 24 & 13 & 9 & 35 \end{pmatrix} \xrightarrow[{-3r_1 + r_3}]{-2r_1 + r_2} \begin{pmatrix} 5 & 8 & 3 & 2 & 7 \\ 0 & 0 & 1 & 4 & 10 \\ 0 & 0 & 4 & 3 & 14 \end{pmatrix}$

This is nightmare #2—a zero on the diagonal and zero(s) all the way south. We declare that column variable (which is $x_2$) to be a free variable.

As you, my Reader, correctly predicted three pages ago, "**I know you are eventually going to tell me what to do with the free variable.**" This would be a good time to do that since the solution to problem 5, which is just two inches (or five centimeters, if you prefer metric) north of this line, already has a free variable.

$\begin{pmatrix} 5 & 8 & 3 & 2 & 7 \\ 0 & 0 & 1 & 4 & 10 \\ 0 & 0 & 4 & 3 & 14 \end{pmatrix}$ corresponds to the system $\begin{cases} 5x_1 + 8x_2 + 3x_3 + 2x_4 = 7 \\ \qquad\qquad x_3 + 4x_4 = 10 \\ \qquad\qquad 4x_3 + 3x_4 = 14 \end{cases}$

where $x_2$ has been declared a free variable.

### So what do you do with a free variable?

First, let me finish the Gaussian elimination.

$\begin{pmatrix} 5 & 8 & 3 & 2 & 7 \\ 0 & 0 & 1 & 4 & 10 \\ 0 & 0 & 4 & 3 & 14 \end{pmatrix} \xrightarrow{-4r_2 + r_3} \begin{pmatrix} 5 & 8 & 3 & 2 & 7 \\ 0 & 0 & 1 & 4 & 10 \\ 0 & 0 & 0 & -13 & -26 \end{pmatrix}$

I'm free!

When you have finished a Gaussian elimination process, each row will have more zeros on the left than the previous row.  Such a matrix is in **echelon form**.

We have to stop for a second and do an English lesson.  Let me put it all in a box.

**Reduced row-echelon** matrices were the result of Gauss-Jordan elimination.  They have 1s along the diagonal (if you mentally eliminate columns that have free variables, as I mentioned on page 94).  They have zeros above and below those ones.  Any all-zero rows are at the bottom of the matrix.

$$\text{example:} \quad \begin{pmatrix} 1 & 7 & 0 & 0 & 5 & 2 & 4 \\ 0 & 0 & 1 & 0 & 6 & 8 & 1 \\ 0 & 0 & 0 & 1 & 3 & 0 & 8 \\ 0 & 0 & 0 & 0 & 0 & 0 & 0 \\ 0 & 0 & 0 & 0 & 0 & 0 & 0 \end{pmatrix}$$

**Echelon** matrices are simpler.  As usual, stick the all-zero rows at the bottom.  The only other requirement is that, as you look at the rows that are above those all-zero rows, each row has more zeros on the left than the row above it.  (English is harder than math.)

$$\text{example:} \quad \begin{pmatrix} 3 & 7 & 8 & 3 & 0 & 9 & 4 \\ 0 & 0 & 1 & 0 & 7 & 9 & 1 \\ 0 & 0 & 0 & 4 & 5 & 3 & 8 \\ 0 & 0 & 0 & 0 & 0 & 2 & 3 \\ 0 & 0 & 0 & 0 & 0 & 0 & 0 \end{pmatrix}$$

The first nonzero element in a row of a matrix that is in echelon form is called a **distinguished element**.  The variables that correspond to distinguished elements are called **pivots** or **leading variables**.

So it all boils down to the fact that you're either a leading variable or you are a free variable.

**But I, your reader, need to know.  Would you please tell me what to do with those free variables?**

Two pages ago, before we had to stop for an English lesson, we had

$$\begin{pmatrix} 5 & 8 & 3 & 2 & 7 \\ 0 & 0 & 1 & 4 & 10 \\ 0 & 0 & 0 & -13 & -26 \end{pmatrix}$$ which corresponds to $$\begin{cases} 5x_1 + 8x_2 + 3x_3 + 2x_4 = 7 \\ x_3 + 4x_4 = 10 \\ -13x_4 = -26 \end{cases}$$

$x_2$ is the free variable.  What to do with it?  You transpose those free variables to the other side of the equation.

Viz., $$\begin{cases} 5x_1 + 3x_3 + 2x_4 = 7 - 8x_2 \\ x_3 + 4x_4 = 10 \\ -13x_4 = -26 \end{cases}$$

We are virtually done.  You want a solution to the original system of linear equations?  Just name a value for the free variable $x_2$.

**Okay.  I say $x_2$ is –40.**

Then we get $$\begin{cases} 5x_1 + 3x_3 + 2x_4 = 7 + 320 \\ x_3 + 4x_4 = 10 \\ -13x_4 = -26 \end{cases}$$

which can be solved by back-substitution.  That will give us a **particular solution**.

If you want another particular solution, then just name another value for $x_2$.  There is an infinite number of solutions, depending on what value you give for the free variable(s).

That is why this chapter is named $\mathsf{Ax = b}$ ☺☺☺ . . .

**I see how to write particular solutions. That's easy. But how do you neatly write all the solutions to the system?**

In order to avoid saying, "First take the system and assign any value you like to the free variable(s), and then back-substitute to find the corresponding values of the leading variables (pivots)," we must resurrect Mr. Jordan.

*Did you think you were done with me?*

$$\begin{pmatrix} 5 & 8 & 3 & 2 & 7 \\ 0 & 0 & 1 & 4 & 10 \\ 0 & 0 & 0 & -13 & -26 \end{pmatrix}$$

> Recall: Gauss-Jordan puts 1s and 0s in all the pivot columns.

$(-1/13)r_3$
$$\begin{pmatrix} 5 & 8 & 3 & 2 & 7 \\ 0 & 0 & 1 & 4 & 10 \\ 0 & 0 & 0 & 1 & 2 \end{pmatrix}$$

$-2r_3 + r_1$
$-4r_3 + r_2$
$$\begin{pmatrix} 5 & 8 & 3 & 0 & 3 \\ 0 & 0 & 1 & 0 & 2 \\ 0 & 0 & 0 & 1 & 2 \end{pmatrix}$$

> I don't like doing this stuff. I've always imagined that people who delight in Gauss-Jordan elimination are the same people who *enjoy* balancing their checkbooks.

$-3r_2 + r_1$
$$\begin{pmatrix} 5 & 8 & 0 & 0 & -3 \\ 0 & 0 & 1 & 0 & 2 \\ 0 & 0 & 0 & 1 & 2 \end{pmatrix}$$

$(1/5)r_1$
$$\begin{pmatrix} 1 & 1.6 & 0 & 0 & -0.6 \\ 0 & 0 & 1 & 0 & 2 \\ 0 & 0 & 0 & 1 & 2 \end{pmatrix}$$

$$\begin{cases} x_1 + 1.6x_2 & = -0.6 \\ x_3 & = 2 \\ x_4 & = 2 \end{cases}$$

So, after transposing, the **general solution** could be written as $x_1 = -0.6 - 1.6x_2$, $x_3 = 2$, $x_4 = 2$.

Often, in order to emphasize that $x_2$ can take on any value (that's called a **parameter**), it is replaced by r (or s or t):

$x_1 = -0.6 - 1.6r$, $x_2 = r$, $x_3 = 2$, $x_4 = 2$.

Another way to write the general solution is in terms of column vectors. From the previous page, $x_1 = -0.6 - 1.6r$,  $x_2 = r$,  $x_3 = 2$,  $x_4 = 2$.

As a column vector this could be written

$$\begin{pmatrix} x_1 \\ x_2 \\ x_3 \\ x_4 \end{pmatrix} = \begin{pmatrix} -0.6 - 1.6r \\ r \\ 2 \\ 2 \end{pmatrix} = \begin{pmatrix} -0.6 \\ 0 \\ 2 \\ 2 \end{pmatrix} + \begin{pmatrix} -1.6r \\ r \\ 0 \\ 0 \end{pmatrix} = \begin{pmatrix} -0.6 \\ 0 \\ 2 \\ 2 \end{pmatrix} + r\begin{pmatrix} -1.6 \\ 1 \\ 0 \\ 0 \end{pmatrix}$$

Saying that in words: The general solution $\begin{pmatrix} x_1 \\ x_2 \\ x_3 \\ x_4 \end{pmatrix}$ is a combination of

$\begin{pmatrix} -0.6 \\ 0 \\ 2 \\ 2 \end{pmatrix}$ and any multiple of $\begin{pmatrix} -1.6 \\ 1 \\ 0 \\ 0 \end{pmatrix}$.

---

*Your Turn to Play*

1. On page 98, we began with $\begin{pmatrix} 5 & 8 & 3 & 2 & 7 \\ 10 & 16 & 7 & 8 & 24 \\ 15 & 24 & 13 & 9 & 35 \end{pmatrix}$

and showed that it was row-equivalent to $\begin{pmatrix} 1 & 1.6 & 0 & 0 & -0.6 \\ 0 & 0 & 1 & 0 & 2 \\ 0 & 0 & 0 & 1 & 2 \end{pmatrix}$

Write the systems of linear equations that each of these two matrices represents.

2. (Continuing the previous problem) Are the solutions to the two systems of equations the same?

3. We noted that the general solution $\begin{pmatrix} x_1 \\ x_2 \\ x_3 \\ x_4 \end{pmatrix}$ is a combination of $\begin{pmatrix} -0.6 \\ 0 \\ 2 \\ 2 \end{pmatrix}$

and any multiple of $\begin{pmatrix} -1.6 \\ 1 \\ 0 \\ 0 \end{pmatrix}$.

---

Specifically, $\begin{pmatrix} x_1 \\ x_2 \\ x_3 \\ x_4 \end{pmatrix} = \begin{pmatrix} -0.6 \\ 0 \\ 2 \\ 2 \end{pmatrix} + r \begin{pmatrix} -1.6 \\ 1 \\ 0 \\ 0 \end{pmatrix}.$

The general solution is the sum of the first matrix plus any real number times the second matrix.

It's pretty easy to see that the first matrix $\begin{pmatrix} -0.6 \\ 0 \\ 2 \\ 2 \end{pmatrix}$ is a solution to either system of equations in answer 1 below.

The challenge is to figure out what system $\begin{pmatrix} -1.6 \\ 1 \\ 0 \\ 0 \end{pmatrix}$ is a solution to.

## .......COMPLETE SOLUTIONS.......

1. $\begin{cases} 5x_1 + 8x_2 + 3x_3 + 2x_4 = 7 \\ 10x_1 + 16x_2 + 7x_3 + 8x_4 = 24 \\ 15x_1 + 24x_2 + 13x_3 + 9x_4 = 35 \end{cases}$    $\begin{cases} x_1 + 1.6x_2 \quad\quad\quad = -0.6 \\ \quad\quad\quad x_3 \quad\quad = 2 \\ \quad\quad\quad\quad x_4 = 2 \end{cases}$

2. That was one of the most important facts about elementary row operations: they leave the solution set unchanged.

3. If you put $\begin{pmatrix} -1.6 \\ 1 \\ 0 \\ 0 \end{pmatrix}$ into the two systems of equations in answer 1, you get

$\begin{cases} 5x_1 + 8x_2 + 3x_3 + 2x_4 = 0 \\ 10x_1 + 16x_2 + 7x_3 + 8x_4 = 0 \\ 15x_1 + 24x_2 + 13x_3 + 9x_4 = 0 \end{cases}$    $\begin{cases} x_1 + 1.6x_2 \quad\quad\quad = 0 \\ \quad\quad\quad x_3 \quad\quad = 0 \\ \quad\quad\quad\quad x_4 = 0 \end{cases}$

If you have gotten this far in mathematics, it is hoped that you would like an occasional theorem to spice things up.

**Theorem**: If $y$ is any particular solution to $Ax = b$ and $z$ is any solution to $Ax = 0$ and if r is any scalar, then $y + rz$ is also a solution to $Ax = b$.

Proof: If $y$ is any particular solution to $Ax = b$, then $Ay = b$. (That was easy.)

If $z$ is any solution to $Ax = 0$, then $Az = 0$.

In matrix algebra, all the "regular" laws of high school algebra apply, except for the commutative law of multiplication of matrices.

Multiply both sides of $Az = 0$ by the real number r and get $r(Az) = r0$.

$0$ is the matrix consisting of all zeros. To multiply a matrix by a scalar r is to multiply every entry of the matrix by r. (We defined scalar multiplication on page 51.) So $r0 = 0$.

$r(Az) = r0$ becomes $r(Az) = 0$.

By the associative law, $(rA)z = 0$.

Scalar multiplication is commutative. (It's only matrix multiplication that doesn't commute.) $(Ar)z = 0$.

Associative law, $A(rz) = 0$.

From the first line of the proof we had $Ay = b$.

Adding those two equations together, $Ay + A(rz) = b$.

Distributive law, $A(y + rz) = b$.

Q.E.D.

And since we are in a very mathematical mood, here is a definition.

Definition: A linear system of the form $Ax = 0$ is called **homogeneous**.

Actually, there are two laws of high school algebra that don't apply to matrix multiplication. (Both of these I mentioned on page 56.) The first is that the commutative law for matrix multiplication does not hold. Usually $AB$ does not equal $BA$.

The second is that $AB = 0$ does not necessarily mean that either $A = 0$ or $B = 0$. Back on page 56 we gave the example

$$\begin{pmatrix} 1 & 1 \\ 2 & 2 \end{pmatrix}\begin{pmatrix} -1 & 1 \\ 1 & -1 \end{pmatrix} = \begin{pmatrix} 0 & 0 \\ 0 & 0 \end{pmatrix}$$

Now $\begin{pmatrix} 1 & 1 \\ 2 & 2 \end{pmatrix}$ certainly isn't the zero matrix. But it is *the next best thing* to being the zero matrix. It is **singular**. You can think of singular matrices as square matrices that are somehow defective when it comes to matrix multiplication.

**Mr. Author, I, your reader, would like to make a small observation: That's a pretty crummy definition of singular matrices. Couldn't you do better than saying that they are "somehow defective"?**

I was just getting to that. You can spot a singular matrix by putting it in echelon form. If there is one or more rows of zeros *at the bottom of a square matrix*, then it is **singular**.

Here's a second definition: An n×n matrix is **singular** if its rank is less than n.

Before you, my reader, object, here is the definition of the **rank of a matrix**: The rank of a matrix is the number of nonzero rows it has after it has been put into echelon form.

Putting $\begin{pmatrix} 1 & 1 \\ 2 & 2 \end{pmatrix}$ into echelon form (using $-2r_1 + r_2$) we get $\begin{pmatrix} 1 & 1 \\ 0 & 0 \end{pmatrix}$

and $\begin{pmatrix} 1 & 1 \\ 0 & 0 \end{pmatrix}$ has one nonzero row. Its rank equals 1.

---

*Your Turn to Play*

1. What is the rank of $\begin{pmatrix} 3 & 7 & 81 & 0.3 \\ 5 & 2 & 34 & 69 \\ 11 & 16 & 196 & 69.6 \end{pmatrix}$ ?

2. Suppose we have a system of linear equations $\mathsf{Ax = b}$ where the rank of $\mathsf{A}$ is equal to 7. What can be said about the rank of the augmented matrix $\begin{pmatrix} \mathsf{A} & \mathsf{b} \end{pmatrix}$? Before you look at the answer, please spend a minute or two thinking about this question. Maybe even doodle a little on a piece of paper. It is the ruminating, not the memorizing of facts, that will help you become a mathematician.

3. Suppose we have a system of linear equations $\mathsf{Ax = b}$ where the rank of $\mathsf{A}$ is equal to 7 and the rank of the augmented matrix $\begin{pmatrix} \mathsf{A} & \mathsf{b} \end{pmatrix}$ is equal to 8. Which of the four nightmares have we encountered? (The nightmares were listed in the boxes on pages 94–96.)

4. Suppose we have a system of linear equations $\mathsf{Ax = b}$ where the rank of $\mathsf{A}$ is equal to 7. Could the rank of the augmented matrix $\begin{pmatrix} \mathsf{A} & \mathsf{b} \end{pmatrix}$ be equal to 6?

---

5. Look at the results of the previous two questions and formulate the theorem that must be true.

### . . . . . . . COMPLETE SOLUTIONS . . . . . . .

1.

$(-5/3)r_1 + r_2$
$(-11/3)r_1 + r_3$
$$\begin{pmatrix} 3 & 7 & 81 & 0.3 \\ 0 & -9\tfrac{2}{3} & -101 & 68.5 \\ 0 & -9\tfrac{2}{3} & -101 & 68.5 \end{pmatrix}$$

and then

$-r_2 + r_3$
$$\begin{pmatrix} 3 & 7 & 81 & 0.3 \\ 0 & -9\tfrac{2}{3} & -101 & 68.5 \\ 0 & 0 & 0 & 0 \end{pmatrix}$$    The rank is equal to 2.

You might have noticed that the third row is a linear combination of the first two rows. Twice the first row plus the second row equals the third row. Adding $-2r_1$ and $-r_2$ to row 3, gives a row of zeros immediately.

2. If the rank of A is equal to 7, then there are 7 nonzero rows after putting A into echelon form. Using those same elementary row operations on

$\left( \text{A} \quad \text{b} \right)$ will leave at least 7 nonzero rows.

3. The only way that the rank of A is less than the rank of the augmented

matrix $\left( \text{A} \quad \text{b} \right)$ is if, after putting it into echelon form, there is a row

of $\left( \text{A} \quad \text{b} \right)$ that has all zeros except for the last column.

This is nightmare #4: the overdetermined system of equations that has no solution. $\text{Ax} = \text{b}$ ☹.

4. Only when pigs fly. If you used elementary row operations on $\left( \text{A} \quad \text{b} \right)$ to put it into echelon form and got 6 nonzero rows, then those same elementary row operations would turn A into an echelon matrix with at most 6 nonzero rows. But we were given that the rank of A is equal to 7.

5. There are several ways you might state it. You might say **Theorem**: If

the rank of $\left( \text{A} \quad \text{b} \right)$ is greater than the rank of A, then $\text{Ax} = \text{b}$ has no

solutions. Or you might write it as **Theorem**: If $\text{Ax} = \text{b}$ has at least one

solution, then the rank of $\left( \text{A} \quad \text{b} \right)$ must be equal to the rank of A.

### *Intermission*

Back in high school geometry, you might have studied a little logic and learned about *contrapositives*.

Suppose you start with an implication: "If it is a zebra, then it has stripes."

The contrapositive is, "If it doesn't have stripes, then it is not a zebra."

In symbols, the contrapositive to "If P then Q" is the statement "If not Q then not P."

An implication and its contrapositive are *logically equivalent*—one is true if and only if the other one is true.

Even more symbolically, P ⇨ Q iff ¬Q ⇨ ¬P.

The first statement of the theorem in answer 5 might be written rank(A  b) > rank(A) ⇨ Ax = b has no solution.  Its contrapositive would be, "If it's not the case that Ax = b has no solution ⇨ it's not the case that rank(A  b) > rank(A)."

This last sentence, if you clean up the English and you remember from question 4 that rank(A  b) < rank(A) is impossible, is the second statement of the theorem.

### *Second Intermission*

Sometimes you may accidentally run into someone who has not read the *Life of Fred* series. When you mention "contrapositive" they will give you a blank stare.

These individuals may mix up contrapositives with *inverses*.

If we have the implication P ⇨ Q, then
. . . its contrapositive is  ¬Q ⇨ ¬P
. . . its Inverse is  ¬P ⇨ ¬Q.

An implication and its inverse are *not* logically equivalent.

True implication:  "If it is a zebra, then it has stripes."

False inverse: "If it's not a zebra, then it doesn't have stripes."

And just for the sake of completeness—the **converse** of  P ⇨ Q is

Q ⇨ P.

We are now at the point where we know that if $Ax = b$ is going to have any solutions at all, the rank of $\begin{pmatrix} A & b \end{pmatrix}$ must be equal to the rank of $A$.

The question is how do we distinguish between the case in which we have one solution ($Ax = b$ ☺) and the case in which we have an infinite number of solutions ($Ax = b$ ☺☺☺. . .)?

**Hey, that's easy. Whenever a free variable pops up you have $Ax = b$ ☺☺☺. . . .**

You are right. And, since you are bringing up old material, how do you distinguish the cases in terms of $A^{-1}$?

**If you are given $Ax = b$ and you can find $A^{-1}$, then using that pretty**

| double-bordered box |

**on page 61, you could instantly go from $Ax = b$ to the single solution $x = A^{-1}b$.**

You, my reader, have scored 100% two times in a row.

**I guess that makes me 200% correct.** [*]

Would you like to go for 300%?

**No. You talk for a while.**

Okay. A third way to tell whether you have exactly one solution or an infinite number of solutions is to look at the **dimensions** of $A$. If $A$ is an m × n matrix (which means m rows and n columns), then $Ax = b$ is a system of m equations and n unknowns. Again, assuming we know that we do have at least one solution—i.e., that the rank of the augmented matrix $\begin{pmatrix} A & b \end{pmatrix}$ equals the rank of $A$—then m < n will automatically mean that we will have one or more free variables and an infinite number of solutions. $Ax = b$ will be an **underdetermined linear system**.

**My turn to talk. Wouldn't that mean that if we know we have one solution, and we know that m = n, then there has gotta be exactly one solution?**

Nope. Some of those m equations might be worthless (as I called them back on page 95). After doing Gaussian elimination you might get

---

[*] Being more than 100% right is an interesting concept. Can you be better than perfect? I would guess that the answer is yes, given the words of the Preamble to the United States Constitution: *We, the People of the United States, in Order to form a more perfect Union. . . .*

rows of zeros in the matrix. In problem 1 of the previous *Your Turn to Play*
we started out with the linear system
$$\begin{cases} 3x_1 + 7x_2 + 81x_3 = 0.3 \\ 5x_1 + 2x_2 + 34x_3 = 69 \\ 11x_1 + 16x_2 + 196x_3 = 69.6 \end{cases}$$
which has three equations (m = 3) and three unknowns (n = 3).
Elementary row operations gave us
$$\begin{pmatrix} 3 & 7 & 81 & 0.3 \\ 0 & -9\tfrac{2}{3} & -101 & 68.5 \\ 0 & 0 & 0 & 0 \end{pmatrix}$$

     It turned out that the third equation was worthless. The third row
was linearly dependent on the other two rows. The rank of the matrix was
equal to 2.

     It's as if the linear system at the top of this page were bragging,
"Hey, I got three equations," but like most braggers, he was full of hot air.
That's why you don't marry someone after dating them only once. You
need some time to find out important information.

     When you are "dating" a linear system of 14 equations and 14
unknowns, you want to ask the question, "What is your rank?" If the
answer comes back, "I'm embarrassed to admit this, but my rank is only
12," then you know there are two worthless equations in the 14. You
know that if you do Gaussian elimination, two of the rows will be all
zeros.

     First appearances can be deceiving. These all look alike, but . . .

$$\begin{cases} 2x_1 + 3x_2 + 4x_3 = 5 \\ 3x_1 + 4x_2 + 5x_3 = 6 \\ 5x_1 + 7x_2 + 8x_3 = 9 \end{cases}$$      has one solution      ☺

$$\begin{cases} 2x_1 + 3x_2 + 4x_3 = 5 \\ 3x_1 + 4x_2 + 5x_3 = 6 \\ 5x_1 + 7x_2 + 9x_3 = 9 \end{cases}$$      has no solution      ☹

$$\begin{cases} 2x_1 + 3x_2 + 4x_3 = 5 \\ 3x_1 + 4x_2 + 5x_3 = 6 \\ 5x_1 + 7x_2 + 8x_3 = 11 \end{cases}$$      has an infinite number of solutions ☺☺☺ . . .

# *Handy Guide to Dating Systems of Linear Equations*
## Ax = b

*First, ask if the rank of the augmented matrix* $\begin{pmatrix} A & b \end{pmatrix}$ *is larger than the rank of* **A**. *If so, this system of equations will never have a solution. End the date immediately and go on to Chapter 3.*

*Second, if the system has passed question number one, then you glance at the number of equations and see if it is less than the number of unknowns . If so, there is an infinite number of solutions.*

*Third, ask the intimate question, "What is your rank?"*

> *If the rank of* **A** *is less than the number of unknowns, there are an infinite number of solutions.*
>
> *If the rank of* **A** *is equal to the number of unknowns, then there is a unique solution.*
>
> *If the one you are dating brags that his or her rank is greater than the number of unknowns, you are dating a liar.* \*

---

\* Ask anyone who has read Chapter 2½. They know that the rank of **A**—which is the dimension of the row space of **A**—is equal to the dimension of the column space of **A**. Of course, if you haven't read Chapter 2½ yet, you may not have the faintest idea what I'm talking about.

**Elsie**

1. Before the start of each choir season, Pat holds a choir party at her house. Four years ago, 1 soprano, 2 altos, 4 tenors, and 3 basses that attended ate a total of 32 slices of Pat's famous cheesecake.

(At choir parties, we will assume that sopranos always each eat $x_1$ slices of cheesecake. Altos, $x_2$, etc.)

     So, four years ago, $x_1 + 2x_2 + 4x_3 + 3x_4 = 32$.

     Three years ago, 3 sopranos, 3 altos, 8 tenors, and 3 basses consumed 51 slices of Pat's cheesecake.

     Two years ago, $4x_1 + 5x_2 + 8x_3 + 4x_4 = 63$.

     Last year, $6x_1 + 6x_2 + 4x_3 + 4x_4 = 58$.

     Find out how many slices each choir member eats.

2. In problem 1, we solved Ax = b. What is the rank of A and the rank of the augmented matrix $\left( A \ \ b \right)$?

3. What does the answer to the previous question tell us in the light of the theorem at the bottom of page 106? So you don't have to turn back to that page, that theorem was 𝕿𝖍𝖊𝖔𝖗𝖊𝖒: If Ax = b has at least one solution, then the rank of $\left( A \ \ b \right)$ must be equal to the rank of A.

4. What we would love to know is: *If the rank of* $\left( A \ \ b \right)$ *is equal to the rank of* A, *then* Ax = b *has at least one solution.*

     How is this implication related to the theorem in the previous question? Use a word from the logic lesson given on page 107.

5. Prove: *If the rank of* $\left( A \ \ b \right)$ *is equal to the rank of* A, *then* Ax = b *has at least one solution.*

*answers*

1.
$$\begin{pmatrix} 1 & 2 & 4 & 3 & 32 \\ 3 & 3 & 8 & 3 & 51 \\ 4 & 5 & 8 & 4 & 63 \\ 6 & 6 & 4 & 4 & 58 \end{pmatrix} \text{ by Gaussian elimination } \begin{pmatrix} 1 & 2 & 4 & 3 & 32 \\ 0 & -3 & -4 & -6 & -45 \\ 0 & 0 & -4 & -2 & -20 \\ 0 & 0 & 0 & 4 & 16 \end{pmatrix}$$

By back-substitution, $x_4 = 4$, $x_3 = 3$, $x_2 = 3$, $x_1 = 2$.

2. Both $\begin{pmatrix} 1 & 2 & 4 & 3 \\ 0 & -3 & -4 & -6 \\ 0 & 0 & -4 & -2 \\ 0 & 0 & 0 & 4 \end{pmatrix}$ and $\begin{pmatrix} 1 & 2 & 4 & 3 & 32 \\ 0 & -3 & -4 & -6 & -45 \\ 0 & 0 & -4 & -2 & -20 \\ 0 & 0 & 0 & 4 & 16 \end{pmatrix}$ have 4 nonzero rows.

3. In question 2, we learne*d that the rank of* A *and the rank of* $\begin{pmatrix} A & b \end{pmatrix}$ *are equal. The theorem states If* $Ax = b$ *has at least one solution, then the rank of* $\begin{pmatrix} A & b \end{pmatrix}$ *must be equal to the rank of* A. The theorem tells us . . . nothing!

4. It is the converse of the theorem. The theorem was in the form P $\Rightarrow$ Q and this is in the form Q $\Rightarrow$ P.

5. We are given that the rank of $\begin{pmatrix} A & b \end{pmatrix}$ is equal to the rank of A. By the definition of rank, which we gave on page 105, if we put A and $\begin{pmatrix} A & b \end{pmatrix}$ into echelon form, they will both have the same number of nonzero rows.

Suppose both A and $\begin{pmatrix} A & b \end{pmatrix}$ had five nonzero rows after they are put in echelon form. The fifth row of $\begin{pmatrix} A & b \end{pmatrix}$ would be equivalent to a linear equation $a_{51}x_1 + a_{52}x_2 + \ldots = b_5$ where not all the a's are equal to zero (since the rank of A is five). Suppose the first nonzero coefficient in that row is $a_{57}$ which would make $x_7$ the pivot variable (also known as the leading variable, which we defined on page 99) for that row. Assign any values you like to the free variables, which might be $x_8$, $x_9$, etc. What you will have then is an equation $a_{57}x_7 = b_5$ where $a_{57} \neq 0$. We can then solve for $x_7$.

Then, on row four, we can use the values we have found for $x_7$, $x_8$, $x_9$, etc., and we have the same situation: $a_{41}x_1 + a_{42}x_2 + \ldots = b_4$ with one pivot variable and maybe some new free variables that weren't on row 5. We assign values to those new free variables and solve for the value of the pivot variable on row 4.

We repeat this solving and back-substituting up through the rows to obtain the general solution. ∎

The purpose of a proof is to *convince* the reader of the truth of something. How can I say this delicately? Some proof writers want to *impress* you more than convince you. They would have rewritten the proof with something like: *Let the rank of* A *be r. After row reduction, the*

*$r^{th}$ row of* $\left( A \quad b \right)$ *would correspond to the equation* $\sum_{i=1}^{n} a_{ri}x_i = b_r$. *Let $a_{rj}$ be such that $a_{rk} = 0$ for all $k < j$ and $a_{rj} \neq 0$.* etc.

The critics will be impressed by the precision of such proofs. The ordinary reader might have a different response.

---

**Gandy**

---

1. You are a door-to-door salesman selling 𝐅𝐫𝐞𝐝'𝐬 𝐅𝐨𝐮𝐧𝐭𝐚𝐢𝐧 𝐏𝐞𝐧𝐬. You knock on the first door and announce to the lady of the house that she can purchase the 𝐒𝐮𝐛𝐮𝐫𝐛𝐚𝐧 𝐂𝐨𝐥𝐥𝐞𝐜𝐭𝐢𝐨𝐧 which consists of 3 fine point pens, 8 medium point pens and 4 broad point pens for only $370. Or she can buy the 𝐑𝐚𝐧𝐜𝐡𝐞𝐫'𝐬 𝐂𝐨𝐥𝐥𝐞𝐜𝐭𝐢𝐨𝐧 which consists of 6 fine point pens, 5 medium point pens and 44 broad point pens for only $1185.

Have you given her enough information so that she can determine the prices of the individual pens?

2. She looks at you and asks, "How come the 𝐑𝐚𝐧𝐜𝐡𝐞𝐫'𝐬 𝐂𝐨𝐥𝐥𝐞𝐜𝐭𝐢𝐨𝐧 has so many broad point pens in it?"

No one had ever asked you that question before. You answer, "If you give me a minute, I'll look in my 𝐈𝐧𝐤𝐛𝐨𝐨𝐤 [the reference manual that 𝐅𝐫𝐞𝐝'𝐬 𝐅𝐨𝐮𝐧𝐭𝐚𝐢𝐧 𝐏𝐞𝐧𝐬 gives to each employee]."

This sentence is an implication, an *if— then—* sentence. State the contrapositive, the inverse, and the converse of, "If you give me a minute, I'll look in my 𝐈𝐧𝐤𝐛𝐨𝐨𝐤."

3. The statement, "If you give me a minute, I'll look in my 𝐈𝐧𝐤𝐛𝐨𝐨𝐤" is true. The converse is also true. If she said, "Never mind about why the 𝐑𝐚𝐧𝐜𝐡𝐞𝐫'𝐬 𝐂𝐨𝐥𝐥𝐞𝐜𝐭𝐢𝐨𝐧 has so many broad point pens in it," you wouldn't haul out your 𝐈𝐧𝐤𝐛𝐨𝐨𝐤 and start rummaging through it. In this particular case, the implication and its converse are both true. Give an example (any example you like) of an *if— then—* sentence that is true where the converse is false.

In case you are curious about what is in the **Inkbook** concerning ranchers and broad point pens, we offer you this footnote.[*]

4.  Give the general solution to the system of linear equations of question one. Your answer will look like what we worked out on the bottom of page 101.

### answers

1.  The easy answer is that since the number of equations is less than the number of unknowns, there is an infinite number of solutions. (From *The Handy Guide to Dating* (page 110): *. . . then you glance at the number of equations and see if it is less than the number of unknowns . If so, there is an infinite number of solutions.*)

   The more sophisticated answer would first consider whether the system of equations is **consistent**—whether there is any solution at all. You would have to check whether *the rank of the augmented matrix* $\begin{pmatrix} A & b \end{pmatrix}$ *is larger than the rank of* A. If it were larger, then there would be no solution, and, of course, she then couldn't determine the prices of the individual pens.

   So in either case—no solution or many solutions—she couldn't determine the prices of the individual pens.

2.    Contrapositive: If I don't look in my **Inkbook**, then you didn't give me a minute.

   Inverse: If you don't give me a minute, then I won't look in my **Inkbook**.

   Converse: If I look in my **Inkbook**, then you gave me a minute."

3.  Your example will probably differ from mine.

   My favorite is, *If it's two o'clock, then the big hand is on the twelve.* That is certainly true.    But the converse isn't true.

---

[*] In the **Inkbook** (written by Prof. Eldwood, 1847 edition) is the explanation: Ranchers are often bothered by deer hunters who accidentally mistake their favorite cow for a deer. Fred's Fountain Pens recommends that ranchers use our broad point pen on their cows.

4. Using Gauss-Jordan elimination,

$$\begin{pmatrix} 3 & 8 & 4 & 370 \\ 6 & 5 & 44 & 1185 \end{pmatrix} \Rrightarrow \begin{pmatrix} 3 & 8 & 4 & 370 \\ 0 & -11 & 36 & 445 \end{pmatrix} \Rrightarrow \begin{pmatrix} 1 & 2.66667 & 1.33333 & 123.333 \\ 0 & 1 & -3.2727 & -40.455 \end{pmatrix}$$

$$\Rrightarrow \begin{pmatrix} 1 & 0 & 10.06 & 231.21 \\ 0 & 1 & -3.27 & -40.455 \end{pmatrix}$$

$x_1 = \$231.21 - \$10.06x_3$     $x_2 = -\$40.46 + \$3.27x_3$ (rounding our answers to the nearest cent)

Some notes:

♪#1: We used ⟹ between the matrices. The matrices are not equal to each other. They are row equivalent. (Row-equivalent was first mentioned on page 65.) Two matrices are equal if they have the same dimensions and all corresponding entries are equal.

♪#2: $x_3$ is the free variable.

♪#3: $x_1$ is the price of a fine point pen. $x_2$ is the price of a medium point pen. $x_3$ is the price of a broad point pen. We are going to do a little detective work. $x_1$ must be a positive number since it's the price of the fine point pen.

From the general solution and $x_1 > 0$,

we have $\$231.21 - \$10.06x_3 > 0$.

So $x_3 < \$22.9831$ and since pens aren't sold for a fraction of penny, $x_3 < \$22.99$.    (or $x_3 \le \$22.98$)

From the other part of the general solution,

we have $-\$40.46 + \$3.27x_3 > 0$.

So $x_3 > \$12.373$ and since pens aren't sold for a fraction of penny, $x_3 > \$12.37$.    (or $x_3 \ge \$12.38$)

Our detective work tells us that the price of a broad point pen is somewhere between $12.37 and $22.99.

If you played with it a little more by trying various prices for the broad point pen, you would discover that $x_3$ equal to $20 would seem to make a lot of sense. Then $x_1$ would be $30, and $x_2$ would be $25.

---

**Radford**

---

1. Lee loved shopping at Mack's Fishing Store. Her whole life revolved around going to that store. Many people write in their diaries thoughts about their love life or their aspirations. Lee wrote about the things she bought at Mack's:

*Monday: Today I bought 2 fishing hooks, 1 hat, and 3 nets. The bill came to $19 (before sales tax).*

*Tuesday: 4 fishing hooks, 2 hats, and 8 nets. The bill was $46.*
*Wednesday: 6 hooks, 5 hats, 2 nets. The bill was $39.*

     All of the fishing hooks were alike. All of the hats were alike. All of the nets were alike. Assume the prices did not change during those three days. Find out how much each cost using Gaussian elimination. You will hit one of the four Nightmares (mentioned on pages 94–96).

2. Suppose her diary had been:

*Monday: Today I bought 2 fishing hooks, 1 hat, and 3 nets. The bill came to $19*
*Tuesday: 4 fishing hooks, 5 hats, and 8 nets. The bill was $61.*
*Wednesday: 8 hooks, 4 hats, 12 nets. The bill was $76.*
*Thursday: 6 hooks, 3 hats, 5 nets. The bill was $41.*

     Find out how much each cost using Gaussian elimination. Describe which of the four Nightmares (mentioned on pages 94–96) you encounter.

3. Had Lee made a mistake if her diary had been:

*Monday: Today I bought 2 fishing hooks, 1 hat, and 3 nets. The bill came to $19*
*Tuesday: 4 fishing hooks, 2 hats, and 6 nets. The bill was $43.*
*Wednesday: 8 hooks, 4 hats, 12 nets. The bill was $76.*
*Thursday: 6 hooks, 3 hats, 5 nets. The bill was $41.*

4. One last look at Lee's diary:

*Monday: Today I bought 2 fishing hooks, 1 hat, and 3 nets. The bill came to $19*
*Tuesday: 4 fishing hooks, 5 hats, and 8 nets. The bill was $61.*
*Wednesday: 8 hooks, 4 hats, 12 nets. The bill was $76.*
*Thursday: 6 hooks, 3 hats, 9 nets. The bill was $57.*

     You know what to do: Use Gaussian elimination and state which Nightmare applies.

***answers***

1.
$$\begin{pmatrix} 2 & 1 & 3 & 19 \\ 4 & 2 & 8 & 46 \\ 6 & 5 & 2 & 39 \end{pmatrix} \Rightarrow \begin{pmatrix} 2 & 1 & 3 & 19 \\ 0 & 0 & 2 & 8 \\ 0 & 2 & -7 & -18 \end{pmatrix}$$

This is Nightmare #1—a zero on the diagonal. Interchange rows 2 and 3.

$$\Rightarrow \begin{pmatrix} 2 & 1 & 3 & 19 \\ 0 & 2 & -7 & -18 \\ 0 & 0 & 2 & 8 \end{pmatrix}$$

And then solving by back-substitution, we obtain $x_3 = \$4$, $x_2 = \$5$, and $x_1 = \$1$.

2.

| 2 | 1 | 3 | 19 |
|---|---|---|---|
| 4 | 5 | 8 | 61 |
| 8 | 4 | 12 | 76 |
| 6 | 3 | 5 | 41 |

| 2 | 1 | 3 | 19 |
|---|---|---|---|
| 0 | 3 | 2 | 23 |
| 0 | 0 | 0 | 0 |
| 0 | 0 | -4 | -16 |

This is Nightmare #3, a row with all zeros. Eliminate that row.

| 2 | 1 | 3 | 19 |
|---|---|---|---|
| 0 | 3 | 2 | 23 |
| 0 | 0 | -4 | -16 |

Solving by back substitution, $x_3 = 4$, $x_2 = 5$, $x_1 = 1$

3. The first step in the Gaussian elimination would be

$$\begin{pmatrix} 2 & 1 & 3 & 19 \\ 0 & 0 & 0 & 5 \\ 0 & 0 & 0 & 0 \\ 0 & 0 & -4 & -16 \end{pmatrix}$$

The second row is Nightmare #4—a row with all zeros except for the last column. This is an impossible situation. Most probably, Lee made a mistake in entering the numbers in her diary.

4.

| 2 | 1 | 3 | 19 |
|---|---|---|---|
| 4 | 5 | 8 | 61 |
| 8 | 4 | 12 | 76 |
| 6 | 3 | 9 | 57 |

| 2 | 1 | 3 | 19 |
|---|---|---|---|
| 0 | 3 | 2 | 23 |
| 0 | 0 | 0 | 0 |
| 0 | 0 | 0 | 0 |

Eliminate the bottom two rows and declare $x_3$ to be a free variable.

| 2 | 1 | 3 | 19 |
|---|---|---|---|
| 0 | 3 | 2 | 23 |

Since we have a free variable, we have to call on Mr. Jordan.

| 2 | 0 | 2.33333 | 11.3333 |
|---|---|---|---|
| 0 | 3 | 2 | 23 |

$\Rightarrow$

| 1 | 0 | 1.16667 | 5.66667 |
|---|---|---|---|
| 0 | 1 | 0.66667 | 7.66667 |

$\Rightarrow$

$x_1 = 5.66667 - 1.16667x_3$

$x_2 = 7.66667 - 0.66667x_3$

---

**Danville**

---

       Robin was taking her employment test at *Harry's Hamburgers*. The last question on the test was a giant system of linear equations. It had 6 equations and 13 unknowns. The question asked, "Find one particular solution to this system of equations." She couldn't believe that such a question was on an employment test for flipping hamburgers.

1. Robin was wondering if this was a trick question that had "This system has no solution" as the correct answer. She flipped open her *Life of Fred: Linear Algebra* book and found two theorems that might apply to this situation:

    ✔ the theorem at the bottom of page 106

    ✔ the theorem of question 5 on page 111

Which of these should she use?

2. The two theorems in the previous question are converses of each other. Both of them are true. One of them is in the form P ⇨ Q and the other is in the form Q ⇨ P. Combine the statement of these two theorems into one theorem using "iff " (if and only if ). Your answer will begin, "The rank of $\left( A \; b \right)$ is equal to. . . ."

3. Robin didn't want to go through the work of solving 6 equations and 13 unknowns. She decided to cheat.

       She looked at someone else's paper. Their solution was $x_1 = 7$, $x_2 = 7$, $x_3 = 7$, $x_4 = 7$. . . . Robin knew that if she just copied that answer, she might be accused of cheating. She looked at another paper and saw the answer $x_1 = 1$, $x_2 = 2$, $x_3 = 3$, $x_4 = 4$. . . .

       She decided to add together these two answers. She wrote, $x_1 = 8$, $x_2 = 9$, $x_3 = 10$, $x_4 = 11$. . . .

       Assuming the other two people had correct answers, would Robin's answer be correct?

4. If she knew that $x_1 = 7$, $x_2 = 7$, $x_3 = 7$, $x_4 = 7$ . . . was a correct solution to Ax = b, and she knew that $x_1 = 8$, $x_2 = 9$, $x_3 = 10$, $x_4 = 11$ . . . was a correct solution to the homogeneous system Ax = 0, would her answer of $x_1 = 8$, $x_2 = 9$, $x_3 = 10$, $x_4 = 11$. . . have been correct (using the theorem at the bottom of page 103)?

5. Robin turned in her test. All three people who took the test were hired. Robin asked why that last question was on the test. The fellow who was grading the tests said, "We here at *Harry's* want to find out who is a quitter when faced with a difficult problem. It's hard work here at *Harry's Hamburgers*, and those people who don't even attempt the last problem on the test aren't hired."

"Hard work?" Robin thought to herself. "That's not for me." She walked out.

The other two test takers were discussing the theorem at the bottom of page 103 which reads: If $y$ is any particular solution to $Ax = b$ and $z$ is any solution to $Ax = 0$ and if $r$ is any scalar, then $y + rz$ is also a solution to $Ax = b$.

"Of course," said one of them, "that theorem doesn't mean that if $y$ is any particular solution to $Ax = b$, that *every* solution $w$ to $Ax = b$ can be expressed as $y + z$ where $z$ is some solution of $Ax = 0$."

"Yeah," said the other test taker. "But that's also true."

Prove it.

### answers

1. This is not an easy question to answer. Robin is working with a linear system $Ax = b$. In order to determine which of the two theorems to use, she first has to find out whether the rank of $\left( A \; b \right)$ is equal to the rank of $A$.

    If they are equal, then use the theorem of question 5 on page 111.

    If they aren't, then use the theorem at the bottom of page 106.

2. The rank of $\left( A \; b \right)$ is equal to the rank of $A$ iff $Ax = b$ has at least one solution.

3. It looks like Robin won't have much chance of flipping burgers. Suppose $y$ and $z$ are solutions to $Ax = b$. Viz., $Ay = b$ and $Az = b$ are both true. Adding these two equations: $Ay + Az = 2b$

    Using the distributive property: $A(y + z) = 2b$

    That means that $y + z$ is the correct solution to $Ax = 2b$, and not to $Ax = b$.

4. Yes. This was a perfect use of that theorem.

5.  Given y is a particular solution to $Ax = b$. Then we are told that w is some solution to $Ax = b$. We need to find a z that is some solution to $Ax = 0$ so that $w = y + z$.

    From the given we have $Aw = b$ and $Ay = b$.

    Subtract one equation from the other: $Aw - Ay = 0$

    Distributive law: $A(w - y) = 0$

    Let $z = w - y$. Then z is a solution to $Ax = 0$.

    Then $w = y + z$.    ■

---

## Joes

    The den meetings for Cub Scouts were taking a heavy toll on Madison's apartment. Too many things were getting broken. She decided to stop giving them Sluice.

    She thought that teaching the boys a little high school algebra might calm them down.

    When the boys arrived, she stood in front of them and wrote $5x = 20$ on a big piece of paper. The boys seemed restless. One of them, Michael, said, "Aw, that's too easy. Everybody knows that's 4."

    Madison knew she needed to crank things up a bit. "Okay boys. I was just warming you up. Let's look at $ax = b$ where a and b are numbers. We solve it by finding $a^{-1}$ and multiplying both sides of the equation by it. She wrote on the paper:

$$ax = b$$
$$a^{-1}ax = a^{-1}b$$
$$x = a^{-1}b.$$

    "Yeah," said Michael. "Like for $5x = 20$, $5^{-1}$ is the same as 1/5. It always works unless you can't invert the a in $ax = b$."

    The other boys started chanting, "You can't invert the a! You can't invert the a! You can't invert the a! You can't invert the a! You can't invert the a! You can't invert the a!" One of the boys tipped over a chair.

    "And when can't you invert the a?" Madison asked.

Three of the boys ran into the other room. Four of the others continued yelling, "You can't invert the a! You can't invert the a! You can't invert the a! You can't invert the a! You can't invert the a! You can't invert the a!"

Michael answered, "$a^{-1}$ is not defined when a equals zero."

1. $ax = b$ looks a lot like $Ax = b$, a system of linear equations. Everyone knows that $Ax = b$ doesn't have a solution if the rank of $\begin{pmatrix} A & b \end{pmatrix}$ is greater than the rank of $A$. We gave a name to an n×n matrix $A$ where the rank of $A$ was less than n. We said that such matrices were *the next best thing* to being a zero matrix. What did we call such matrices?

If $A^{-1}$ is a square matrix such that $A^{-1}A = I$, then $A$ is called **invertible**. You could also say that $A$ is **nonsingular**. If You can't invert the A!, then, of course, you would say that $A$ is singular.

The boys' chant changed from "You can't invert the a!" to "We want Sluice! We want Sluice! We want Sluice! We want Sluice!" The three boys came in from the other room when they heard Sluice! One of them had Madison's goldfish in his hands.

Madison attempted to continue her lecture. "Everyone knows that $A^{-1}A = I$, where I is the identity matrix*."

2. In question 3, below, we are going to show that if $A^{-1}A = I$, then $AA^{-1}$ must also equal I, but first, we are going to need a lemma.** Prove that for an invertible matrix $C$, if $CC = C$, then $C$ is the identity matrix I.
3. Prove that if $A^{-1}A = I$, then $AA^{-1}$ must also equal I. (This is one of the rare cases in which matrix multiplication does commute.) This is not an especially easy proof.

---

✱ In problems 2 and 4 on page 66, we introduced the identity matrix I. It had the property that for any matrix $B$, it was always true that $BI = IB = B$.

✱✱ In geometry, you might have learned that a lemma is a theorem you prove first in order to get ready to prove a more major theorem. It makes the proof of the major theorem easier.

***answers***

1. If $A$ is an $n \times n$ matrix whose rank is less than n, $A$ is called a **singular** matrix.

2. We are given $CC = C$. *On the left side* of each side of the equation, multiply by $C^{-1}$. (Since matrix multiplication is not commutative, if we multiply *on the left* one side of an equation, we must multiply *on the left* the other side of the equation.)

$CC = C$ becomes $\qquad\qquad C^{-1}(CC) = C^{-1}C$

Associative law $\qquad\qquad (C^{-1}C)C = C^{-1}C$

By definition of $C^{-1}$ $\qquad\qquad I\ C = I$

By definition of I $\qquad\qquad C = I$

3. $A^{-1}A = I$ is given. *On the left side* of each side of the equation, multiply by $A$.

So $A^{-1}A = I$ becomes $\qquad\qquad A(A^{-1}A) = AI.$

By the associative law, $\qquad\qquad (AA^{-1})A = AI$

Since I is the identity matrix, $\qquad\qquad (AA^{-1})A = A$

Multiply both sides on the right

by $A^{-1}$, $\qquad\qquad (AA^{-1})(AA^{-1}) = AA^{-1}$

By the lemma of the previous problem, where $(A A^{-1})$ is $C$, we can assert that $AA^{-1}$ is I.

---

## Kelseyville

1. You are an element in a matrix in echelon form. To the left of you, all the entries are zero. You might not be a distinguished element. Why not? (In symbols, this question might read: If $C$ is a echelon matrix, and if for some particular i and j it is true that $(C)_{ik} = 0$ for all $k < j$, then why is it not necessarily true that $(C)_{ij}$ is a distinguished element?)

The definition of a distinguished element was given in the box on page 99.

2. $Ax = b$ represents a system of linear equations that has 5 leading variables and 9 free variables. $A$ is an $m \times n$ matrix. What can you say about m and n?

3. Jackie asked the police quartermaster what kind of ammo to use.

"There are three different brands," the quartermaster explained. "There's FlavorFire, Wack'em, and DeadOn."

Without hesitation, Jackie said, "Give me a box of each."

The quartermaster put a box of each on the counter and wrote $41 in the department budget ledger.

Next week, Jackie asked for 2 boxes of FlavorFire, 1 box of Wack'em, and 4 boxes of DeadOn. The quartermaster wrote $98.

The week after that, it was 4 boxes of FlavorFire, 1 box of Wack'em, and 1 box of DeadOn. The cost was $77.

First read problem 4, below, and then decide which of the five methods of solving
$$\begin{cases} x_1 + x_2 + x_3 = 41 \\ 2x_1 + x_2 + 4x_3 = 98 \\ 4x_1 + x_2 + x_3 = 77 \end{cases}$$
will make your work easiest. Use that method and solve the system.

4. For the first three weeks of every month, Jackie always placed the same orders:

$1^{st}$ week: one box of each.

$2^{nd}$ week: 2 boxes of FlavorFire, 1 box of Wack'em, and 4 boxes of DeadOn.

$3^{rd}$ week: 4 boxes of FlavorFire, 1 box of Wack'em, and 1 box of DeadOn.

This month, the ammo manufacturers changed the prices, so the quartermaster wrote:

$1^{st}$ week: $46.

$2^{nd}$ week: $113.

$3^{rd}$ week: $85.

Find the new prices of the three brands of ammo.

FlavorFire   Wack'em   DeadOn

5. Here's a different way of looking at Ax, the product of a matrix times a column vector. Fill in the blanks:

$$Ax = \begin{pmatrix} a_{11} \; a_{12} \; a_{13} \\ \\ a_{33} \end{pmatrix} \begin{pmatrix} x_1 \\ \\ \end{pmatrix} = \begin{pmatrix} a_{11}x_1 + \\ \\ \end{pmatrix} = \begin{pmatrix} a_{11}x_1 \\ \\ \end{pmatrix} + \begin{pmatrix} a_{12}x_2 \\ \\ \end{pmatrix} + \begin{pmatrix} \\ \\ \end{pmatrix}$$

$$= \begin{pmatrix} a_{11} \\ \\ \end{pmatrix} x_1 + \begin{pmatrix} a_{12} \\ \\ \end{pmatrix} x_2 + \begin{pmatrix} a_{13} \\ \\ \end{pmatrix} x_3$$

which shows that Ax can be represented as a linear combination of the *column vectors* of A.

6. Consider the implication P ⇨ Q.

Fill in the blanks with either *contrapositive, inverse,* or *converse.*

    The inverse of the converse is the _____.

    The contrapositive of the inverse is the _____.

**answers**

1. The entry in the second row, second column of  $\begin{pmatrix} 3 & 4 & 8 & 2 & 7 \\ 0 & 0 & 5 & 8 & 6 \\ 0 & 0 & 0 & 4 & 4 \\ 0 & 0 & 0 & 0 & 0 \end{pmatrix}$
has all zeros to its left, but
it isn't a distinguished element, because
it is itself a zero.  Distinguished elements
are nonzero.

2. Leading variables and free variables correspond to entries in a matrix in echelon form.  So first place A in echelon form using elementary row operations.

    Every leading variable corresponds to a distinguished element which is the first nonzero element in a row of the matrix.  So there is a nonzero row in A for every one of the 5 leading variables.  Other rows of A will be all-zero rows.  So $5 \leq m$.

    Every column of A (in echelon form) will correspond either to a leading (pivot) variable or to a free variable.  So $n = 5 + 9$.

3. Given that we are solving Ax = b twice—for two different values of b, the *LU*-decomposition  method is the easiest.

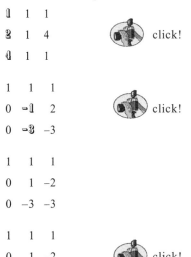

$$\begin{array}{ccc} 1 & 1 & 1 \\ 2 & 1 & 4 \\ 4 & 1 & 1 \end{array} \qquad \text{click!}$$

$$\begin{array}{ccc} 1 & 1 & 1 \\ 0 & -1 & 2 \\ 0 & -3 & -3 \end{array} \qquad \text{click!}$$

$$\begin{array}{ccc} 1 & 1 & 1 \\ 0 & 1 & -2 \\ 0 & -3 & -3 \end{array}$$

$$\begin{array}{ccc} 1 & 1 & 1 \\ 0 & 1 & -2 \\ 0 & 0 & -9 \end{array} \qquad \text{click!}$$

$$\begin{matrix} 1 & 1 & 1 \\ 0 & 1 & -2 \\ 0 & 0 & 1 \end{matrix}$$

So A is decomposed into LU.  $A = LU = \begin{pmatrix} 1 & 0 & 0 \\ 2 & -1 & 0 \\ 4 & -3 & -9 \end{pmatrix} \begin{pmatrix} 1 & 1 & 1 \\ 0 & 1 & -2 \\ 0 & 0 & 1 \end{pmatrix}$

$Ax = b$ becomes $LUx = b$ becomes $Ly = b$  (when we replace $Ux$ with $y$).

Solving $Ly = b$, which is

$$\begin{cases} y_1 & = 41 \\ 2y_1 - y_2 & = 98 \\ 4y_1 - 3y_2 - 9y_3 = 77 \end{cases}$$  by forward-substitution gives $y_1 = 41$

$y_2 = -16$

$y_3 = 15$

Solving $Ux = y$, which is

$$\begin{cases} x_1 + x_2 + x_3 = 41 \\ x_2 - 2x_3 = -16 \\ x_3 = 15 \end{cases}$$  by back-substitution gives $x_3 = 15$

$x_2 = 14$

$x_1 = 12$

4.  The nice thing about *LU*-decomposition is that we don't have to go through the decomposition again.  Only b has changed.

Solving $Ly = b$, which is

$$\begin{cases} y_1 & = 46 \\ 2y_1 - y_2 & = 113 \\ 4y_1 - 3y_2 - 9y_3 = 85 \end{cases}$$  by forward-substitution gives $y_1 = 46$

$y_2 = -21$

$y_3 = 18$

Solving $Ux = y$, which is

$$\begin{cases} x_1 + x_2 + x_3 = 46 \\ x_2 - 2x_3 = -21 \\ x_3 = 18 \end{cases}$$  by back-substitution gives $x_3 = 18$

$x_2 = 15$

$x_1 = 13$

5.  $Ax =$

$$\begin{pmatrix} a_{11} & a_{12} & a_{13} \\ a_{21} & a_{22} & a_{23} \\ a_{31} & a_{32} & a_{33} \end{pmatrix} \begin{pmatrix} x_1 \\ x_2 \\ x_3 \end{pmatrix} = \begin{pmatrix} a_{11}x_1 + a_{12}x_2 + a_{13}x_3 \\ a_{21}x_1 + a_{22}x_2 + a_{23}x_3 \\ a_{31}x_1 + a_{32}x_2 + a_{33}x_3 \end{pmatrix} = \begin{pmatrix} a_{11}x_1 \\ a_{21}x_1 \\ a_{31}x_1 \end{pmatrix} + \begin{pmatrix} a_{12}x_2 \\ a_{22}x_2 \\ a_{32}x_2 \end{pmatrix} + \begin{pmatrix} a_{13}x_3 \\ a_{23}x_3 \\ a_{33}x_3 \end{pmatrix}$$

$$= \begin{pmatrix} a_{11} \\ a_{21} \\ a_{31} \end{pmatrix} x_1 + \begin{pmatrix} a_{12} \\ a_{22} \\ a_{32} \end{pmatrix} x_2 + \begin{pmatrix} a_{13} \\ a_{23} \\ a_{33} \end{pmatrix} x_3$$

6. The inverse of the converse is the _____ .

Start with $P \Rightarrow Q$. Its converse is $Q \Rightarrow P$. The inverse of that is not-Q $\Rightarrow$ not-P. This is the <u>contrapositive</u> of $P \Rightarrow Q$.

The contrapositive of the inverse is the _____ .

Start with $P \Rightarrow Q$. Its inverse is not-P $\Rightarrow$ not-Q. The contrapositive of that is $Q \Rightarrow P$. This is the <u>converse</u> of $P \Rightarrow Q$.

# Chapter Two and a Half
## Vector Spaces

It was noon. Betty and Alexander arrived at the picnic grounds on the Great Lawn and found the picnic table that they had reserved. They saw the cans of picnic rice scattered on the ground and broken jars of mustard smashed against the tree trunks.

"I wonder where Fred is?" Alexander said. "He's usually very punctual." Alexander was about to set the picnic basket on the table.

"Don't!" Betty warned Alexander. "It's got Sluice all over it."

Alexander put the basket on the ground and picked up the table. "I'll just go and dip it in the Great Lake."

Putting a picnic basket on the ground at the Great Lawn was a dangerous thing to do. Immediately, it was surrounded by bunnies.

Alexander carried the table over his head as he walked toward the lake. He found Fred waist-deep in the water doing the Twist surrounded by dead fish. Fred grinned at Alexander but didn't offer an explanation. He was too embarrassed.

Alexander plunged the table into the water and shook it. More fish died.

As man and boy walked back to the picnic site, Alexander was the first to spot the bunnies. When you are six feet tall, you can see a lot farther than when you are three feet tall. "Six," he said to himself.

Fred didn't know what Alexander was talking about, but he didn't want to be left out of the conversation, so he expatiated on that idea, "You know, there are people that have such good memories that they can remember when the act of counting was considered theoretical math."

"Yeah," Alexander said. "Going from 'a bunch of bunnies' to 'six bunnies' is a big step for a two-year-old."

Fred's heart started to beat faster. They were "talking math." Fred drifted into his expository mode, "Yes, and then after they learned counting, we made it more theoretical by introducing addition. And then more theoretical with 6x + 5y. And more theoretical with sin x. And then we abstracted all of these and called them functions*."

**I, your reader, can guess what is coming. We're going to climb to the heights again. Given the title of the chapter,** *Vector Spaces* **, I bet we are going to abstract the idea of vectors.**

You could almost write this book without my help.

**Hardly. Before I started this book, the only vectors I knew where the geometrical ones which were arrows.**

Yes. Back in *Life of Fred: Calculus* we introduced vectors with these words:

Here is a picture of a vector named **A** and his

another view
from the heights

bigger brother next to him. His bigger brother is three

times as long as **A** and points in the same direction as **A**.

We'll call him 3**A**. (This is what it means to multiply a vector by a scalar. It's called

**scalar multiplication**.)

**And in this linear algebra book, back on page 32, you said a vector was a matrix that has only one row or only one column. Later, you called them skinny matrices.**

Alexander set the table down with a thud, and the bunnies scattered. "And the whole point of this abstracting," said Alexander, "is that anything we say about functions will be true about counting, about 6x + 5y, and about sin x. It's labor-saving."

---

★ Given any two sets, a **function** is a rule which associates to each element of the first set exactly one element of the second set. The first set is called the **domain** and the second set is called the **codomain**. The set of all elements of the codomain that are the **image** of at least one element of the domain using the function is the **range** of the function. The range is a subset of the codomain.

Addition is a function in which the first set is the set of all ordered pairs of numbers and the second set is the set of all numbers. The ordered pair (5, 7) is mapped by the addition function to 12.

Even *counting* is a function. The codomain is the set of **whole numbers** = {0, 1, 2, 3, 4. . .}. A little more than a hundred years ago Giuseppe Peano wrote five axioms (the Peano Axioms) about the whole numbers and proved:

**Theorem**: 1 + 1 = 2.

In the Peano Axioms 2–4, he described the **successor function**. The successor function applied to 4779 is 4780.

Betty set the picnic basket on the table and joined the conversation, "Yes, and anything we say about vector spaces will be true about every example of a vector space."

"It can be tricky creating a new definition," Fred commented. "You look at a bunch of different things and ask, 'What do they have in common?' For example, when you were a kid sitting at the kitchen table, your mother wanted to teach you the concept of *three*. She may have pointed to  and you might have said, 'bananas.'

Then she pointed to and then she pointed out the window to the

pets you have in the yard and then she asked you,

'What do these all have in common?'"

Alexander smiled and said, "I would have said that you could eat them."

Betty punched Alexander playfully on the arm and told him, "Let Fred continue."

And Fred continued, "And then your mother would have pointed to

and she would have asked you, 'What do the bananas, the cherries, the chickens, and the chairs have in common?'"

Alexander couldn't resist, "I would have said that you could sit on them."

Betty rolled her eyes.

Fred persisted, "And then she pointed out the window to some towers and asked you, 'What do the bananas, the cherries, the chickens, the chairs, and the towers have in common?"

Alexander: "I could own them all."

Fred: "And she pointed to three stars."

Betty interrupted their play, "Okay boys. Enough." She looked at Fred and asked, "Is this how you really learned about the concept of *three*?"

Fred blushed a little. "Well, actually, I had read about it in a book and when my mother pointed to the bananas and the cherries, I told her that the sets { 🍌 } and { 🍒 } had the same cardinality."

We are about to abstract the "vector spaceyness" out of skinny matrices and geometric arrows.

First, we could call our vector space $\mathcal{V}$ (or $\mathcal{U}$ or $\mathcal{W}$). We will let the vectors in $\mathcal{V}$ be letters like b, u, v, w, or x.

We are going to need scalar multiplication (multiplying a real number times a vector). We will let *r, s,* and *t* stand for real numbers.

So the multiplication of real number *t* times vector u would be written as *t*u.

**Hold it! Stop the show! I, your reader, don't have all those fancy computer type fonts in my pencil. If I write tu, I can't tell the difference between the real number t and the vector u.**

Teachers at the blackboard have the same problem. The traditional solution is to write a tiny arrow over each vector.
**That will take forever! You expect me to write $\overrightarrow{u}$ every time you write u?**

You can if you like, but most people aren't so artistic. A simple little "half-arrow" will do. $\vec{u}$ = u.

If we are going to describe a vector space $\mathcal{V}$, we need to talk about two things: vector addition and scalar multiplication.

When I write "vector addition," u + v, you should think:

skinny matrices

$$(5 \quad 4 \quad 8) + (2 \quad 3 \quad 1) = (7 \quad 7 \quad 9)$$

geometric arrows

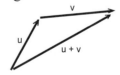

If this head-to-tail addition of arrows is new to you, just look at the skinny matrix addition instead.

When I write "scalar multiplication," $t$u, you should think:

skinny matrices

$$2(5 \quad 4 \quad 8) = (10 \quad 8 \quad 16)$$

geometric arrows

In vector space $\mathcal{V}$, vector addition, u + v, will have four properties, and scalar multiplication, $t$u, will have four properties.

Vector addition is ① commutative, ② associative, ③ has an identity, and ④ has inverses. What that means is:

① Commutative: For every u and v in $\mathcal{V}$, u + v = v + u.

② Associative: For every u, v, and w in $\mathcal{V}$, (u + v) + w = u + (v + w).

③ Identity: There is a vector 0 (called the **zero vector**) with the property that for every vector v in $\mathcal{V}$, v + 0 = 0 + v = v.

④ Inverses: For every v in $\mathcal{V}$, there is a w in $\mathcal{V}$ such that v + w = 0. (Notation: if w is the inverse of v, then we write w as –v. So ④ could be written: For every v in $\mathcal{V}$, there is a –v in $\mathcal{V}$ such that v + –v = 0.)

Chant with me . . .

*Vector addition is commutative, associative, identity, and inverses.*
*Vector addition is commutative, associative, identity, and inverses.*
*Vector addition is commutative, associative, identity, and inverses.*
*Vector addition is commutative, associative, identity, and inverses.*
*Vector addition is commutative, associative, identity, and inverses.*
*Vector addition is commutative, associative, identity, and inverses.*

In terms of skinny matrices, you are imagining . . .

Commutative: $(2 \quad 3 \quad 9) + (8 \quad 8 \quad 4) = (8 \quad 8 \quad 4) + (2 \quad 3 \quad 9)$

Associative: $((2 \quad 3 \quad 9) + (8 \quad 8 \quad 4)) + (7 \quad 1 \quad 5) = (2 \quad 3 \quad 9) + ((8 \quad 8 \quad 4) + (7 \quad 1 \quad 5))$

Identity: $(2 \quad 3 \quad 9) + (0 \quad 0 \quad 0) = (2 \quad 3 \quad 9)$

Inverses: The inverse of $(2 \quad 3 \quad 9)$ is $(-2 \quad -3 \quad -9)$ since
$$(2 \quad 3 \quad 9) + (-2 \quad -3 \quad -9) = (0 \quad 0 \quad 0)$$

---

*Intermission*
## Overlapping Algebras

In high school, there were beginning algebra and advanced algebra. In both courses you were probably taught how to factor $x^2 - 5x + 6$ into $(x - 2)(x - 3)$.

In college, there are two more algebra courses: linear algebra and abstract algebra. They also overlap a little.

You know what linear algebra is about: Fred going on a picnic with Betty and Alexander, solving systems of linear equations, vector spaces, etc.

Abstract algebra deals with algebraic structures: groups, rings, fields, integral domains, etc. Abstract algebra uses words like commutative, associative, identity, and inverses.

If you did your chanting on the previous page, you might be able to guess where some of the overlap between linear algebra and abstract algebra is.

---

## A Very Short Course in Abstract Algebra

Start with some nonempty set S and some binary operation ∘. That's a <u>groupoid</u>. (A binary operation is a function whose domain is ordered pairs.) Example of a groupoid: S is the whole numbers and ∘ is defined by a∘b = a + 5b. So 7∘8 equals 47. This particular groupoid is not associative. (7∘8)∘2 doesn't equal 7∘(8∘2).

A groupoid that is associative is called a <u>semigroup</u>.

A semigroup that has an identity element is called a <u>monoid</u>.

A monoid in which where every element has an inverse is called a <u>group</u>.

A group that is commutative (a∘b = b∘a for every a and b in S) is called an <u>abelian group</u>.

Start with a nonempty set S and two binary operations ∘ and •. If (S, ∘) is an abelian group and (S, •) is a semigroup, and if the left distributive law holds—a•(b∘c) = (a•b)∘(a•c)—and the right distributive property holds—(a∘b)•c = (a •c)∘(b •c)—then (S, ∘, •) is called a <u>ring</u>.

A <u>field</u> is, roughly speaking, a ring with commutativity for both ∘ and • thrown in, plus an identity for •, plus inverses under • for almost all the elements. The most important example of a field is the real numbers under addition and multiplication.

On the first time through, most people get lost at about the monoid level of abstraction. It gets a lot easier to understand when examples are given and you have weeks to study it.

1. Vector addition is commutative, associative, has an identity and inverses for every element. That makes vector addition a(n)_____.

2. Is $(Q, +)$ an abelian group? $Q$ is the standard abbreviation for the rational numbers. (The letter Q was chosen to stand for *quotient*.) The rational numbers, informally speaking, are the fractions. Formally, $Q = \{a/b \mid a$ and $b$ are integers and b ≠ 0$\}$.

3. Is $(Q, -)$ an abelian group? In other words, are the rational numbers under subtraction an abelian group?

4. Is $(Q, ×)$ an abelian group? In other words, are the rational numbers under multiplication an abelian group?

## . . . . . . . COMPLETE SOLUTIONS . . . . . . .

1. Vector addition is a nonempty set with a binary operation, so it's a groupoid. Since vector addition is associative, it is also a semigroup. Since $v + 0 = 0 + v = v$ for every vector $v$, it is also a monoid. Since inverses exist for every vector $v$, vector addition is a group. Since vector addition is commutative, it is an abelian group.

2. Certainly, addition of fractions is a binary operation.
Certainly, it's associative. $(½ + ¼) + ⅓ = ½ + (¼ + ⅓)$
Zero is a rational number, and it is a nice identity: $⅓ + 0 = 0 + ⅓ = ⅓$.
Inverses exist. The inverse of $⅓$ is $–⅓$.
Addition is commutative: $⅓ + ½ = ½ + ⅓$.
It all checks out. $(Q, +)$ is an abelian group.

3. $(Q, -)$ is not even a semigroup. Subtraction isn't associative.
$$(½ - ¼) - ⅓ \neq ½ - (¼ - ⅓)$$

4. $×$ is a binary operation.
$×$ is associative.
$(Q, ×)$ has an identity, viz., 1 since $⅝ × 1 = 1 × ⅝ = ⅝$.
$(Q, ×)$ has inverses for almost every element of $Q$. For example, the inverse of 2/7 is 7/2 since $2/7 × 7/2 = 1$. But there is one element of $(Q, ×)$ that doesn't have an inverse. There is no inverse for the rational number zero. $0 × ? = 1$. $(Q, ×)$ is a monoid, but not a group.

Since vector addition is a group, all the theorems about groups that you have learned/are learning/will learn in abstract algebra applies to vector addition. E.g., Theorem: The identity is unique. (So in linear algebra we know: If $0 + v = v$ and $0' + v = v$, then $0$ equals $0'$.) E.g., Theorem: The inverse of an element of a group is unique. (So in linear algebra we know: If $v + w = 0$ and $v + u = 0$, then $w = u$.)

Back to linear algebra. We were describing a vector space $\mathcal{V}$ and had given the four properties of vector addition: ① commutative, ② associative, ③ has an identity, and ④ has inverses.

To finish the job, we name the four properties of scalar multiplication (multiplying a real number times a vector) $sv$. (We are letting *r, s,* and *t* stand for real numbers, and letters like b, u, v, w, or x be vectors.)

Scalar multiplication ① is associative, ② has identity, ③ is distributive, and ④ is distributive. What this means is:

① Associative: for every pair of real numbers *s* and *t*,
    and every v in $\mathcal{V}$, $(st)v = s(tv)$.
② Identity: multiplying by the number one offers no surprises: $1v = v$.

③ Distributive: for every scalar *s* and every pair of vectors v and w, scalar multiplication distributes over vector addition: $s(v + w) = sv + sw$.

④ Distributive: for every pair of scalars *s* and *t* and every vector v, scalar multiplication distributes over scalar addition: $(s + t)v = sv + tv$.


It's time to chant . . .
*Scalar multiplication—associative, identity, and double distributive.*
*Scalar multiplication—associative, identity, and double distributive.*
*Scalar multiplication—associative, identity, and double distributive.*
*Scalar multiplication—associative, identity, and double distributive.*


Now when anyone stops you on the street and asks you what the definition of a vector space $\mathcal{V}$ is, you can say that vector spaces consist of vectors with an addition that has four properties: commutative, associative, has an identity, and has inverses. These vectors can be multiplied by real numbers, where that multiplication has four properties: associative, has an identity and is double distributive.

What you have given is the definition of a vector space over the real numbers ($\mathbb{R}$). You could also define vector spaces over the rational numbers ($\mathbb{Q}$), and vector spaces over the complex numbers ($\mathbb{C}$).*

If you had already studied abstract algebra, we could define a more general concept: a vector space $\mathcal{V}$ over any field.** But you really don't lose anything if you keep thinking about the real numbers when we talk about scalars.

---

*Intermission*

# Cancellation

Back in high school algebra, one trick you learned was that you could go from x + 5 = y + 5 to x = y. At that time, your teacher said something like, "You add –5 to each side of the equation, and when equals are added to equals, the results are equal."

And you accepted that. It sounded good.

A week later, you went from 7x = 7y to x = y. Your teacher said, "You multiply both sides by $\frac{1}{7}$ and when equals are multiplied by equals the results are equal."

And you accepted that. It sounded good.

Now, after all these years, it is time to give you the real reason for cancellation. You are old enough now to hear the truth. Do you remember when your parents sat down with you and gave you the birds-and-the-bees talk? When they told you that the stork really doesn't bring babies?

---

* The complex numbers $\mathbb{C}$ are any numbers that can be written as $s + ti$ where $s$ and $t$ are real numbers and $i = \sqrt{-1}$. Here are some complex numbers: 5 + 7i, 3 – 4i, 8, ⅝ + 39079723.3i, and my favorite, $\pi$i.

** And if you would want to really go over the edge, substitute *ring* for *field* and what you get is a **module**. For those of you who have read *Gone With the Wind*, we paraphrase one of Scarlett O'Hara's famous lines, "I'll think about that tomorrow."

### *Intermission* (*continued*)

All those years (before you got hair under your arms) they told you about the stork.

And you accepted that. It sounded good.

Now sit down and get a firm grip on this book, and I'll tell you the real reason why you can go from x + 5 = y + 5 to x = y.

It is because addition *is a function*. (I mentioned that in the footnote on page 128.) By definition of function, if you add 2 + 3 on Tuesday, you will get the same answer if you add 2 + 3 on Wednesday. And on Thursday, if you add 8/4 + 3 you will still get the same answer. Addition associates to each ordered pair of numbers *exactly one answer* in the codomain.

R. I. P.

Stork

So if I start with x + 5 and add –5 to it, I will get exactly the same answer as when I add –5 to y + 5. The x + 5 and the y + 5 play the role of the 2 in the previous paragraph, and the –5 plays the role of the 3.

Because multiplication is a function, you could go from 7x = 7y to x = y. Multiplying 7x by $\frac{1}{7}$ gives you the same answer as multiplying 7y by $\frac{1}{7}$.

In advanced algebra, when you went from $\sqrt{x}$ = 4 to x = 16, you could do so because squaring is a function.

In trig, you went from θ = 30° to sin θ = sin 30° by taking the sine of both sides of the equation. You could do that because sine is a function.

Now the upshot of all this is that vector addition is also a function. Because of that, the cancellation law applies.

If you know V + W = U + W, then you can write that V = U.

You will use that in the *Your Turn to Play* which is coming very soon.

1. Suppose we are looking at $r\mathsf{v}$ (where $r$ is a scalar and $\mathsf{v}$ is a vector). Is $r\mathsf{v}$ a scalar, a vector, or neither?

2. Show that for any vector $\mathsf{v}$, $0\mathsf{v} = \mathsf{0}$, where $0$ is the real number zero, and $\mathsf{0}$ is the zero vector. If you were writing this by hand, it might look like: $0\vec{v} = \vec{0}$.

    This is not exactly easy to show. Use any of the eight properties of vector spaces plus anything you know about the real numbers plus, of course, cancellation.

    In doing this problem, you will really learn the eight axioms for vector spaces. In just looking at the answer, you will learn about 5% of what you would have learned otherwise. Showing that $0\mathsf{v} = \mathsf{0}$ is a puzzle to be played with. Solving it is what makes math fun. The other way to "do math" is to memorize definitions, memorize theorems, and memorize the proofs of theorems, and that's no fun at all. Don't cheat yourself.

3. How would you show that for any vector $\mathsf{v}$, $1\mathsf{v} = \mathsf{v}$?

4. For real numbers $r$ and $s$, and vector $\mathsf{u}$, which kind of addition does each of the plus signs in $(r + s)\mathsf{u} = r\mathsf{u} + s\mathsf{u}$ represent?

**. . . . . . . COMPLETE SOLUTIONS . . . . . . .**

1. $r\mathsf{v}$ is a vector. The result of multiplying a scalar times a vector is a vector.

2. *From what we know about the real numbers*     $0 + 0 = 0$

Take your choice: $\begin{cases} \textit{Multiply both sides by } \mathsf{v} \\ \textit{Scalar multiplication is a function} \end{cases}$   $(0 + 0)\mathsf{v} = 0\mathsf{v}$

*One of the two distributive*
*properties for scalar multiplication*     $0\mathsf{v} + 0\mathsf{v} = 0\mathsf{v}$

*Identity property for* $\mathsf{0}$ *(defined on page 131)*     $0\mathsf{v} + 0\mathsf{v} = 0\mathsf{v} + \mathsf{0}$
*Cancellation*     $0\mathsf{v} = \mathsf{0}$

3. That's one of the four properties of scalar multiplication. On page 134 we wrote: "② Identity: multiplying by the number one offers no surprises: $1\mathsf{v} = \mathsf{v}$."

4. In $(r + s)\mathsf{u} = r\mathsf{u} + s\mathsf{u}$, the first plus sign is the addition of real numbers. The second plus sign is vector addition.

We have now arrived at $\mathcal{V}$.

When you hear "vector space $\mathcal{V}$" your brain jumps in two directions

## Properties of $\mathcal{V}$

A nonempty set with a binary operation of vector addition that is an abelian group . . .

For every $\mathsf{u}$, $\mathsf{v}$, and $\mathsf{w}$ in $\mathcal{V}$:

① commutative    $\mathsf{u} + \mathsf{v} = \mathsf{v} + \mathsf{u}$

② associative    $(\mathsf{u} + \mathsf{v}) + \mathsf{w} = \mathsf{u} + (\mathsf{v} + \mathsf{w})$

③ identity    there is a vector 0 in $\mathcal{V}$ such that $\mathsf{v} + 0 = 0 + \mathsf{v} = \mathsf{v}$ for every $\mathsf{v}$

④ inverses    for every $\mathsf{v}$ in $\mathcal{V}$ there is a $-\mathsf{v}$ such that $\mathsf{v} + -\mathsf{v} = 0$

. . . together with a binary operation called scalar multiplication which multiplies a real number $s$ times a vector $\mathsf{v}$ to obtain a vector $s\mathsf{v}$ with four properties:

For any real numbers $s$ and $t$ and any vectors $\mathsf{v}$ and $\mathsf{w}$:

1 associative      $(st)\mathsf{v} = s(t\mathsf{v})$
2 identity        $1\mathsf{v} = \mathsf{v}$
3 distributive    $s(\mathsf{v} + \mathsf{w}) = s\mathsf{v} + s\mathsf{w}$
4 distributive    $(s + t)\mathsf{v} = s\mathsf{v} + t\mathsf{v}$

## Examples of $\mathcal{V}$

row vectors     $(2 \quad 5 \quad -4 \quad 8)$

column vectors    $\begin{pmatrix} 3 \\ -92 \\ \pi \\ 14 \end{pmatrix}$

geometric arrows    $\nwarrow \mathsf{u}$

and for variety, you could substitute $\mathbb{Q}$ (the rationals) or $\mathbb{C}$ (the complex numbers) for $\mathbb{R}$ (the real numbers)

Does it look like the left column (the theory) outweighs the right column (the examples)? So much machinery. So little meat.

| | |
|---|---|
| B-B-Q spice shaker | 6.00 |
| Bottle of Filet Sauce | 8.49 |
| Filet Sauce dauber | 5.00 |
| Chef's hat | 18.00 |
| Oak cutting board | 36.00 |
| Hecks Kitchen Grill | 1800.00 |
| 500-gallon propane tank | 350.00 |
| subtotal | 2223.49 |
| tax & gratuity | 276.51 |
| total | 2500.00 |

(In *Life of Fred: Geometry*, Fred was talked into spending $2500 so that he could cook some weenies.)

What else can be considered vectors in a vector space?

One of the most surprising examples of a vector space is the set of all 2×3 matrices. Suppose we call $\begin{pmatrix} 5 & 7 & 1 \\ 3 & 8 & 2 \end{pmatrix}$ and $\begin{pmatrix} 3 & 1 & 6 \\ 4 & 0 & 1 \end{pmatrix}$ vectors. We can add two matrices if they have the same dimensions. We can multiply a matrix by a scalar: $5\begin{pmatrix} 3 & 2\pi & \sqrt{381} \\ 6 & -7 & \ln 8 \end{pmatrix} = \begin{pmatrix} 15 & 10\pi & 5\sqrt{381} \\ 30 & -35 & 5\ln 8 \end{pmatrix}$

The four properties for vector addition hold. Matrix addition is commutative and associative. The identity is $\begin{pmatrix} 0 & 0 & 0 \\ 0 & 0 & 0 \end{pmatrix}$ and the inverse of

$\begin{pmatrix} 9 & \sin 29^\circ & -7 \\ e & 5-\sqrt{3} & 2 \end{pmatrix}$ is $\begin{pmatrix} -9 & -\sin 29^\circ & 7 \\ -e & -5+\sqrt{3} & -2 \end{pmatrix}$

The four properties for scalar multiplication hold. It is associative. It has an identity and is double distributive.

So anything we prove or define about vector spaces will now be true for any set of matrices that all are of the same dimension.

Another example of a vector space is the set of all polynomials of degree 5 or less. Here are two vectors in that vector space: $189 + 3x^2 - 97x^4$ and $888x^4 + x^5$.

Notice that if you add any two polynomials of degree 5 or less, you get a polynomial of degree 5 or less. The set of degree 5 or less polynomials is said to be **closed** under addition. (You can't, for example, add two fifth degree polynomials and get a sixth degree polynomial answer.)

Recall that π and e are real numbers.

π is approximately equal to
3.141592653589793238462643383279502884197169399375105820974944592307816406286208998628034825342117067982148086513282306647093844609550582231725359408128481117450284102701938521105559644622948954930381964428810975665933446128475648233786783165271201909145648566923460348610454326648213393607260249141273724587006606315588174881520920962829254091171536436789259036001133053054882046652138414695194151160943305727036575959195309218611738193261179310511854807446237996274956735188575272489122793818301194912983367336244065664308602139494639522473719070217986094370277053921717629317675238467481846766940513200056812714526356082778577134275778960917363717872146844090122495343014654958537105079227968925892354201995611212902196086403441815981362977477130996051870721134999999837297804995105973173281609631859502445945534690830264252230825334468503526193118817101000313783875288658753320838142061717766914730359825349042875546873115956286388235378759375195778185778053217122680661300192781109590216420198938090525720106548586327886593615338182796823010019520353018529689995773622599413891249721775283479131515174857242454159069595082953311686172785588907509838175463746493931925506040092770167113900984824012858361603563707660104710181942955996174696784075451679989632066920403145513015359112523824300355876402474964732639141992726042699227967823547816360093417216412199245863150302861829745557067498385054945881

e was defined in calculus as
$\lim_{y \to \infty} (1 + 1/y)^y$

Scalar multiplication for polynomials of degree 5 or less is exactly what you would expect. If I multiply $7 + 30x - 100x^4 + 4x^5$ by 6, I get $42 + 180x - 600x^4 + 24x^5$.

It is straightforward to check that the four axioms for vector addition (commutative, associative, identity, and inverse), and the four axioms for scalar multiplication (associative, identity, and double distributive) are all true for polynomials of degree 5 or less.

Of course, the set of all 11×23 matrices is also a vector space, and the set of all polynomials of degree 400 or less is a vector space.

*Your Turn to Play*

1. What about the set of *all* polynomials. Is it a vector space?
2. What about the set of all matrices. Is it a vector space?
3. What about the set of all polynomials of degree 5. Is it a vector space?

**. . . . . . . COMPLETE SOLUTIONS . . . . . . .**

1. For the set of all polynomials, we have a binary operation of addition that satisfies commutative, associative, identity (it's the zero polynomial), and inverses. We have a binary operation of multiplying a real number times a polynomial that is associative, identity (the real number 1) and double distributive. So the set of all polynomials over the real numbers is a vector space.
2. We don't have a binary operation for addition of any two matrices. We have no way to add two matrices if, for example, one of them is a 2×3 and the other is a 5×4.

Let's be explicit about what a **binary operation** is. A binary operation is a function which maps ordered pairs $(a, b)$ into a set C. A binary operation is **well-defined** if for every $a$ and every $b$, there is exactly one element of C that $(a, b)$ is mapped to.

For the binary operation of matrix addition, $a$ and $b$ are both matrices. Unless they have the same dimensions (the same number of rows and the same number of columns), there wouldn't be an answer. Matrix addition is not well-defined for most pairs of matrices.

For the binary operation of scalar multiplication, the ordered pair $(a, b)$ consists of all real numbers for $a$, and all vectors for $b$. The ordered pair is mapped into C, which in this case is the set of all vectors.

3. In order to determine whether something is a vector space, there are actually 11 items to check.

First, the set must be nonempty.    (one item)

Second, the binary operations for vector addition and scalar multiplication are well-defined.    (two items)

Third, vector addition has the usual four properties (commutative, associative, identity, and inverses).    (four items)

Fourth, scalar multiplication has the usual four properties (associative, identity and double distributive).    (four items)

Let's see for which of the 11 items the set of all fifth degree polynomials fails.

It certainly isn't nonempty. $17 + 43.2x^5$ is a fifth degree polynomial.

Vector addition (in this case, polynomial addition) is *not* closed. Here are two fifth degree polynomials that do not add to a fifth degree polynomial: $9 + 2x^5$ and $6x^2 - 2x^5$.

Scalar multiplication is also not closed. If you multiply a fifth degree polynomial by zero, the answer is not a fifth degree polynomial.

Polynomial addition is commutative and associative, but there is no identity. What *fifth degree* polynomial would you add to $2x^4 + 8x^5$ in order to get $2x^4 + 8x^5$?

Since there is no identity for addition in fifth degree polynomials, the idea of finding inverses doesn't even make sense. In a vector space $\mathcal{V}$, the inverse of v is another element of $\mathcal{V}$ so that their sum is equal to the identity.

Here are five vector spaces that are all different, and yet, are all the same underneath.

***First vector space***: all 1×2 matrices such as (3    9) and (2    7). When you add them, you get (5    16).

***Second vector space***: all ordered pairs of numbers such as (3, 9) and (2, 7). If you add them, you get (5, 16). This second vector space can be considered as the set of all points in a plane. This is sometimes called $\mathbb{R}^2$.

***Third vector space***: all vectors whose tails are at the origin such as ![(3, 9), (2, 7)] and when you add the vectors you get ![(5, 16), (3, 9), (2, 7)]

***Fourth vector space***: all complex numbers such as 3 + 9i and 2 + 7i. If you add them, you get 5 + 16i.

***Fifth vector space***: all polynomials of degree 1 or less such as 3 + 9x and 2 + 7x. If you add them, you get 5 + 16x.

Now if we look at the vector space $\mathbb{R}^3$—the set of all ordered triples of real numbers—we could match it up with the vector space of all 1×3 matrices. We could also match it up with the set of all vectors in three dimensions with tails at the origin.

And we could match up $\mathbb{R}^3$ with all polynomials of degree 2 or less. I have trouble figuring out how to match up $\mathbb{R}^3$ with the complex number example. When we go from $\mathbb{R}^2$ to $\mathbb{R}^3$ we lose one of five examples.

*a very brief Your Turn to Play*

1. When we go to $\mathbb{R}^4$ which of the systems that match up with $\mathbb{R}^3$ do we lose?
.......**COMPLETE SOLUTIONS**.......

1. We still have 1×4 matrices. We still have all polynomials of degree 3 or less. What I can't figure out is how to draw vectors in four dimensions.

The dimension of $\mathbb{R}^3$ is 3. The set of all 4-tuples (that's what they are called) of real numbers is $\mathbb{R}^4$, and the dimension of $\mathbb{R}^4$ is 4. $\mathbb{R}^n$, which is the set of all **n-tuples** of real numbers, has dimension n. But I can't talk about the dimension of a vector space until I've defined linear combinations and linear independence, so please forget that I mentioned *dimension* until later in this chapter.

It is fun to think about $\mathbb{R}^{\infty}$. A vector in $\mathbb{R}^{\infty}$ might look like (3, –4, 9376962, 5½, 0.07, 8, 101, 43, 9, ⅝, 10000, 92, –π, ln 6, $\sqrt{387}$, 7e, . . . ). I'm just not allowed to tell you that $\mathbb{R}^{\infty}$ has an infinite dimension yet. Remember, I didn't bring up the topic of dimension yet.

Another unmentionable topic is the idea of matching up. On the previous page, $\mathbb{R}^3$ was matched up with all polynomials of degree 2 or less. The official word for matching up is **isomorphic**. If I take two vectors in $\mathbb{R}^3$ such as (3, 4, 5) and (8, –3, 2) and add them, I get (11, 1, 7). If I take the corresponding polynomials, $3 + 4x + 5x^2$ and $8 – 3x + 2x^2$, and add them, I get $11 + x + 7x^2$, which matches up with the (11, 1, 7). $\mathbb{R}^3$ and the set of all polynomials of degree 2 or less are isomorphic. In abstract algebra they use the word *isomorphic* a lot. They even define what isomorphic means.*

Here is another example of a vector space. It doesn't look like any of the other examples we have seen. And if this were an abstract algebra book, we would note that this example is not isomorphic to any of the previous examples.

The vectors in this vector space will be functions. In the footnote on page 128, a function was defined as a rule that associates to each element of the domain exactly one element of the codomain. If the

---

✶ In English, set A with a binary operation ∘ is isomorphic to set B with a binary operation •, if there is a 1-1 correspondence between A and B such that the 1-1 correspondence preserves the operation.

In symbols, (A, ∘) is isomorphic to (B, •) iff there exists a 1-1 and onto function f:A→B, such that for every $a_1$ and $a_2 \in$ A, $f(a_1 \circ a_2) = f(a_1) \bullet f(a_2)$.

In abstract algebra, they will, hopefully, give a ton of examples which makes the definition of isomorphism easier to understand.

function were named f, the domain named A, and the codomain named B, then the notation is f:A→B.

The vectors in this vector space $\mathcal{V}$ will be real-valued functions (which could be written f:ℝ→ℝ).

We define vector addition as the addition of functions.

In symbols, if f, g ∈ $\mathcal{V}$, then f + g is defined as (f + g)(r) = f(r) + g(r) for every real number r.

We define scalar multiplication as the multiplication of a function by a real number.

For example, if f(x) = 2 + 30x⁴, then 5f(x) is 10 + 150x⁴.

And for variety, you can, as always, replace ℝ by ℚ or by ℂ to get a different vector space.

> Here are some examples of real-valued functions whose domains are the real numbers:
> $f(x) = x^{25}$
> $f(x) = \sin x$
> $f(x) = \ln (x^2 + 3)$
> $f(x)$ = the digit in the thousands place when x is expressed as a decimal number.
>
> These are *not* examples of real-valued functions whose domains are the real numbers:
> $f(x) = \sqrt{x}$   since –2 is not in the domain.
> $f(x) = \ln x$   since x ≤ 0 is not in the domain.
> $f(x) = 5 + xi$   since f(3) is not a real number.
> $f(x) = \pm 2x$   since f is not a function.

We now have the definition of a vector space along with many examples. When someone mentions vector spaces to Fred, he immediately thinks of the 11 items in the definition (set not empty; the two binary operations well-defined; vector addition is commutative, associative, and has identity and inverses; scalar multiplication is associative, has an identity and double distributive), and he thinks of all the various examples including substituting ℚ and ℂ for ℝ.

But you and I are not six-year-olds who are full professors of math at a university. In order to keep from going crazy when someone mentions vector space, it might be best to just think of one example of a vector space, such as ℝ³.

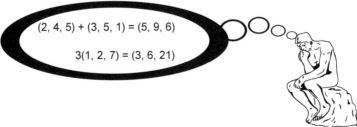

(2, 4, 5) + (3, 5, 1) = (5, 9, 6)

3(1, 2, 7) = (3, 6, 21)

Betty took things out of the picnic basket and set them on the table. A whole-wheat bagel, a slice of turkey, and a peach was her lunch. Alexander started to look a little sad.

Betty hopped off the picnic table bench and headed back to the car to get the boys' lunch. She carried an extra-large pizza box from PieOne Pizza.

"What kind is it?" Alexander shouted.

"A linear combo," Betty answered. "It's a new kind of combination pizza that Stanthony has just invented."

When Alexander opened the box, the smell of melted cheese and all the ingredients of a combo pizza* sent Alexander into seventh heaven.

The toppings were arranged in a single line across the pizza. That's why Stanthony named it the linear combo.

Fred, as usual, wasn't much interested in food. He thought of what a **linear combination** would be for vector spaces. You take some of the

---

✶ Stanthony, the owner of PieOne, was famous for his combo pizzas. His standard American combo was pepperoni, mushrooms, sausage, green pepper, onion, and extra cheese.

The India combo pizza had pickled ginger, minced mutton, and a form of cottage cheese called paneer.

For the veggie combo, he would throw on alfalfa sprouts, artichoke hearts, avocado, baby leeks, broccoli, capers, capicolla, carrots, cherry tomatoes, eggplant, green peppers, olives, lettuce, mushrooms, onions, peas, both porcini and portobello mushrooms, red onions, red peppers, roasted cauliflower, roasted garlic, scallions, shallots, snow peas, spinach, sun dried tomatoes, sweet corn, watercress, yellow peppers, yellow squash, and zucchini, and then season it with basil, bay leaf, chili powder, chives, dill, garlic, jalapeno peppers, laurel, marjoram, fenugreek, oregano, parsley, pepper, rosemary, and cardamon.

For the meaty combo, Stanthony decorated the pizza with pieces of bacon, beef, chicken, chorizo, duck, honey-cured ham, meatballs, pepperoni, salami, sausage, turkey, venison, American bison, steak, veal, yak, hare, rabbit, kangaroo, opossum, goat, ibex, pork, reindeer, moose, antelope, giraffe, squirrel, whale, bear, duck, dove, quail, mutton, ostrich, pheasant, grouse, partridge, pigeon, woodcock, goose, frog, anchovy, cod, eel, halibut, salmon, shark, tuna, abalone, clam, conch, oyster, scallop, crayfish, lobster, prawns, salmon, shrimp, squid, and tuna.

A customer once ordered the meaty combo and complained that he couldn't taste the ostrich. Stanthony rummaged around in the pizza for several minutes and found three pieces of ostrich meat.

"You were kinda light on the ostrich, weren't you?" the customer moaned. Stanthony shrugged his shoulders. The meaty combo is three inches tall.

vectors, like $v_1$, $v_2$, and $v_3$, and you take some scalars like $r_1$, $r_2$, and $r_3$ and combine them: $r_1v_1 + r_2v_2 + r_3v_3$.*

Some linear combinations are short: $r_1v_1$.

And some are long: $r_1v_1 + r_2v_2 + r_3v_3 + r_4v_4 + r_5v_5 + r_6v_6 + r_7v_7 + r_8v_8$ $+ r_9v_9 + r_{10}v_{10} + r_{11}v_{11} + r_{12}v_{12} + r_{13}v_{13} + r_{14}v_{14}$. (All are of finite length.)

Fred liked to think of linear combinations as $\sum_{i=1}^{n} r_iv_i$, but that is a little too compact for most people's tastes.

Betty began to slice her whole-wheat bagel. Alexander picked up the whole pizza and began to eat it. He didn't want to wait till Betty was done with the knife.

And Fred sat on the picnic table bench and thought of linear combinations in vector spaces.

---

*Your Turn to Play*

1. Fred first thought Suppose I start with a couple of vectors in $\mathbb{R}^3$ such as $(1, 3, -1)$ and $(5, 3, 2)$. I wonder if it's possible to find a linear combination of these two vectors that would equal $(27, 33, 1)$?

That would mean that there were real numbers r and s such that $r(1, 3, -1) + s(5, 3, 2) = (27, 33, 1)$.

Or, I could use row vectors and eliminate the commas:
$r(1 \quad 3 \quad -1) + s(5 \quad 3 \quad 2) = (27 \quad 33 \quad 1)$.
This turns into $(r \quad 3r \quad -r) + (5s \quad 3s \quad 2s) = (27 \quad 33 \quad 1)$.
And using vector addition, $(r + 5s \quad 3r + 3s \quad -r + 2s) = (27 \quad 33 \quad 1)$.

Since r and s are the unknowns, I'll use $x_1$ and $x_2$ in place of them.
$(x_1 + 5x_2 \quad 3x_1 + 3x_2 \quad -x_1 + 2x_2) = (27 \quad 33 \quad 1)$.

Now it is your turn to play. Express this equation in the form $Ax = b$ and solve it for $x_1$ and $x_2$.

2. Will the procedure in question 1, above, always work? In other words, does it matter which two vectors in $\mathbb{R}^3$ that we start with? Will we always be able to express $(27, 33, 1)$ as a linear combination of those two vectors?

3. Fred then thought Suppose I start with three vectors in $\mathbb{R}^3$ such as

---

★ If you followed my suggestion from two pages ago, you might be thinking of a linear combination like $3(1, 8, 9) + 4(3, 7, 6) + 5(2, 0, -7)$.

$(1, 5, 2), (-3, 8, 5)$ and $(5, 3, -1)$. I wonder if it's possible to find a linear combination of these three vectors that would equal $(1, 53, 24)$?

(Hint: You can go through all the steps that Fred went through in question one—That would mean that there were real numbers r and s and t such that $r(1, 5, 2) + s(-3, 8, 5) + t(5, 3, -1) = (1, 53, 24)$—or you can, if you see the pattern, spring immediately to $Ax = b$ and solve the system of linear equations. It's your choice.)

4. Will the procedure in question 3, above, always work? In other words, does it matter which three vectors in $\mathbb{R}^3$ that we start with? Will we always be able to express $(1, 53, 24)$ as a linear combination of those three vectors?

5. On the other hand, is it possible to find three vectors in $\mathbb{R}^3$ so that *any* vector in $\mathbb{R}^3$ can be written as a linear combination of those three vectors?

**. . . . . . . COMPLETE SOLUTIONS . . . . . . .**

1. $(x_1 + 5x_2 \quad 3x_1 + 3x_2 \quad -x_1 + 2x_2) = (27 \quad 33 \quad 1)$ would become

$$\begin{pmatrix} 1 & 5 \\ 3 & 3 \\ -1 & 2 \end{pmatrix} \begin{pmatrix} x_1 \\ x_2 \end{pmatrix} = \begin{pmatrix} 27 \\ 33 \\ 1 \end{pmatrix}$$

The augmented matrix is $\begin{pmatrix} 1 & 5 & 27 \\ 3 & 3 & 33 \\ -1 & 2 & 1 \end{pmatrix}$

Using Gaussian elimination, after several elementary row operations, we arrive at $\begin{pmatrix} 1 & 5 & 27 \\ 0 & 1 & 4 \\ 0 & 0 & 0 \end{pmatrix}$

So $x_2 = 4$. Back-substituting into the first row, $x_1 + 5(4) = 27$. $x_1 = 7$.

So $7(1, 3, -1) + 4(5, 3, 2) = (27, 33, 1)$. Or, in English, $(27, 33, 1)$ is a linear combination of $(1, 3, -1)$ and $(5, 3, 2)$.

2. I hope that you gave some thought to this question and are not just reading the questions and then reading the answers. I know that it's a lot easier to just read the questions and read the answers, but in doing that, you won't learn linear algebra. The *Your Turn to Play* sections are meant to

be the transition stage between reading the text and being entirely on your own for the questions in the Cities at the end of each chapter.

Reading the text is like jogging in the park.

The *Your Turn to Play* is like running three miles.

The Cities are like a marathon.

Skip the steps involved in working out for yourself the solutions in the *Your Turn to Play* and you'll die when you hit the Cities.

Now to the question of whether we can take any two vectors in $\mathbb{R}^3$ and will always be able to express (27, 33, 1) as a linear combination of them. Suppose our vectors are (0, 1, 0) and (0, 0, 1). No matter what real numbers $r_1$ and $r_2$ we choose, $r_1(0, 1, 0) + r_2(0, 0, 1)$ will never generate a 27 as the first coordinate of the answer.

3. I'll go slowly and do all the steps. If you spring ahead, you could just flip each of the row vectors into column vectors and get $Ax = b$ instantly.

Start with $\qquad r(1, 5, 2) + s(-3, 8, 5) + t(5, 3, -1) = (1, 53, 24)$

Change to row vectors $\quad r(1\ 5\ 2) + s(-3\ 8\ 5) + t(5\ 3\ -1) = (1\ 53\ 24)$

Scalar multiplication $\ (r\ 5r\ 2r) + (-3s\ 8s\ 5s) + (5t\ 3t\ -t) = (1\ 53\ 24)$

Vector addition $\quad (r\ -3s + 5t \quad 5r + 8s + 3t \quad 2r + 5s - t) = (1\ 53\ 24)$

Change to $Ax = b$
where $x_1 = r$, $x_2 = s$
and $x_3 = t$

$$\begin{pmatrix} 1 & -3 & 5 \\ 5 & 8 & 3 \\ 2 & 5 & -1 \end{pmatrix} \begin{pmatrix} x_1 \\ x_2 \\ x_3 \end{pmatrix} = \begin{pmatrix} 1 \\ 53 \\ 24 \end{pmatrix}$$

Augmented matrix
$$\begin{pmatrix} 1 & -3 & 5 & 1 \\ 5 & 8 & 3 & 53 \\ 2 & 5 & -1 & 24 \end{pmatrix} \quad \overset{\substack{-5r_1 + r_2 \\ -2r_1 + r_3}}{\Longrightarrow}$$

$$\begin{pmatrix} 1 & -3 & 5 & 1 \\ 0 & 23 & -22 & 48 \\ 0 & 11 & -11 & 22 \end{pmatrix} \overset{\substack{1\ r_2 r_3 \\ (1/11)\ r_3}}{\Longrightarrow} \begin{pmatrix} 1 & -3 & 5 & 1 \\ 0 & 1 & -1 & 2 \\ 0 & 23 & -22 & 48 \end{pmatrix} \overset{-23r_2 + r_3}{\Longrightarrow} \begin{pmatrix} 1 & -3 & 5 & 1 \\ 0 & 1 & -1 & 2 \\ 0 & 0 & 1 & 2 \end{pmatrix}$$

$x_3 = 2$.

Back-substituting, $x_2 - 2 = 2$. $x_2 = 4$.

Back-substituting, $x_1 - 3(4) + 5(2) = 1$. $x_1 = 3$.

So (1, 53, 24) can be written as a linear combination of the three given vectors: 3(1, 5, 2) + 4(–3, 8, 5) + 2(5, 3, –1) = (1, 53, 24).

4. Starting with (0, 0, 1), (0, 0, 2), and (0, 0, 9), you will never find real numbers $r_1$, $r_2$, $r_3$ such that $r_1$(0, 0, 1) + $r_2$(0, 0, 2) + $r_3$(0, 0, 9) = (1, 53, 24).

5. That's a cinch. My favorite three vectors would be (1, 0, 0), (0, 1, 0), and (0, 0, 1). Lots of people like these three vectors. They are given the special names **i**, **j**, and **k**. This saves a little bit of writing. Instead of having to write (0, 1, 0), you can abbreviate it as **j**.

Now if we start with any old vector in $\mathbb{R}^3$ such as (3.944, $\pi$, $\log_3 77$), we can express it as a linear combination of **i**, **j**, and **k**:

(3.944, $\pi$, $\log_3 77$) = 3.944**i** + $\pi$**j** + $\log_3 77$**k**.

Or, if you don't like abbreviations,

(3.944, $\pi$, $\log_3 77$) = 3.944(1, 0, 0) + $\pi$(0, 1, 0) + $\log_3 77$(0, 0, 1).

In $\mathbb{R}^4$ we start to run out of letters for (1, 0, 0, 0), (0, 1, 0, 0), (0, 0, 1, 0), and (0, 0, 0, 1), and use instead the letters $\mathbf{e}_1$, $\mathbf{e}_2$, $\mathbf{e}_3$, and $\mathbf{e}_4$.

In $\mathbb{R}^{7208}$ this becomes especially handy: $\mathbf{e}_1$, $\mathbf{e}_2$, $\mathbf{e}_3$, . . . , $\mathbf{e}_{7208}$.

Take your choice: either write $\mathbf{e}_{5893}$ or write (0, 0, 0, 0, 0, 0, 0, 0, 0, 0, 0,
0, 0, 0, 0, 0, 0, 0, 0, 0, 0, 0, 0, 0, 0, 0, 0, 0, 0, 0, 0, 0, 0, 0, 0, 0, 0, 0, 0, 0, 0, 0, 0, 0, 0, 0, 0, 0, 0, 0, 0, 0, 0, 0, 0, 0, 0, 0, 0, 0, 0, 0, 0, 0, 0, 0, 0, 0, 0, 0, 0, 0, 0, 0, 0, 0, 0, 0, 0, 0, 0, 0, 0, 0, 0, 0, 0, 0, 0, 0, 0, 0, 0, 0, 0, 0, 0, 0, 0, 0, 0, 0, 0, 0, 0, 0, 0, 0, 0, 0, 0, 0, 0, 0, 0, 0, 0, 0, 0, 0, 0, 0, 0, 0, 0, 0, 0, 0, 0, 0, 0, 0, 0, 0, 0, 0, 0, 0, 0, 0, 0, 0, 0, 0, 0, 0, 0, 0, 0, 0, 0, 0, 0, 0, 0, 0, 0, 0, 0, 0, 0, 0, 0, 0, 0, 0, 0, 0, 0, 0, 0, 0, 0, 0, 0, 0, 0, 0, 0, 0, 0, 0, 0, 0, 0, 0, 0, 0, 0, 0, 0, 0, 0, 0, 0, 0, 0, 0, 0, 0, 0, 0, 0, 0, 0, 0, 0, 0, 0, 0, 0, 0, 0, 1, 0, 0, 0, 0, 0, 0, 0, 0, 0, 0, 0, 0, 0, 0, 0, 0, 0, 0, 0, 0, 0, 0, 0, 0, 0, 0, 0, 0, 0, 0, 0, 0, 0, 0, 0,
0, 0, 0, 0, 0, 0, 0, 0, 0, 0, 0, 0, 0, 0, 0, 0, 0, 0, 0, 0, 0, 0, 0, 0, 0, 0, 0, 0, 0, 0, 0, 0, 0, 0, 0, 0, 0, 0, 0, 0, 0, 0)

*note*

In $\mathbb{R}^3$, we say that any vector is in the **span** of (1, 0, 0), (0, 1, 0), and (0, 0, 1). Or we say that any vector in $\mathbb{R}^3$ is in the span of **i**, **j**, and **k**. Or we say that the whole space $\mathbb{R}^3$ is spanned by **i**, **j**, and **k**.

Or we write $\mathbb{R}^3$ = span {(1, 0, 0), (0, 1, 0), (0, 0, 1)}.

Or we write $\mathbb{R}^3$ = span {**i**, **j**, **k**}.

**Hey! I, your reader, think that you, Mr. Author, forgot something. You forgot to mention what the spam of something is.**

The word is *span*, not *spam*.

**Big deal. Now define it!**

I have a better idea.

*Your Turn to Play*

1. You define it. I'll start the definition. You finish it. *A vector* v *is in the span of vectors* $v_1$, $v_2$, . . . , $v_n$ *if*. . . .

2. On page 144, we defined the vector space $\mathcal{V}$ of real-valued functions. Two vectors in $\mathcal{V}$ are $f(x) = \sin x$ and $g(x) = \cos x$. Is the identity function, $h(x) = x$ in the span of f and g?

3. If $f(x) = \sin x$ and $g(x) = \cos x$, is $h(x) = \sin 2x$ in the span of f and g?

4. If $f(x) = \sin x$ and $g(x) = \cos x$, is $h(x) = 5 \sin x$ in the span of f and g?

## . . . . . . . COMPLETE SOLUTIONS . . . . . . .

1. *A vector* v *is in the span of vectors* $v_1$, $v_2$, . . . , $v_n$ *if* v can be expressed as a linear combination of $v_1$, $v_2$, . . . , $v_n$.

2. "*Is h(x) = x in the span of f(x) = sin x and g(x) = cos x*" is equivalent to asking if we can find real numbers $r_1$ and $r_2$ so that $r_1 \sin x + r_2 \cos x = x$. I can't think of how to do that.

3. That is equivalent to asking if we can find real numbers $r_1$ and $r_2$ so that $r_1 \sin x + r_2 \cos x = \sin 2x$.

Some of my readers don't remember all of the multiple angle formulas from trig. Some have forgotten, for example, $\cos^5 x = (5/8)\cos x + (5/16)\cos 3x + (1/16)\cos 5x$, which was given in *Life of Fred: Trigonometry*. The formula you are trying to remember for this question is

$$\sin 2\theta = 2 \sin \theta \cos \theta$$

So, we are asking if we can find real numbers $r_1$ and $r_2$ so that $r_1 \sin x + r_2 \cos x = 2 \sin x \cos x$. No matter what values of $r_1$ and $r_2$ I think of, I can't make that equation true *for all x*. In English, $\sin 2x$ is not in the span of $\sin x$ and $\cos x$.

4. Finally, we get an easy question. Let $r_1 = 5$ and $r_2 = 0$. Then $5f + 0g = h$.

On page 143, we mentioned the vector space of all polynomials of degree 2 or less. One example of a **spanning set** for that vector space would be $\{1, x, x^2\}$. Every polynomial of degree 2 or less can be represented as a linear combination of 1, x, and $x^2$.

Then, $\{1, x, x^2, 3x + 892x^2\}$ would also be a spanning set for the vector space of all polynomials of degree 2 or less. For example, if I took some arbitrary vector like $94 + 555x + 2207x^2$, I could express it as a

linear combination of those four elements of the spanning set. Viz.,
$94 + 555x + 2207x^2 = $ **94** + **555**x + **2207**x^2 $+ $ **0**$(3x + 892x^2)$.

   That last element of the spanning set, $3x + 892x^2$, is—how can we say this politely?  That last element is worthless.  The set $\{1, x, x^2\}$ works just fine as a spanning set for the vector space.  When you toss in the extra element $3x + 892x^2$, we get a set that is **linearly dependent**.

   $\{1, x, x^2, 3x + 892x^2\}$ is linearly dependent.

**Definition**: A set of vectors $\{v_1, v_2, v_3, \ldots, v_n\}$ is linearly dependent if one of the vectors is in the span of the other vectors.

---

*Your Turn to Play*

1.  Complete this definition without using the word *span*: A set of vectors $\{v_1, v_2, v_3, \ldots, v_n\}$ is linearly dependent if one of the vectors. . . .

2.  In $\mathbb{R}^3$ is the set $\{(1, 0, 0), (0, 3, 0), (52, -9, 0)\}$ linearly dependent?

3.  In vector space $\mathcal{V}$, is the set $\{0, v_1, v_2, v_3, v_4\}$ linearly dependent?

4.  Suppose you have a vector space $\mathcal{V}$ in which for some real numbers $r_1$, $r_2, r_3, \ldots, r_n$, you know that $r_1v_1 + r_2v_2 + r_3v_3 + \ldots + r_nv_n = 0$.  Must the set $\{v_1, v_2, v_3, \ldots, v_n\}$ be linearly dependent?  (Assume $n \geq 2$.)

5.  In the answer to the previous question, we showed that, "If $r_1v_1 + r_2v_2 + r_3v_3 + \ldots + r_nv_n = 0$, and not all the $r_i$ are zero, then the set $\{v_1, v_2, \ldots, v_n\}$ must be linearly dependent."  That is true.

   What about the statement: "If $\{v_1, v_2, \ldots, v_n\}$ is linearly dependent, then it is possible to find real numbers $r_1, r_2, r_3, \ldots, r_n$, not all of them equal to zero, such that $r_1v_1 + r_2v_2 + r_3v_3 + \ldots + r_nv_n = 0$."  Is that also true?  (Of course, if all the $r_i$ equal zero, then $r_1v_1 + r_2v_2 + \ldots + r_nv_n = 0$ is trivially true.)

6.  [A question for English majors—and those mathematicians who want to learn to convey their thoughts to the outside world]  Combine the results of the previous two questions into a single "iff" theorem.  (Recall that *iff* stands for "if and only if.")

 **. . . . . . . COMPLETE  SOLUTIONS . . . . . . .**

1.  A set of vectors $\{v_1, v_2, v_3, \ldots, v_n\}$ is linearly dependent if one of the vectors can be expressed as a linear combination of the other vectors.

---

2. $(52, -9, 0) = \mathbf{52}(1, 0, 0) + (-3)(0, 3, 0)$, and therefore the set is linearly dependent.

3. Some linear algebra books might be tempted to make that into a theorem by saying that any set which contains the zero vector, $0$, must be linearly dependent. Note that $0$ is in the span of $\{v_1, v_2, v_3, v_4\}$ since $0 = 0v_1 + 0v_2 + 0v_3 + 0v_4$.

But *it is not true that* any set which contains the zero vector, $0$, must be linearly dependent. There is one counterexample. There is one set which contains $0$ which is not linearly dependent, namely $\{0\}$.

Any **singleton** (a set containing only one element) is not linearly dependent because you can't find an element that is a linear combination of *the other elements* of that set.

4. Many people might say yes and give the argument:

Start with $r_1v_1 + r_2v_2 + r_3v_3 + \ldots + r_nv_n = 0$

Add $-r_nv_n$ to both sides: $r_1v_1 + r_2v_2 + r_3v_3 + \ldots + r_{n-1}v_{n-1} = -r_nv_n$

Divide both sides by $-r_n$:

$(r_1/-r_n)v_1 + (r_2/-r_n)v_2 + (r_3/-r_n)v_3 + \ldots + (r_{n-1}/-r_n)v_{n-1} = v_n$

And then they would say, "Look, I've expressed one of the vectors,

$v_n$, as a linear combination of the other vectors."

But you and I see the error in that argument. As in many fallacious algebraic arguments, division by zero messes things up. What if $r_n$ is zero? The true statement is . . .

**Theorem**: If $r_1v_1 + r_2v_2 + r_3v_3 + \ldots + r_nv_n = 0$, *and not all the $r_i$ are zero*, then the set $\{v_1, v_2, v_3, \ldots, v_n\}$ must be linearly dependent.

5. Suppose the set $\{v_1, v_2, \ldots, v_n\}$ is linearly dependent.

By definition, one of the vectors, say $v_i$, is in the span of the other vectors. If $v_i$ isn't $v_n$, then rename $v_i$ as $v_n$ and $v_n$ as $v_i$. Now $v_n$ is in the span of the other vectors.* (This renaming makes what I'm going to write next a lot easier to read.)

By definition of span, $v_n$ is a linear combination of $\{v_1, v_2, \ldots, v_{n-1}\}$.

---

★ Instead of switching the names of the two vectors, some authors will write something like, "We know that $v_i$, for some i, is in the span of the other vectors. For definiteness, assume $v_n$ is in the span of the other vectors."

By definition of linear combination, there exist real numbers $r_1$, $r_2$, $r_3$, $r_4$, . . . , $r_{n-1}$ such that $r_1v_1 + r_2v_2 + r_3v_3 + . . . + r_{n-1}v_{n-1} = v_n$.

Add $-v_n$ to each side of the equation, and we have

$$r_1v_1 + r_2v_2 + r_3v_3 + . . . + r_{n-1}v_{n-1} - v_n = 0,$$

which could be written    $r_1v_1 + r_2v_2 + r_3v_3 + . . . + r_{n-1}v_{n-1} + (-1)v_n = 0$.

We have found real numbers $r_i$ such that $\sum_{i=1}^{n} r_iv_i = 0$ where not all of the $r_i$ are equal to zero.

6. **Theorem**: A set of vectors $\{v_1, v_2, . . . , v_n\}$ is linearly dependent iff there exist real numbers $r_1$, $r_2$, $r_3$, $r_4$, . . . , $r_n$, not all equal to zero, such that $r_1v_1 + r_2v_2 + r_3v_3 + . . . + r_nv_n = 0$.

---

### *Intermission*

Do you remember back a hundred years ago when you studied geometry in high school? Way back then, many textbooks presented proofs in Statement–Reason form.

Given the figure as marked, prove that $\angle E \cong \angle W$.

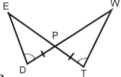

| *S* | *R* |
|---|---|
| *1.* $\overline{DP} \cong \overline{TP}$, $\angle D \cong \angle T$ | *1.* Given |
| *2.* $\angle DPE \cong \angle TPW$ | *2.* Vertical Angle Thm. |
| *3.* $\triangle DPE \cong \triangle TPW$ | *3.* ASA |
| *4.* $\angle E \cong \angle W$ | *4.* Def of $\cong$ $\triangle$ |

We could have done that with the solution to question 5 in the *Your Turn to Play*. It would have looked a little like this:

| *S* | *R* |
|---|---|
| *1.* $\{v_1, v_2, . . . , v_n\}$ is linearly dependent. | *1.* Given |
| *2.* $v_n$ is in the span of the other vectors. | *2.* Def of lin. dep. |
| *3.* $v_n$ is a lin. comb. of $\{v_1, v_2, . . . , v_{n-1}\}$. | *3.* Def of span |
| *4.* there exist real numbers $r_1$, $r_2$, $r_3$, $r_4$, . . . , $r_{n-1}$ such that $r_1v_1 + r_2v_2 + r_3v_3 + . . . + r_{n-1}v_{n-1} = v_n$. | *4.* Def of lin. comb. |
| *5. etc.* | |

Next year, this linear algebra will seem as simple to you as geometry does now. It's the nature of math: The current stuff is hard, and the stuff you have already learned looks easy.

Sometimes it's pretty easy to see if a set of vectors is linearly dependent. In question 2 of the *Your Turn to Play* that we just did, it was not hard to see that (52, –9, 0) could be expressed as a linear combination of (1, 0, 0) and (0, 3, 0).

Sometimes you need to do a little computation in order to see if a set of vectors is linearly dependent.

Betty reached back into the picnic basket and took out a fancy bottle and poured some of the yellow liquid into a glass.

Fred watched her. Alexander kept working on his linear combo pizza. Fred looked at the back of the bottle:

Antelope essence

Antelope essence

**Nutrition Facts**

Amount Per Serving

| | |
|---|---|
| Fiber | 1 g |
| Fructose | 30 g |
| Fat | 6 g |
| Folacin | 2 mg |

Then Betty pulled a second bottle from the picnic basket and poured some purple liquid into her glass. She swirled the glass and the liquid fizzed.

Beatle juice

Beatle juice

**Nutrition Facts**

Amount Per Serving

| | |
|---|---|
| Fiber | 20 g |
| Fructose | 60 g |
| Fat | 10 g |
| Folacin | 80 mg |

Then a splash of Croton oil. The mixture turned brown.

Croton oil

Croton oil

**Nutrition Facts**

Amount Per Serving

| | |
|---|---|
| Fiber | 12 g |
| Fructose | 6 g |
| Fat | 42 g |
| Folacin | 30 mg |

And a dribble of Dogwood.

Dogwood

Dogwood

**Nutrition Facts**

Amount Per Serving

| | |
|---|---|
| Fiber | 9 g |
| Fructose | 157 g |
| Fat | 38 g |
| Folacin | 23 mg |

Alexander and Fred both staring at Betty as she put a straw into the now muddy-brown liquid.

Betty explained, "The math department secretary, Belinda, sold me these four bottles. She told me that it's the newest thing in liquid nutrition. You get all four major food groups—fiber, fructose, fat, and folacin—in just the right amounts if you use these four bottles. She said it was invented by a doctor so it was medically approved. And it was only $29 for each bottle."

She took a bite of her turkey-bagel sandwich and then took a slurp of her health drink.

Fred headed to the mathematics. Here were the four vectors:

$a = (1, 30, 6, 2)$
$b = (20, 60, 10, 80)$
$c = (12, 6, 42, 30)$
$d = (9, 157, 38, 23)$.

Fred's question was:

*Is one of the bottles unnecessary?*

Which translates into:

*Is one of the vectors a linear combination of the other three vectors?*

Which translates into:

*Is one of the vectors in the span of the other three vectors?*

Which translates into:

*Is the set {a, b, c, d} linearly dependent?*
(By the definition on page 151.)

Which translates into:

*Do there exist real numbers $r_1$, $r_2$, $r_3$, and $r_4$, not all of which are zero, such that $r_1a + r_2b + r_3c + r_4d = 0$?*
(By the theorem two pages ago.)

Alexander smiled. He knew what was going on. All the facts fit together: (1) Betty had a playful spirit. (2) Belinda was a known booby. She was the one who got Fred drafted into the army (in *Life of Fred: Beginning Algebra*). (3) Betty wouldn't spend $116 (= 4 × $29) on health drinks. (4) Everyone knows that the four major food groups are sugar, salt, saturated fat, and caffeine. And finally, (5) croton oil is a drastic purgative. In other words, Betty was just putting the boys on.

Which translates into finding a nonzero solution to:

$r_1(1, 30, 6, 2) + r_2(20, 60, 10, 80) + r_3(12, 6, 42, 30) + r_4(9, 157, 38, 23) = 0$

Using row vectors:

$r_1(1\ \ 30\ \ 6\ \ 2) + r_2(20\ \ 60\ \ 10\ \ 80) + r_3(12\ \ 6\ \ 42\ \ 30) + r_4(9\ \ 157\ \ 38\ \ 23) = 0$

Using scalar multiplication:

$(r_1\ \ 30r_1\ \ 6r_1\ \ 2r_1) + (20r_2\ \ 60r_2\ \ 10r_2\ \ 80r_2) + (12r_3\ \ 6r_3\ \ 42r_3\ \ 30r_3)$
$\quad + (9r_4\ \ 157r_4\ \ 38r_4\ \ 23r_4) = (0\ \ 0\ \ 0\ \ 0)$

Using vector addition:

$(r_1 + 20r_2 + 12r_3 + 9r_4\quad 30r_1 + 60r_2 + 6r_3 + 157r_4\quad 6r_1 + 10r_2 + 42r_3 + 38r_4\quad 2r_1 + 80r_2 + 30r_3 + 23r_4) = (0\ \ 0\ \ 0\ \ 0)$

Using x instead of r, since r is the unknown:

$(x_1 + 20x_2 + 12x_3 + 9x_4\quad 30x_1 + 60x_2 + 6x_3 + 157x_4\quad 6x_1 + 10x_2 + 42x_3 + 38x_4\quad 2x_1 + 80x_2 + 30x_3 + 23x_4) = (0\ \ 0\ \ 0\ \ 0)$

Which in terms of matrices translates into:

$$\begin{pmatrix} 1 & 20 & 12 & 9 \\ 30 & 60 & 6 & 157 \\ 6 & 10 & 42 & 38 \\ 2 & 80 & 30 & 23 \end{pmatrix} \begin{pmatrix} x_1 \\ x_2 \\ x_3 \\ x_4 \end{pmatrix} = \begin{pmatrix} 0 \\ 0 \\ 0 \\ 0 \end{pmatrix}$$

Which translates hee hee hee into hə hə hə something ho ho ho that we can solve by any of the methods we've covered so far.

**Wait! Stop! I, your reader, get the feeling that something has been going on. What's all this** hee hee hee  hə hə hə  ho ho ho **business?**

Well, I, your author, was feeling kind of left out. Betty was having all the fun playing with Fred's mind. I wanted to have some fun too.

**You don't mean . . . were you putting me on? Were you fooling with me?**

Well hee hee hee look at all this stuff hə hə hə on the top half of this page. It's certainly all true, but look at where we started ho ho ho on the previous page with those four vectors

a = (1, 30, 6, 2)

b = (20, 60, 10, 80)

c = (12, 6, 42, 30)

d = (9, 157, 38, 23)  and then look at the matrix equation we arrived at. You never have to go through all that "which translates into" stuff. It's all true, but it's also all unnecessary. It is a one-step procedure.

*Your Turn to Play*

1. [For English majors and for those mathematicians who wish to communicate with the outside world] Describe in English how you would determine whether a set of row vectors is linearly dependent. (As usual, no fair looking at the answer until you have figured it out on your own.)

2. We did it in English. Now we will do it in symbols. Complete the following: *In order to determine whether the row vectors*
$u_1 = (u_{11}, u_{12}, \ldots, u_{1n}), u_2 = (u_{21}, \ldots, u_{2n}), \ldots, u_m = (u_{m1}, u_{m2}, \ldots, u_{mn})$
*are linearly dependent, see if there is a nonzero solution to* $Ax = 0$ *where* $A$ *is defined by* $(A)_{ij} = \ldots$.

3. Is one of Betty's bottles unnecessary? (In golf, you have par 5 holes. This is a par 12-minute question.)

### . . . . . . . COMPLETE SOLUTIONS . . . . . . .

1. Take the row vectors and transpose them into column vectors. (*Transposing* matrices was defined on page 96.) These are the columns of matrix $A$. Solve $Ax = 0$ to see if there is a nonzero solution. If there is, then the row vectors are linearly dependent.

2. We want the first column of $A$ to be $u_1$. We want the last column of $A$ to be $u_m$. We want the row 5, column 8 element of $A$ to be the $5^{th}$ element of $u_8$. So $(A)_{ij} = u_{ji}$.

3. On the previous page, we got as far as wondering whether this system has a nonzero solution:

$$\begin{pmatrix} 1 & 20 & 12 & 9 \\ 30 & 60 & 6 & 157 \\ 6 & 10 & 42 & 38 \\ 2 & 80 & 30 & 23 \end{pmatrix} \begin{pmatrix} x_1 \\ x_2 \\ x_3 \\ x_4 \end{pmatrix} = \begin{pmatrix} 0 \\ 0 \\ 0 \\ 0 \end{pmatrix}$$

(On page 104, we called this a homogeneous equation.)

The augmented matrix is $\begin{pmatrix} 1 & 20 & 12 & 9 & 0 \\ 30 & 60 & 6 & 157 & 0 \\ 6 & 10 & 42 & 38 & 0 \\ 2 & 80 & 30 & 23 & 0 \end{pmatrix}$

We use the three elementary row operations:
  ① *Interchange any two rows.*
  ② *Multiply any row by a nonzero number.*
  ③ *Take any row,*
      *multiply it by any number,*
      *and add it to another row.*

$$\begin{pmatrix} 1 & 20 & 12 & 9 & 0 \\ 30 & 60 & 6 & 157 & 0 \\ 6 & 10 & 42 & 38 & 0 \\ 2 & 80 & 30 & 23 & 0 \end{pmatrix} \quad \substack{\text{becomes} \\ \text{by Gaussian} \\ \text{elimination}} \quad \begin{pmatrix} 1 & 20 & 12 & 9 & 0 \\ 0 & 40 & 6 & 5 & 0 \\ 0 & 0 & 1 & 0.1667 & 0 \\ 0 & 0 & 0 & 0 & 0 \end{pmatrix}$$

The matrix is now in echelon form. (Your answer may be different than mine, because echelon form is not unique. When I did it, I interchanged rows 2 and 4 to make the arithmetic easier.)

$x_4$ is a free variable. (Nightmare #2 on page 94.)

We can set $x_4$ equal to any number we wish.

We set $x_4$ equal to some nonzero number (like 4.03869921532, or 1) and find the values of $x_3$, $x_2$, and $x_1$ by back-substitution.

Since we have a solution in which not all of the $x_i$ are equal to zero, we may assert that:

  ✓ The set {a, b, c, d} is linearly dependent.

  ✓ One of the vectors is in the span of the other three.

  ✓ One of the bottles is unnecessary.

We now know how to check if we have *too many* bottles (take the row vectors, turn them into column vectors of A and solve Ax = 0, looking for a zero on the diagonal and zeros below that entry. (Nightmare #2.)

The other question is whether we have *too few* bottles. In other words, using the four bottles, can we mix up *any* particular drink, say, 5g of fiber, 6 g of fructose, 7 g of fat, and 8 mg of folacin?

In other words, do the four bottles span the space of all possible drinks?

Antelope
essence

Beatle
juice

Croton oil

Dogwood

In the language of vector spaces, if **a**, **b**, **c**, and **d** are elements of a vector space $\mathcal{V}$, and **v** is any element of $\mathcal{V}$, is it always possible to express **v** as a linear combination of **a**, **b**, **c**, and **d**?

In other words, is every **v** in the span of **a**, **b**, **c**, and **d**?

Or, is {**a**, **b**, **c**, **d**} a spanning set for $\mathcal{V}$?

---

*Your Turn to Play*

1. Name three vectors that span $\mathbb{R}^3$.
2. Name 27 vectors that fail to span $\mathbb{R}^3$.

.......**COMPLETE SOLUTIONS**.......

1. We did this back on page 149 when we wrote that $\mathbb{R}^3$ = span {(1, 0, 0), (0, 1, 0), (0, 0, 1)} which we abbreviated as $\mathbb{R}^3$ = span {**i**, **j**, **k**}.

There are many other possibilities for spanning sets for $\mathbb{R}^3$. For example, {7**i**, 3**j**, –9**k**} will also work. In a moment we will show that any three vectors of $\mathbb{R}^3$ will work as long as they are not linearly dependent.

2. Your answer may differ from mine. The 27 vectors I choose are: (1, 0, 0), (2, 0, 0), (3, 0, 0), . . . , (26, 0, 0), (27, 0, 0). No linear combination of these vectors will ever give me (0, 0, 88).

---

Back to Betty's bottles. We are given that:

**a** = (1, 30, 6, 2)
**b** = (20, 60, 10, 80)
**c** = (12, 6, 42, 30)
**d** = (9, 157, 38, 23)

and we want to know if, for any arbitrary vector **v** = ($v_1$, $v_2$, $v_3$, $v_4$), will we always be able to find scalars $r_1$, $r_2$, $r_3$, and $r_4$ such that

$$r_1\mathbf{a} + r_2\mathbf{b} + r_3\mathbf{c} + r_4\mathbf{d} = \mathbf{v}.$$

This is all going to look familiar—very much like what we did three pages ago. If I put it in small type, you can skip over it and make this whole process into a one-step procedure.

Which translates into finding any ~~nonzero~~ solution to:

$r_1$(1, 30, 6, 2) + $r_2$(20, 60, 10, 80) + $r_3$(12, 6, 42, 30) + $r_4$(9, 157, 38, 23) = **v**

Using row vectors:

$r_1(1 \ 30 \ 6 \ 2) + r_2(20 \ 60 \ 10 \ 80) + r_3(12 \ 6 \ 42 \ 30) + r_4(9 \ 157 \ 38 \ 23) = \mathsf{V}$

Using scalar multiplication:

$(r_1 \ 30r_1 \ 6r_1 \ 2r_1) + (20r_2 \ 60r_2 \ 10r_2 \ 80r_2) + (12r_3 \ 6r_3 \ 42r_3 \ 30r_3) + (9r_4 \ 157r_4 \ 38r_4 \ 23r_4) = (v_1 \ v_2 \ v_3 \ v_4)$

Using vector addition:

$(r_1 + 20r_2 + 12r_3 + 9r_4 \quad 30r_1 + 60r_2 + 6r_3 + 157r_4 \quad 6r_1 + 10r_2 + 42r_3 + 38r_4 \quad 2r_1 + 80r_2 + 30r_3 + 23r_4) = (v_1 \ v_2 \ v_3 \ v_4)$

Using x instead of r, since r is the unknown:

$(x_1 + 20x_2 + 12x_3 + 9x_4 \quad 30x_1 + 60x_2 + 6x_3 + 157x_4 \quad 6x_1 + 10x_2 + 42x_3 + 38x_4 \quad 2x_1 + 80x_2 + 30x_3 + 23x_4) = (v_1 \ v_2 \ v_3 \ v_4)$

Which, in terms of matrices, translates into the question of whether we can always find a solution to:

$$\begin{pmatrix} 1 & 20 & 12 & 9 \\ 30 & 60 & 6 & 157 \\ 6 & 10 & 42 & 38 \\ 2 & 80 & 30 & 23 \end{pmatrix} \begin{pmatrix} x_1 \\ x_2 \\ x_3 \\ x_4 \end{pmatrix} = \begin{pmatrix} v_1 \\ v_2 \\ v_3 \\ v_4 \end{pmatrix}$$

## Handy Summary

**Dependent**: A set of row vectors is linearly dependent if $\mathsf{Ax} = \mathsf{0}$ has a nonzero solution. (The row vectors are the columns of $\mathsf{A}$.)

**Span**: A set of row vectors spans the vector space if $\mathsf{Ax} = \mathsf{v}$ always has a solution. (Again, the row vectors are the columns of $\mathsf{A}$.)

## Discussion of the Handy Summary

**Dependent**: $\mathsf{Ax} = \mathsf{0}$ will have a nonzero solution if, as we solve the augmented matrix, we get at least one zero on the diagonal with zeros below it. Then we have at least one free variable, which we can set equal to a nonzero number.

**Span**: We want a solution—any solution—to $\mathsf{Ax} = \mathsf{v}$.

What we don't want is Nightmare #4 (which we talked about back on page 96). We don't want to be working with Gaussian elimination and find one of the rows looking like $\begin{pmatrix} 0 & 0 & 0 & 0 & v_3 \end{pmatrix}$ which has no solution.

1. As Alexander munched on his linear combo pizza, he thought back to his sixteenth birthday party. Besides the usual gifts of shirts, socks, a car, and some ties, he received three special gifts:

A Stanthony's Seaside pizza in which every square inch of the pizza contained 1 gram of salt, 1 anchovy, and 3 pieces of kelp.

A Stanthony's Salt Flat pizza in which every square inch of the pizza contained 3 grams of salt and 3 pieces of kelp.

A Stanthony's Ocean King pizza in which every square inch of the pizza contained 2 grams of salt and 1 anchovy.

For your convenience, these may be summarized as:
(1, 1, 3), (3, 0, 3), and (2, 1, 0).

In order to get the right flavor, Alexander might take a large bite of one of the pizzas, a small bite of one of them, and a medium bite of the third.

Your question: Was one of the pizzas unnecessary? I.e., could taking the appropriate size bites of two of the pizzas equal a bite of the third?

2. (Continuing the previous problem) Were the three types of pizza sufficient for Alexander to get any flavor combination of salt, anchovy and kelp that he desired?

### . . . . . . . C O M P L E T E   S O L U T I O N S . . . . . . .

1. $Ax = 0$ for checking linear dependence.

$$\begin{pmatrix} 1 & 3 & 2 & 0 \\ 1 & 0 & 1 & 0 \\ 3 & 3 & 0 & 0 \end{pmatrix} \implies \begin{pmatrix} 1 & 3 & 2 & 0 \\ 0 & -3 & -1 & 0 \\ 0 & -6 & -6 & 0 \end{pmatrix} \implies \begin{pmatrix} 1 & 3 & 2 & 0 \\ 0 & -3 & -1 & 0 \\ 0 & 0 & -4 & 0 \end{pmatrix}$$

We're looking for a zero on the diagonal and zeros below that entry.

The only solution to this homogeneous set of equations is the **trivial solution** (in which all variables equal zero). No pizza was unnecessary. The three row vectors are **linearly independent**. (Please note how we snuck two definitions into the answer to this question.)

Also please notice that we do not put equal signs between the matrices. The matrices are row-equivalent. They are not equal. Two matrices are equal only if they have the same dimensions and all corresponding entries are equal.

2. $Ax = v$ for checking to see if the three pizzas span the space of all flavor combinations.

$$\begin{pmatrix} 1 & 3 & 2 & v_1 \\ 1 & 0 & 1 & v_2 \\ 3 & 3 & 0 & v_3 \end{pmatrix} \implies \begin{pmatrix} 1 & 3 & 2 & v_1' \\ 0 & -3 & -1 & v_2' \\ 0 & -6 & -6 & v_3' \end{pmatrix} \implies \begin{pmatrix} 1 & 3 & 2 & v_1'' \\ 0 & -3 & -1 & v_2'' \\ 0 & 0 & -4 & v_3'' \end{pmatrix}$$

where the $v_i'$ are linear combinations of the $v_i$, and where the $v_i''$ are linear combinations of the $v_i'$.

$v_2'$, for example, is $(-1)v_1 + v_2$.

No nightmare of $0 \quad 0 \quad 0 \quad v_i''$ has occurred. We can use back-substitution and find a solution for $v_3$, $v_2$, and $v_1$. The three pizzas span the space of all flavors.

Let's make things a little easier. In checking for linear dependence (as we did in question 1 in the *Your Turn to Play*), why drag along the last column of zeros? All we are interested in is whether there is a zero on the diagonal with zeros below it.

In checking whether a set of vectors spans the space (as we did in question 2 in the *Your Turn to Play*), why drag along the $v_i$? All we were interested in is whether there was a row of zeros.

### New Quicker Summary

**Dependent**: A set of row vectors is linearly dependent if A in echelon form has a zero on the diagonal with zeros below that entry. (The row vectors are the columns of A.)

**Span**: A set of row vectors spans the vector space if A in echelon form doesn't have a row of zeros. (Again, the row vectors are the columns of A.)

*Echelon matrices* were defined in the middle of page 99.

**Proposition**: (which is a light-weight theorem): Three row vectors will never span $\mathbb{R}^4$.

Proof: Suppose our three row vectors were:
$(1\ \ 2\ \ 3\ \ 4), (3\ \ 8\ \ 5\ \ 6)$, and $(4\ \ 5\ \ 9\ \ 1)$.

Then A would be
$$\begin{pmatrix} 1 & 3 & 4 \\ 2 & 8 & 5 \\ 3 & 5 & 9 \\ 4 & 6 & 1 \end{pmatrix}$$

Using elementary row operations, the fourth row is doomed to be all zeros. The 1 in the first row will chew out zeros below it:

$$\begin{pmatrix} 1 & 3 & 4 \\ 0 & 2 & -3 \\ 0 & -4 & -3 \\ 0 & -6 & -15 \end{pmatrix}$$

Then the 2 in the second row will chew out zeros below it, and the distinguished element (page 99) in the third row will chew out a zero below it. The result will be all zeros in the last row. By the New Quicker Summary, on the previous page, these three row vectors do not span $\mathbb{R}^4$. ∎

Some notes on this proof:

♪#1: Suppose, after the 1 in the first row had chewed out zeros below it, we found that there was a zero in the second row–second column position instead of the 2. Then we couldn't use it to chew out zeros below it. Instead, interchange that row with any row below it that contains a nonzero element in the second-column position. In the worst case, there are only zeros below it. But the worst case is not so bad. All the work of creating zeros in that column is already done, and we can move on to the next column.

♪#2: There are some people who are more fastidious than you or I. They would complain, "What kind of a stupid proof was that? Giving one example doesn't make a proof!" They might even draw nasty pictures of me:  I would have to respond: (1) I am not that handsome; and (2) I would tell them that I wasn't done with the proof. I was just warming up. Replace $(1\ \ 2\ \ 3\ \ 4)$, $(3\ \ 8\ \ 5\ \ 6)$, and $(4\ \ 5\ \ 9\ \ 1)$ with $(a_{11}, a_{12}, a_{13}, a_{14})$, $(a_{21}, a_{22}, a_{23}, a_{24})$, and $(a_{31}, a_{32}, a_{33}, a_{34})$. Replace "the fourth row is doomed to be all zeros" by "it will follow that the elements of the fourth row will become all zeros."

Replace "the 1 in the first row will chew out zeros below it" with "$a_{11}$ will create zeros in the entries below it."

With those changes, you would have a proof that you could print in a book.

**Proposition**: Five row vectors in $\mathbb{R}^4$ will always be linearly dependent.

Proof: The five row vectors will create an **A** that is $4 \times 5$. It will have four rows and five columns. After Gaussian elimination we will have four distinguished elements. (Or less, if some of the rows are all zero.) We will therefore have at most four pivots (also known as leading variables, page 99). Thus, we will have at least one free variable. (As we used to say, "It all boils down to the fact that you're either a leading variable or you are a free variable.")

We have at least one free variable. Set it to a nonzero value. We now have a non-trivial solution to **Ax = 0**. Thus any five row vectors in $\mathbb{R}^4$ will always be linearly dependent. ∎

Put these two propositions together—

☞ Three (or fewer) row vectors will never span $\mathbb{R}^4$.

☞ Five (or more) row vectors in $\mathbb{R}^4$ will always be linearly dependent—and even your baby brother could come up with the conclusion that:

☞ *If a bunch of vectors is going to span $\mathbb{R}^4$ and is going to be linearly independent, there have to be exactly four of them.*

Any fewer than four vectors, and they won't span $\mathbb{R}^4$.

Any more than four vectors, and they'll be dependent.

**Definition**: A **basis** for a vector space is a set of vectors that both spans the vector space and is linearly independent.

*Your Turn to Play*

1. Will every set of four vectors in $\mathbb{R}^4$ be a basis for $\mathbb{R}^4$?
2. Can you find four vectors in $\mathbb{R}^4$ that span $\mathbb{R}^4$, but are linearly dependent?
3. Can you find four vectors in $\mathbb{R}^4$ that are linearly independent, but which fail to span $\mathbb{R}^4$? Please consider this a puzzle to solve. Take some time to

find a proof rather than just looking at the answer. I know it's easier just to look at the answer, but our goal is not just to learn linear algebra. At this point in your life, you are well on your way to becoming a mathematician. An essential part of being a mathematician is being able to find truth—to prove things.

Today, you'll prove whether there are four linearly independent vectors in $\mathbb{R}^4$ which fail to span $\mathbb{R}^4$. The fact that most mathematicians know whether that's true is immaterial. It's new to you.

Month after month, year after year, as a mathematician-in-training you will prove stuff that other people have already proved. Then one day, you will prove something that no one else has ever proved. The transition will be almost imperceptible. It will feel no different than all the previous proofs you have done, since in each case you will be proving something that is new to you. The only difference is that you will have given the world a gift. You will have pushed back a little bit the dark forest of the unknown that surrounds us.

# . . . . . . . COMPLETE SOLUTIONS . . . . . . .

1. No. Here is an example of a set of four vectors that fail to span $\mathbb{R}^4$: $(1, 0, 0, 0)$, $(5, 0, 0, 0)$, $(-4, 0, 0, 0)$, and $(\pi, 0, 0, 0)$. No linear combination of those vectors will ever give you $(0, 1, 0, 0)$.
Those four vectors also fail to be linearly independent.

2. If $v_1$, $v_2$, $v_3$, and $v_4$ were four linearly dependent vectors that spanned $\mathbb{R}^4$, then one of them, say $v_4$, would be a linear combination of the other three. That means $v_4 = r_1 v_1 + r_2 v_2 + r_3 v_3$ for some real numbers $r_1$, $r_2$, and $r_3$.
For any $w$ in $\mathbb{R}^4$, $w = r_4 v_1 + r_5 v_2 + r_6 v_3 + r_7 v_4$ since $v_1$, $v_2$, $v_3$, and $v_4$ span $\mathbb{R}^4$.
By algebra, for any $w$ in $\mathbb{R}^4$, $w = r_4 v_1 + r_5 v_2 + r_6 v_3 + r_7(r_1 v_1 + r_2 v_2 + r_3 v_3)$.
By more algebra, $w = (r_4 + r_7 r_1)v_1 + (r_5 + r_7 r_2)v_2 + (r_6 + r_7 r_3)v_3$.
This shows that any $w$ in $\mathbb{R}^4$ is in the span of $\{v_1, v_2, v_3\}$ which is impossible by the proposition proved two pages ago. Therefore, four linearly dependent vectors can't span $\mathbb{R}^4$.

3. If $v_1$, $v_2$, $v_3$, and $v_4$ are four linearly independent vectors that do not span $\mathbb{R}^4$, then there is a $w$ in $\mathbb{R}^4$ which is not a linear combination of $v_1$, $v_2$, $v_3$, and $v_4$. Then $v_1$, $v_2$, $v_3$, $v_4$, and $w$ are five linearly independent vectors in $\mathbb{R}^4$. This contradicts the proposition of two pages ago. Thus, four linearly independent vectors in $\mathbb{R}^4$ will always span $\mathbb{R}^4$.

The arguments we have made for $\mathbb{R}^4$ carry over easily to any vector space $\mathcal{V}$.

**Lemma**: Given a vector space $\mathcal{V}$ with a basis $v_1$, $v_2$, . . . , $v_n$. Then any vector $u$ in $\mathcal{V}$ can be expressed as a linear combination of the $v_i$ *uniquely*. Proof: The vector $u$ can certainly be expressed as a linear combination of the $v_i$ in at least one way since the $v_i$ form a basis. (Every basis spans the vector space.)

So the only thing to show is the uniqueness.

Suppose $u$ could be expressed as a linear combination of the $v_i$ in two different ways:

$$u = \sum_{i=1}^{n} r_i v_i \qquad \text{where the } r_i \text{ are scalars*}$$

$$u = \sum_{i=1}^{n} q_i v_i \qquad \text{where the } q_i \text{ are scalars.}$$

Subtracting one equation from the other: $0 = \sum_{i=1}^{n} (r_i - q_i) v_i$.

Since the $v_i$ are a basis, they are linearly independent by definition. Since $0 = \sum_{i=1}^{n} (r_i - q_i) v_i$, the coefficients $r_i - q_i$ must all be zero (by the theorem on page 152). And since $r_i = q_i$ for each i, we have uniqueness.

The $r_i$ are the **coordinates** of $u$ with respect to basis $v_1$, $v_2$, . . . , $v_n$.

A lot of people call the next theorem a central one in linear algebra. Others call it absolutely fundamental or exceedingly important. If you are on a committee interviewing people to teach a course in linear algebra, you might ask them to prove this theorem. We'll compromise and call it a "Nice theorem."

---

✱ If you prefer: $u = r_1 v_1 + r_2 v_2 + \ldots + r_n v_n$

The Nice theorem deserves to have a box put around it.

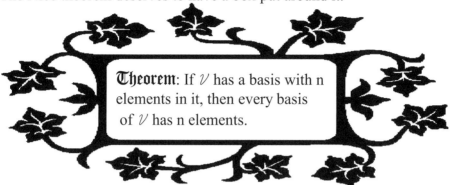

**Theorem**: If $\mathcal{V}$ has a basis with n elements in it, then every basis of $\mathcal{V}$ has n elements.

**Proof**:

Suppose that $v_1, v_2, \ldots, v_n$ and $w_1, w_2, \ldots, w_m$ are two bases for vector space $\mathcal{V}$. We want to show n = m.

First, let's eliminate the possibility that n > m. We assume n > m and look for a contradiction.

Since the $v_i$ span $\mathcal{V}$, each of the $w_i$ can be expressed as a linear combination of the $v_i$. We have m equations with n unknowns—fewer equations than unknowns. As we know from *Dating Systems of Linear Equations,* (page 110), there is an infinite number of solutions.

That translates into, "There is more than one way to express the $w_i$ as linear combinations of the $v_i$." The lemma on the previous page states that it not possible.

*Mutatis mutandis,** we eliminate the possibility that m > n. ∎

Some vector spaces that we have seen do not have bases with a finite number of elements, e.g., the vector space of real-valued functions or the vector space of all polynomials. We call those vector spaces **infinite dimensional**.

(One basis for the set of all polynomials would be $\{1, x, x^2, x^3, x^4, \ldots\}$.)

---

∗ *Mutatis mutandis* even passes spell check on my computer. [mew-TA-tis mew-TAN-dis] A little Latin always adds a little sparkle to one's discourse. *Mutatis mutandis* means "the necessary changes having been made." Normal people say, "similarly," but would you want to be called normal?

Using the Nice theorem we can say that for any vector space that has a finite basis, we may attach a unique number to it—the number of elements in any of its bases. If vector spaces were cows, they might look like:

each branded with a unique number. We call that number the **dimension** of the vector space.

---

1. Suppose we have a cow marked with a seven (a vector space with dimension equal to seven). We know by the Nice theorem that every basis will have seven elements in it.

     Show that no set of six vectors can span this cow. (This is a par 10 question. On the average, this may take about ten minutes of thinking to work through it.)

2. Same cow. Show that every set of eight vectors will be linearly dependent. (A par 15 question.)

#### . . . . . . . COMPLETE SOLUTIONS . . . . . . .

1. Assume the six vectors span the vector space. We will look for a contradiction. First, we note that the six vectors couldn't be linearly independent. If they were, they would be a basis with less than seven elements, which is not permitted by the Nice theorem.

    Since the six vectors are linearly dependent, at least one of them is expressible as a linear combination of the other five. If we eliminate that vector, then the remaining five vectors still span the cow. In symbols: If $v_1$, $v_2$, $v_3$, $v_4$, $v_5$, $v_6$ are linearly dependent, then one of the vectors, say $v_6$, is a linear combination of the other five vectors. $v_6 = r_1v_1 + r_2v_2 + r_3v_3 + r_4v_4 + r_5v_5$. (The $r_i$ are scalars.) If $w$ is any arbitrary vector in the space and since the six $v_i$ span the space by assumption, we can find six scalars such that $w = r_6v_1 + r_7v_2 + r_8v_3 + r_9v_4 + r_{10}v_5 + r_{11}v_6$. Substituting the value of $v_6$ into the second equation, we have $w = r_6v_1 + r_7v_2 + r_8v_3 + r_9v_4 + r_{10}v_5 + r_{11}(r_1v_1 + r_2v_2 + r_3v_3 + r_4v_4 + r_5v_5)$ which shows that the space is spanned by the five vectors after chopping out $v_6$.

---

Are the five remaining vectors linearly independent? If yes, then we contradict the Nice theorem. If no, then we chop out another vector.

Continue chopping. At some point we will have a linearly independent set of six or fewer vectors which spans the space, which contradicts the Nice theorem. (This process of chopping will have to stop at some time since a set consisting of a single vector is always linearly independent. Singletons are always linearly independent.)

If you change "7" to "n" in the above argument, we have established:

**Theorem**: In an n-dimensional vector space, no set of n − 1 vectors will span the space.

2. Let $v_1$, $v_2$, . . ., $v_8$ be any set of eight vectors in a seven-dimensional space. Let $b_1$, $b_2$, . . ., $b_7$ be a basis for that space. We will assume that the $v_i$ are linearly independent and look for a contradiction.

Consider the ordered list $v_1$, $v_2$, . . ., $v_8$, $b_1$, $b_2$, . . ., $b_7$. One by one, cross off that list any of the $b_i$ that are a linear combination of elements earlier on the list. For example, we would cross off $b_3$ if it were a linear combination of $v_1$, $v_2$, . . ., $v_8$, $b_1$, $b_2$.

Notice that not all of the $b_i$ would be crossed off, since that would mean that all of the $b_i$ would be linear combinations of $v_1$, $v_2$, . . ., $v_8$, which would mean that $v_1$, $v_2$, . . ., $v_8$ would span the whole vector space. Since we're assuming the $v_i$ are linearly independent, the $v_i$ would be a basis with 8 elements in it (contradicting the Nice theorem).

We now have a list that would look like $v_1$, $v_2$, . . ., $v_8$, $b_2$, $b_5$. That list is a basis for the vector space*, and it contains more than 8 elements (again, contradicting the Nice theorem). ∎

If you change "7" to "n" in the above argument, we have established:

**Theorem**: In an n-dimensional vector space, no set of n + 1 vectors will be linearly independent.

---

✱ (1) It is linearly independent by the way that we constructed it. (2) Every $b_i$ not on the list is expressible as a linear combination of elements on the list, so each of $b_1$, $b_2$, . . ., $b_7$ is expressible as a linear combination of elements on the list. Hence the list spans the entire vector space.

Betty looked at the turkey in her whole-wheat bagel sandwich. "Oh!" she exclaimed. She opened the sandwich and carefully trimmed off a bit of turkey fat. Alexander continued to munch on his combo pizza. A trickle of oil ran down his arm.

Fred watched his two friends. He, himself, wasn't eating. Betty's work with a knife reminded him of the Chop Down theorem in linear algebra. Alexander made Fred think of the Fill 'er Up theorem. This might be a good time to talk about these theorems.

**Chop Down theorem**: If you have a set of linearly dependent vectors $v_1, v_2, \ldots, v_m$, you can always find a vector $v_i$ that is a linear combination of the earlier ones on the list. Chopping $v_i$ from the list won't affect the span of the list.

Can a forest have too many trees?

Proof: Pick the smallest k such that $v_1, v_2, \ldots, v_k$ are linearly dependent. Certainly, there is such a k since $v_1, v_2, \ldots, v_m$ is linearly dependent to start with.

By definition of linear dependence (page 151, problem 1), one of the $v_1, v_2, \ldots, v_k$ is a linear combination of the other vectors. $v_i = \sum_{\substack{j=1 \\ j \neq i}}^{k} r_j v_j$ *

The coefficient $r_k$ can't be zero since k was chosen to be the smallest k such that $v_1, v_2, \ldots, v_k$ are linearly dependent. That means that we can solve the equation in the footnote below and express $v_k$ as a linear combination of $v_1, v_2, \ldots, v_{k-1}$.

The second half of the proof is to show that chopping $v_i$ from the list won't affect the span of the list. (We did this for a particular case two pages ago, at the bottom of the page.) Now for the general case, if w is any vector in the span of $v_1, v_2, \ldots, v_m$, then $w = \sum_{\alpha=1}^{m} r_\alpha v_\alpha$ and we replace the $v_k$ in that summation by a linear combination of $v_1, v_2, \ldots, v_{k-1}$. ∎

---

✱ This is called the "closed form." No ellipses ( . . . ) are needed. Without the sigma notation it would look like: $v_i = r_1 v_1 + r_2 v_2 + \ldots + r_{i-1} v_{i-1} + r_{i+1} v_{i+1} + \ldots + r_{k-1} v_{k-1} + r_k v_k$.

## 𝔍ill 'er 𝔘p theorem[*]:

If you have a set of linearly independent vectors $v_1, v_2, \ldots, v_m$ in an n-dimensional space $\mathcal{V}$, you can always find vectors to add to that list to create a basis for $\mathcal{V}$ if it isn't a basis already.

Proof: Let $b_1, b_2, \ldots, b_n$ be a basis for $\mathcal{V}$. Consider the ordered list $v_1, v_2, \ldots, v_m, b_1, b_2, \ldots, b_n$.

This set of vectors is linearly dependent by the theorem two pages ago ("In an n-dimensional vector space, no set of n + 1 vectors will be linearly independent.")

By the Chop Down theorem, one of the vectors on this list is a linear combination of the earlier ones on the list. It certainly isn't one of the $v_i$ because we are given that the $v_1, v_2, \ldots, v_m$ are linearly independent. Chop off one of the $b_i$. By the second part of the Chop Down theorem, we know that chopping off one of the $b_i$ won't affect the span of the list.

The original ordered list $v_1, v_2, \ldots, v_m, b_1, b_2, \ldots, b_n$ spans all of $\mathcal{V}$ since the list contains a basis for $\mathcal{V}$, so the list without one of the $b_i$ will still span $\mathcal{V}$.

Repeat this process until the ordered list $v_1, \ldots, v_m, b_1, b_2, \ldots, b_n$ has exactly n elements in it. It spans $\mathcal{V}$. The only remaining question is whether this chopped down list is a basis. I.e., is it linearly independent?

If it weren't, then we could chop off another $b_i$ and the new list would still span $\mathcal{V}$. This is not possible by the theorem two pages ago ("In an n-dimensional vector space, no set of n – 1 vectors will span the space.") ∎

Not eating, Fred sat there writing on a clipboard and saying to himself, "Chop, chop, chop!" and "Time for a fill up!" Betty and Alexander worried about Fred's propensity to skip eating. She sliced off a tiny piece of her sandwich and placed it near his clipboard. Alexander put a pepperoni on top of Betty's offering.

Fred started on a fresh sheet of paper and drew a pair of axes. Then he drew a bunch of arrows with tails at the origin. Living in $\mathbb{R}^2$, he didn't notice the food. As Fred so often exclaimed when he was teaching, "Paper is so well-suited for $\mathbb{R}^2$."

---

✶ A word about the names of these theorems. Until *Life of Fred: Linear Algebra* was published, no one else called them the Chop Down theorem or the Fill 'er Up theorem. Other books call them by less memorable names like Theorem 5.6.1.

Vectors with tails at the origin is one way to illustrate $\mathbb{R}^2$. It's much more visual than looking at ordered pairs like $(5, 7)$. Sometimes it's easier

to work with $(5, 7)$, and sometimes _____ works better—where the vector with tail at the origin has its head at the point $(5, 7)$.

Alexander took a napkin and drew a large vector arrow pointing at the food. Fred didn't notice. Betty remarked, "It looks like Fred is off in his own little world."

My own little world Fred thought to himself. I wonder if I could create a little subspace of $\mathbb{R}^2$? It would be a little vector space inside of $\mathbb{R}^2$—a little place to play inside of big old $\mathbb{R}^2$.

Fred's first try to find a subspace of $\mathbb{R}^2$:

How about just the one vector $(5, 7)$? Would that make a vector space inside of $\mathbb{R}^2$?

He thought for a moment. No, that won't work. Vector addition wouldn't be a binary operation (which was defined on page 140). If I add $(5, 7)$ to itself I get a vector that is not in my subspace.

Fred's second try to find a subspace of $\mathbb{R}^2$:

How about just the zero vector $(0, 0)$? If I add it to itself, I get a unique answer inside the subspace, so vector addition is a binary operation. All the other properties of a vector space are trivially true. It works. (This subspace is called the **zero subspace**.)

## Properties of a vector space

A nonempty set with a binary operation of vector addition that is an abelian group . . .
For every u, v, and w in $\mathcal{V}$:
① commutative        $u + v = v + u$
② associative        $(u + v) + w = u + (v + w)$
③ identity        there is a vector 0 in $\mathcal{V}$ such that $v + 0 = 0 + v = v$ for every v
④ inverses        for every v in $\mathcal{V}$ there is a –v such that $v + -v = 0$

. . . together with a binary operation called scalar multiplication which multiplies a real number $s$ times a vector v to obtain a vector $s$v with four properties:

For any real numbers $s$ and $t$ and any vectors v and w:
① associative        $(st)v = s(tv)$
② identity        $1v = v$
③ distributive        $s(v + w) = sv + sw$
④ distributive        $(s + t)v = sv + tv$

*Your Turn to Play*

1. What about a subspace that contained no vectors at all? Would this be a vector space?

2. In Fred's third try to find a subspace of $\mathbb{R}^2$, he thought of all the vectors in $\mathbb{R}^2$ which have their heads in the first quadrant—all vectors with heads (a, b) where a ≥ 0 and b ≥ 0.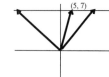
Would this be a subspace of $\mathbb{R}^2$?
(Official definition of subspace: $\mathcal{W}$ is a **subspace** of $\mathcal{V}$ if $\mathcal{W}$ is a subset of $\mathcal{V}$, and $\mathcal{W}$ is a vector space using the same binary operations as $\mathcal{V}$.)

3. Since every set is a subset of itself, Fred's fourth try to find a subspace of $\mathbb{R}^2$ was the set $\mathbb{R}^2$ itself. Is $\mathbb{R}^2$ a subspace of $\mathbb{R}^2$?

4. Fred's fifth try to find a subspace of $\mathbb{R}^2$ consisted of all vectors whose heads were on the line y = 5. That failed in so many categories (vector addition wasn't a binary operation, scalar multiplication wasn't a binary operation, etc.) that Fred quickly tossed that idea in the garbage can.

Fred's sixth try to find a subspace of $\mathbb{R}^2$ consisted of all vectors whose heads were on a line that passed through the origin. Is this a subspace of $\mathbb{R}^2$?

.......**COMPLETE SOLUTIONS**.......

1. Looking at the box on the previous page, the first part of the definition of a vector space is that it is a nonempty set. The zero subspace that Fred found in his second try to find a subspace of $\mathbb{R}^2$, which contains one vector, is the smallest possible subspace.

2. The vector addition is commutative, associative, and has the identity (0, 0). Where it fails is that most vectors don't have an inverse. A vector like (5, 7) doesn't have an inverse *in the subspace*. (–5, –7) isn't in the first quadrant.

3. The zero subspace of $\mathbb{R}^2$ is the smallest possible subspace. $\mathbb{R}^2$ itself is the largest possible subspace.

4. Suddenly, everything works nicely. Add two vectors whose heads are on a line through the origin, and the sum is a vector which is also on the line. (Vector addition is a binary operation.) Multiply a vector whose head is on the line through the origin and the result is a vector which is also on the line. (Scalar multiplication is a binary operation.)

All four vector addition properties (remember the chant?) commutative, associative, identity, and inverses and all four scalar multiplication properties associative, identity, and double distributive are inherited directly from $\mathbb{R}^2$. There is no need to check them.

There is an insight lurking around in that *Your Turn to Play*.
**Insight**: All we have to check to make sure nonempty set $\mathcal{W}$ is a subspace of $\mathcal{V}$ is that vector addition and scalar multiplication in $\mathcal{W}$ are binary operations in $\mathcal{W}$.

Restatement of **Insight**: Nonempty set $\mathcal{W}$ is a subspace of $\mathcal{V}$ if for every pair of vectors, $w_1$ and $w_2$, in $\mathcal{W}$, and every scalar r, both $w_1 + w_2$ and rw are also in $\mathcal{W}$.

Re-restatement of **Insight**: Nonempty set $\mathcal{W}$ is a subspace of $\mathcal{V}$ if $\mathcal{W}$ is **closed under** vector addition and scalar multiplication. (The meaning of "closed under" is given in the previous restatement.)

Re-re-restatement of **Insight**: Nonempty set $\mathcal{W}$ is a subspace of $\mathcal{V}$ if every linear combination $(r_1 w_1 + r_2 w_2)$ of vectors in $\mathcal{W}$ is also in $\mathcal{W}$.

In Fred's six tries to find a subspace of $\mathbb{R}^2$, he had three successes (the zero subspace, a line through the origin and all of $\mathbb{R}^2$). These are the only three possible subspaces of $\mathbb{R}^2$.

*a very short Your Turn to Play*

1. Consider $\mathbb{R}^3$ as the set of all vectors with tails at the origin and heads in three-dimensional space. The vector space $\mathbb{R}^3$ has four possible subspaces. What are they? Please, please, with sugar and chocolate sprinkles on top: *Don't look at the answer until you have given this some thought.*

# ·······COMPLETE SOLUTIONS·······

1. The reason I ask you not to cheat and just look at the answer is that by thinking on your own you most easily and naturally learn linear algebra. If I were to simply list the subspaces of $\mathbb{R}^3$: (1) the zero space; (2) any line through the origin; (3) any plane through the origin; and (4) $\mathbb{R}^3$ itself, then the most predictable response on your part would be Ho-Hum.

If math is presented as just a list of things to memorize, only masochists would be mathematicians. Instead, real math is a playground—a place to invent and explore.

When I was in the tenth grade, my geometry teacher wrote on the board: *Prove that in any triangle, if the angle bisectors of two angles are congruent, then the triangle is isosceles.* He said that the proof wasn't easy, but it was possible to do using regular high school geometry.

I headed home that night and worked on it. Next morning before school I brought him a proof. He found a flaw in the proof. The next morning I brought him another (flawed) proof. Days, weeks, months, and the theorem would not yield.

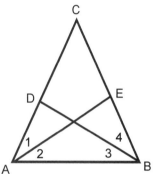

Given: $\angle 1 \cong \angle 2$, $\angle 3 \cong \angle 4$
and AE = BD.
To show: AC = BC

When I was a math teacher at the college level, the proof finally fell into place. A fourteen-year quest was completed. The search was almost as much fun as finding the proof.

Classes in the culinary arts should, by their very nature, be fun. Except for Fred, most everyone really enjoys eating. But suppose that the teacher just stood up in front of the class and read a list of cheeses. How thrilling would it be to hear the enumeration: Acorn, Affidelice au Chablis, Ambert, American Cheese, Aragon, Armenian String, Aromes au Gene de Marc, Asadero, Baguette Laonnaise, Basket Cheese, Bavarian Bergkase, Beaufort, Bierkase, Bleu de Gex, Blue Castello, Boursault, Boursin, Bouyssou, Braudostur, Breakfast Cheese, Brebis du Lochois, Bresse Bleu, Brick, Cashel Blue, Castellano, Castelmagno, Castelo Branco, Castigliano, Cathelain, Cendre d'Olivet, Cerney, Chabichou, Chaource, Charolais, Chaumes, Cheddar, Cheshire, Chevres, Chontaleno, Civray, Coeur de Chevre, Colby, Comte, Coolea, Cooleney, Coquetdale, Corleggy, Cornish Pepper, Cotherstone, Cotija, Cougar Gold, Coulommiers, Coverdale, Cream Havarti, Crema Agria, Crema Mexicana, Creme Fraiche, Crescenza, Croghan, Crottin du Chavignol, Crowdie, Crowley, Cuajada, Cure Nantais, Curworthy, Cypress Grove Chevre. . .

and then be told, "Your assignment is to memorize the list. In the next class meeting, we will start with the cheeses beginning with *D*"? Yuck!

The easiest way to teach is to list facts and have the students memorize them. It is also the most deadly way to teach.

You often can't tell at first glance whether a problem will be easy or difficult. The theorem *In any triangle, if the angle bisectors of two angles are congruent, then the triangle is isosceles* looked pretty easy, but turned out to be hard.

It's obvious (and also true!) that if $\mathcal{W}$ is a subspace of $\mathcal{V}$, then the dimension of $\mathcal{W}$ ≤ the dimension of $\mathcal{V}$. Proving that is probably a par 10 question—one that will take about ten minutes on the average. Proving that will involve one or more of the four recent theorems. It is the natural way to learn these theorems, rather than just sitting down and trying to memorize them.

*Your Turn to Play*

1. Show that if $\mathcal{W}$ is a subspace of $\mathcal{V}$, then the dimension of $\mathcal{W}$ ≤ the dimension of $\mathcal{V}$. Assume dim($\mathcal{V}$) = n. I.e, $\mathcal{V}$ has finite dimension.

Here are the five recent theorems:

**Nice theorem**: If $\mathcal{V}$ has a basis with n elements in it, then every basis of $\mathcal{V}$ has n elements.

**Theorem**: In an n-dimensional vector space, no set of n – 1 vectors will span the space.

**Theorem**: In an n-dimensional vector space, no set of n + 1 vectors will be linearly independent.

**Chop Down theorem**: If you have a set of linearly dependent vectors $v_1$, $v_2$, . . ., $v_m$, you can always find a vector $v_i$ that is a linear combination of the earlier ones on the list. Chopping the $v_i$ from the list won't affect the span of the list.

**Fill 'er Up theorem**: If you have a set of linearly independent vectors $v_1$, $v_2$, . . ., $v_m$ in an n-dimensional space $\mathcal{V}$, you can always find vectors to add to that list to create a basis for $\mathcal{V}$.

**. . . . . . .COMPLETE SOLUTIONS . . . . . . .**

1. $\mathcal{W}$ has at least one vector in it, since it is a vector space. The set consisting of one vector (a singleton) is always linearly independent. By the Fill 'er Up theorem, we can augment that set into a basis for $\mathcal{W}$.

Call that basis $b_1, b_2, \ldots, b_m$. By the Nice theorem, every basis of $\mathcal{W}$ will have m elements in it. The dimension of $\mathcal{W}$ is m. If $m \leq n$, we are done.

Otherwise, $m > n$. By the third theorem in the list of five recent theorems, $b_1, b_2, \ldots, b_m$ are linearly dependent in $\mathcal{V}$. The question at this point is whether $b_1, b_2, \ldots, b_m$ are linearly dependent in $\mathcal{W}$. If they are, we have a contradiction—a linearly dependent set that is also a basis.

Since $b_1, b_2, \ldots, b_m$ are linearly dependent in $\mathcal{V}$, one of the $b_i$ is a linear combination of the other b's.    $b_i = \sum_{\substack{j=1 \\ j \neq i}}^{m} r_j b_j$

Since $\mathcal{W}$ and $\mathcal{V}$ share the same scalar field (the r's), this same equation will hold in $\mathcal{W}$ which shows that $b_1, b_2, \ldots, b_m$ are linearly dependent in $\mathcal{W}$.

Vector arrows in $\mathbb{R}^2$ and $\mathbb{R}^3$ with tails at the origin are good ways to visualize subspaces. It's easy to see the results of multiplying by a scalar. Three times a vector arrow is an arrow three times as long. The inverse of an arrow is an arrow pointing in the opposite direction.

But when we want to look at subspaces of $\mathbb{R}^7$, vectors like (3, 21, 0, 1, 18, 6, 20) are the way to go. Or, better yet, we could eliminate the commas and look at the row vector (3  21  0  1  18  6  20), which is a 1×7 matrix. That saves us the effort of writing the commas.*

Now let us suppose we have some 5×7 matrix $A$ which has that row vector in it:

$$A = \begin{pmatrix} -4 & -28 & 0 & 0 & -20 & 8 & -16 \\ 8 & 56 & 1 & 0 & 46 & 24 & 33 \\ 3 & 21 & 0 & 1 & 18 & 6 & 20 \\ 4 & 28 & 3 & 0 & 38 & 32 & 19 \\ 0 & 0 & 6 & 2 & 42 & 48 & 22 \end{pmatrix}$$

---

✱ Mathematicians are known for finding ways of doing things that require less effort. (My mother had a different way of saying that. She called it "lazy.") In the olden days, when one shepherd had 9 sheep and the other one had 6 sheep, and they wanted to find out how many sheep they both had, they would have to get the sheep all together in one group and count them. When mathematicians came along, they just figured out how to do addition: 9 + 6 = 15.

There are five row vectors in A. Call them $r_1$, $r_2$, $r_3$, $r_4$, $r_5$.
So $r_3 = (3 \quad 21 \quad 0 \quad 1 \quad 18 \quad 6 \quad 20)$.

Consider span$\{r_1, r_2, r_3, r_4, r_5\}$. That is the set of all linear combinations of $r_1$, $r_2$, $r_3$, $r_4$, $r_5$. By the fourth *Insight* back on page 174, span$\{r_1, r_2, r_3, r_4, r_5\}$ is a subspace of $\mathbb{R}^5$.

This subspace is called the **row space** of A.

What if you do some elementary row operations on A? Will it affect the row space of A?

> **Elementary Row Operations**
> ① Interchange any two rows.
> ② Multiply any row by a nonzero number.
> ③ Take any row,
>     multiply it by any number,
>     and add it to another row.

① Interchanging any two rows is the same as asking if the span of $r_1$, $r_2$, $r_3$, $r_4$, $r_5$ is the same as the span of $r_2$, $r_1$, $r_3$, $r_4$, $r_5$. Yes. If you can express a vector of $\mathbb{R}^5$ as a linear combination of $r_1$, $r_2$, $r_3$, $r_4$, $r_5$, you can express it as a linear combination of $r_2$, $r_1$, $r_3$, $r_4$, $r_5$.

② If you multiply any row by a nonzero number, the span is not affected. Suppose, for example, you multiply $r_3$ by 79. If you can express a vector of $\mathbb{R}^5$ as a linear combination of $r_1$, $r_2$, $r_3$, $r_4$, $r_5$, you can express it as a linear combination of $r_1$, $r_2$, $79r_3$, $r_4$, $r_5$.

③ If you take any row, multiply it by any number, and add it to another row, the span is still not affected. Suppose, for example, you multiply $r_3$ by 88 and add it to $r_5$. If you can express a vector of $\mathbb{R}^5$ as a linear combination of $r_1$, $r_2$, $r_3$, $r_4$, $r_5$, you can express it as a linear combination of $r_1$, $r_2$, $r_3$, $r_4$, $88r_3 + r_5$. That's provable using high school algebra.*

---

★ For unbelievers, here are the details. Suppose we have some vector $v$ in $\mathbb{R}^5$ that is in the span of $r_1$, $r_2$, $r_3$, $r_4$, $r_5$. Then, for some real numbers $a_1$, $a_2$, $a_3$, $a_4$, and $a_5$, we have $v = a_1 r_1 + a_2 r_2 + a_3 r_3 + a_4 r_4 + a_5 r_5$. Rename $88r_3 + r_5$ as $r_6$. I have to show that $v$ is in the span of $r_1$, $r_2$, $r_3$, $r_4$, $r_6$.

Solve $88r_3 + r_5 = r_6$ for $r_5$.     $r_5 = r_6 - 88r_3$.

So $v = a_1 r_1 + a_2 r_2 + a_3 r_3 + a_4 r_4 + a_5 r_5$
$= a_1 r_1 + a_2 r_2 + a_3 r_3 + a_4 r_4 + a_5(r_6 - 88r_3)$
$= a_1 r_1 + a_2 r_2 + (a_3 - 88a_5)r_3 + a_4 r_4 + a_5 r_6$

and we now have $v$ as a linear combination of $r_1$, $r_2$, $r_3$, $r_4$, $r_6$.

Sharp-eyed readers have noted that if I multiply $r_3$ by 88 and add it to $r_5$, I *do not decrease* the span—linear combinations of $r_1$, $r_2$, $r_3$, $r_4$, $88r_3 + r_5$ can equal anything in the span of $r_1$, $r_2$, $r_3$, $r_4$, $r_5$.

So elementary row operation ③ (If you take any row, multiply it by any number, and add it to another row) won't decrease the row space of A. But I can go from $r_1$, $r_2$, $r_3$, $r_4$, $88r_3 + r_5$ back to $r_1$, $r_2$, $r_3$, $r_4$, $r_5$ by the elementary row operation of multiplying $r_3$ by –88 and add it to $88r_3 + r_5$.

In the symbolism of set theory, we have shown

span $\{r_1, r_2, r_3, r_4, r_5\}$ ⊂ span $\{r_1, r_2, r_3, r_4, 88r_3 + r_5\}$ ⊂ span $\{r_1, r_2, r_3, r_4, r_5\}$.*

And in set theory, A ⊂ B ⊂ A implies A = B.

Hence, span $\{r_1, r_2, r_3, r_4, r_5\}$ = span $\{r_1, r_2, r_3, r_4, 88r_3 + r_5\}$.

*Mutatis mutandis,* we have **Theorem**: For *any* matrix A, the row space of A is unaffected by elementary row operations.

*Mutatis mutandis,* define the **column space** of any matrix A.**

*Mutatis mutandis,* define **elementary column operations**.

*Mutatis mutandis,* we have Theorem: For *any* matrix A, the column space of A is unaffected by elementary column operations.

Back on page 105, we defined the rank of a matrix A as the number of nonzero rows A has after it has been put in echelon form. For the time being, we will rename that as the row-rank of A. Combining the thoughts of this and the previous page, we can say that the dimension of the row space of A is equal to the row-rank of A.

*Mutatis mutandis,* we define the column-rank of A and note that the column-rank of A is equal to the dimension of the column space of A.

_____

✱ ⊂ means "is a subset of."

✱✱ For those who like it spelled out, let $c_1$, $c_2$, . . ., $c_n$ be the column vectors of a matrix A whose dimensions are m×n. Define the column space of A as the span of $\{c_1, c_2, . . ., c_n\}$.

**I, your reader, am a little bemused. I remember that mutatis mutandis means "similarly." But aren't you laying it on a little thick—all those definitions and theorems lined up in a row. Regular linear algebra books do that all the time, but that's not your style.**

I was just hurrying through the boring stuff. I hate 900-page math books. Maybe since they are all "lined up in a row," I should have drawn some ducks.

**Anything!**

That gives me an idea.

| defintion of row space | row space unaffected by elem ops | definition of column space | definition of elem col ops | column space unaffected by elem ops | dimension of row space = row-rank | dimension of column space = col-rank |

**Tanks! Ducks are a lot more pacific.**[*]

I agree, but in this case, ducks would be misleading. We are about to present the most explosive theorem in linear algebra. It takes most students by complete surprise. Most intelligent readers are completely blown out of the water.

**I'm ready to be blown out of the water. Fire away.**

You asked for it. Here goes.

Start with any m×n matrix A. For concreteness, think of the 5×7 matrix of three pages ago:

$$A = \begin{pmatrix} -4 & -28 & 0 & 0 & -20 & -8 & -16 \\ 8 & 56 & 1 & 0 & 46 & 24 & 33 \\ 3 & 21 & 0 & 1 & 18 & 6 & 20 \\ 4 & 28 & 3 & 0 & 38 & 32 & 19 \\ 0 & 0 & 6 & 2 & 42 & 48 & 22 \end{pmatrix}$$

What is the dimension of its row space?

**I don't know offhand. There might be only one linearly independent row, or all five rows could be linearly independent.**

**In symbols, 1 ≤ row-rank ≤ 5.**

---

[*] When it is not capitalized, *pacific* means tranquil or peaceful.

What is the dimension of its column space?

**I don't know offhand. There might be only one linearly independent column, or all seven columns could be linearly independent. In symbols, 1 ≤ column-rank ≤ 7.**

In a classroom setting, we sometimes have a little contest. I place a new-in-the-box stapler on the front desk as a prize for the first student to create a 5×7 matrix with row-rank = 3 and column-rank = 5. There is a 10-minute time limit.

Some of the students begin by writing

$$A = \begin{pmatrix} ? & ? & ? & ? & ? & 0 & 0 \\ ? & ? & ? & ? & ? & 0 & 0 \\ ? & ? & ? & ? & ? & 0 & 0 \\ 0 & 0 & 0 & 0 & 0 & 0 & 0 \\ 0 & 0 & 0 & 0 & 0 & 0 & 0 \end{pmatrix}$$

and then try to replace each of the question marks with numbers.

Because of the most explosive, surprising[*] theorem of linear algebra, I still own that new-in-the-box stapler.

**Explosive theorem**: For any matrix, the number of linearly independent rows is always equal to the number of linearly independent columns!

Restatement: For any matrix, the dimension of the row space is always equal to the dimension of the column-space.

Re-restatement: For any matrix, the row-rank is equal to the column rank.

The Explosive theorem states, for example, that in any matrix, if there are three linearly independent rows, there must be exactly three linearly independent columns. Except for 🔲 , no one has ever thought that this was obvious.

Why should the row space of a matrix have anything to do with its column space?

---

[*] Except for those who paid attention on page 110, in the Handy Dating Guide, when you called your date a liar.

Proof: The critical part of the proof is first showing that the dimension of the column space of A (the **column-rank**) is unaffected by elementary row operations. Please read that carefully. *Column-rank* unaffected by elementary *row* operations. That will allow us to turn the corner and relate rows and columns.

We have an m×n matrix A with columns $c_1, c_2, c_3, \ldots, c_n$. Some of these columns form a basis for the column space of A. Suppose, for example, that the first six columns form a basis. Then any of the other columns, $c_7, c_8, \ldots, c_n$ can be expressed as a linear combination of $c_1, c_2, c_3, \ldots, c_6$.

For example, $c_8 = \sum_{i=1}^{6} a_i c_i$     for scalars $a_1, a_2, a_3, a_4, a_5,$ and $a_6$.

Now what happens if I interchange two rows of A? Every column vector has two of its entries interchanged. I can't call them $c_1, c_2, c_3, \ldots,$ $c_n$ anymore. They are different vectors. After the row interchange, we'll call the new column vectors $c'_1, c'_2, c'_3, \ldots, c'_n$.

But $c'_8$ is still the same linear combination of the first six columns.

$c'_8 = \sum_{i=1}^{6} a_i c'_i$     for the same scalars $a_1, a_2, a_3, a_4, a_5,$ and $a_6$.

The $a_1, a_2, a_3, a_4, a_5,$ and $a_6$ do not change at all.

And if I multiply the 7th row of A by a nonzero number, all the column vectors will have their 7th entries multiplied by that number. If the new column vectors are $c''_1, c''_2, c''_3, \ldots, c''_n$ we still have $c''_8$ is still the same linear combination of the first six columns.

$c''_8 = \sum_{i=1}^{6} a_i c''_i$     for the same scalars $a_1, a_2, a_3, a_4, a_5,$ and $a_6$.

And likewise, if I add a multiple of one row to another row.

Similarly, we note that the *row-rank* of A will be unaffected by elementary *column* operations.

Now the rest of the proof is smooth sailing. Please turn back two pages and look at the matrix A. If you perform elementary row operations on it using Gauss-Jordan elimination, you have (i) any all-zero rows at the bottom of the matrix, (ii) the first nonzero elements of the other rows (what we call the distinguished elements) are 1s, and (iii) above and below those 1s are all zeros. This was called a reduced row-echelon form defined on page 99.

$$\text{A becomes} \quad \mathsf{B} = \begin{pmatrix} 1 & 7 & 0 & 0 & 5 & 2 & 4 \\ 0 & 0 & 1 & 0 & 6 & 8 & 1 \\ 0 & 0 & 0 & 1 & 3 & 0 & 8 \\ 0 & 0 & 0 & 0 & 0 & 0 & 0 \\ 0 & 0 & 0 & 0 & 0 & 0 & 0 \end{pmatrix}$$

Now take B and use elementary column operations wiping out all the nonzero entries to the right of the distinguished elements.

$$\text{B becomes} \quad \mathsf{C} = \begin{pmatrix} 1 & 0 & 0 & 0 & 0 & 0 & 0 \\ 0 & 0 & 1 & 0 & 0 & 0 & 0 \\ 0 & 0 & 0 & 1 & 0 & 0 & 0 \\ 0 & 0 & 0 & 0 & 0 & 0 & 0 \\ 0 & 0 & 0 & 0 & 0 & 0 & 0 \end{pmatrix}$$

The row-rank and column-rank of C is the same as the row-rank and column-rank of A by the first part of this proof.

The rows of C that contain a 1 form a basis of the row space.*

The columns of C that contain a 1 form a basis of the column space.

So the number of 1s in C is both the dimension of the row space and the column space of C.    ■

We can now just talk about the rank of A. We don't need to say row-rank or column-rank.

In about 57 pages, we have explored the basic ideas of vector spaces—all the way up to this Explosive theorem. Let's do some traveling, as we head off to the Cities.

---

\* Back on page 149, we called those three rows of C—(1 0 0 0 0 0 0), (0 0 1 0 0 0 0), and (0 0 0 1 0 0 0)—by their cute little names $e_1$, $e_3$, and $e_4$.

## Bowlegs

1.  Do the polynomials of degree one (such as 6x, –30x, 5πx, and $\sqrt{382}$ x) form a vector space?

2.  Let {$v_1$, $v_2$, . . ., $v_6$} be a linearly independent set of vectors in a vector space.  Show that every nonempty subset of that set is linearly independent.

3.  Show that (2, 6, –2), (–3, 4, –23), and (5, 1, 23) are linearly dependent.

4.  Show that (3, 4.8, 9.2), (⅝, –19.732, 0), (π, $\sqrt{382}$, 335), and (2, –π, ⅔) are linearly dependent.

5.  If A is a 5×9 matrix with real number entries, and if you interchange rows 2 and 4 in A, will the resulting matrix be equal to A?

*answers*

1.  Several of the parts of the definition of a vector space fail.  For example, vector addition is not closed.  Add 6x and –6x and you get a polynomial that is not of degree one.  Also there is no zero vector.

2.  The definition of {$v_1$, $v_2$, . . ., $v_6$} being linearly independent is that the only solution to $\sum_{i=1}^{6} r_i v_i = 0$ is one in which all the $r_i$ are equal to zero.

   If some subset of {$v_1$, $v_2$, . . ., $v_6$} were linearly dependent, then there exist real numbers, not all equal to zero, such that the linear combination of those vectors, using those coefficients, would equal zero.  Adding to that linear combination $0v_i$ for all the $v_i$ that were not in the subset yields a nontrivial solution to $\sum_{i=1}^{6} r_i v_i = 0$.

   In symbols, the above argument might be rewritten as:

Let set $\mathscr{A}$ = {$v_1$, $v_2$, . . ., $v_6$}.  Let $\mathscr{B} \subset \mathscr{A}$.

Then $v_i \in \mathscr{B}$ iff  i ∈ I, where I ⊂ {1, 2, 3, 4, 5, 6}.  (I is called an index set.)

If there exist $r_i$, for i ∈ I, not all zero, such that $\sum_{i \in I} r_i v_i = 0$, then $\sum_{i=1}^{6} r_i v_i = 0$, and $\mathscr{A}$ is linearly dependent. ∎  (In this proof, we proved the contrapositive.  See page 107 for the definition of contrapositive.)

3. Using the New Quicker Summary on page 162, we first construct A with (2, 6, –2), (–3, 4, –23), and (5, 1, 23) as *columns* of A.

$$A = \begin{pmatrix} 2 & -3 & 5 \\ 6 & 4 & 1 \\ -2 & -23 & 23 \end{pmatrix}$$   Then we put A into echelon form.

$$\longrightarrow \begin{pmatrix} 2 & -3 & 5 \\ 0 & 13 & -14 \\ 0 & -26 & 28 \end{pmatrix} \longrightarrow \begin{pmatrix} 2 & -3 & 5 \\ 0 & 13 & -14 \\ 0 & 0 & 0 \end{pmatrix}$$

and then we note that (in the words of the New Quicker Summary): *A set of row vectors is linearly dependent if A in echelon form has a zero on the diagonal with zeros below that entry*, which is what we have in this case.

4. There is the hard way and the easy way to attack this problem. The hard way is to use the technique of the previous problem. The easier way is to recall the theorem from page 169: *In an n-dimensional vector space, no set of* n + 1 *vectors will be linearly independent.* We're done.

5. At the top of page 162 was the definition of equality between matrices: *Two matrices are equal only if they have the same dimensions and all corresponding entries are equal.* If you interchange rows, the corresponding entries are not equal. A and the resulting matrix are row equivalent. (Row-equivalent was defined on page 65.)

---

**Hot Coffee**

---

1. Thinking geometrically, (1, 0, 0) and (1, 37, 0) are two vectors whose tails are at the origin and whose heads are at (1, 0, 0) and (1, 37, 0). Describe geometrically what span {(1, 0, 0) and (1, 37, 0)} would look like.

2. Next to the Great Lawn on the KITTENS campus, where Fred, Betty, and Alexander are having their picnic, is the Great Lake. It's a fun place to go sailing. Let **i** represent a west wind of 1 mph and **j** a south wind of 1 mph. Suppose the winds on the Great

Lake are always of the form $\mathbf{r}i + \mathbf{j}$ where r is any real number. Do these winds form a vector space?

3. The set of all polynomials of degree 2 or less is a vector space. Are the vectors $2 + 6x - 2x^2$, $-3 + 4x - 23x^2$, and $5 + x + 23x^2$ linearly independent or linearly dependent?

4. Show that a vector space can only have one zero vector. Start by assuming that both $0$ and $0'$ are zero vectors.

*answers*

1. $\mathbf{i}$ is certainly in the span of $\{(1, 0, 0)$ and $(1, 37, 0)\}$. ($\mathbf{i}$, as defined on page 149, is equal to $(1, 0, 0)$.)

$\mathbf{j}$ is in the span of $\{(1, 0, 0)$ and $(1, 37, 0)\}$ since
$$\mathbf{j} = \frac{-1}{37}(1, 0, 0) + \frac{1}{37}(1, 37, 0).$$

✓ The dimension of span$\{(1, 0, 0)$ and $(1, 37, 0)\} = 2$ since $(1, 0, 0)$ and $(1, 37, 0)$ are linearly independent.
✓ $\mathbf{i}$ and $\mathbf{j}$ are in span$\{(1, 0, 0)$ and $(1, 37, 0)\}$.
✓ $\mathbf{i}$ and $\mathbf{j}$ are linearly independent.
✓ $\{\mathbf{i}, \mathbf{j}\}$ is a basis for span$\{(1, 0, 0)$ and $(1, 37, 0)\}$.

Thus, geometrically, span$\{(1, 0, 0)$ and $(1, 37, 0)\}$ is the whole X–Y plane.

2. No, since there is no zero vector.

3. Working with $2 + 6x - 2x^2$, $-3 + 4x - 23x^2$, and $5 + x + 23x^2$ seems so messy. As usual, let's take the easier approach. On pages 142–143 we "matched up" the vector space of all polynomials of degree 2 or less with $\mathbb{R}^3$. So instead of dealing with $2 + 6x - 2x^2$, $-3 + 4x - 23x^2$, and $5 + x + 23x^2$, we'll look at $(2, 6, -2)$, $(-3, 4, -23)$, and $(5, 1, 23)$. We instantly know that these vectors are linearly dependent.*

4. By definition of zero vector,
   (i) $0$ is a zero vector if for every $\mathbf{v}$ in $\mathcal{V}$, $\mathbf{v} + 0 = 0 + \mathbf{v} = \mathbf{v}$.
   (ii) $0'$ is a zero vector if for every $\mathbf{v}$ in $\mathcal{V}$, $\mathbf{v} + 0' = 0' + \mathbf{v} = \mathbf{v}$.
$0$ equals $0 + 0'$ using (ii) where we let $\mathbf{v} = 0$.

$0 + 0'$ equals $0'$ using (i) where we let $\mathbf{v} = 0'$. Thus $0 = 0'$ by the transitive law of equality. (If $a = b$ and $b = c$, then $a = c$)

──────────────────

✶ . . . because we did the work in problem 3 in the previous City (Bowlegs).

| Yell |
| --- |

1. Suppose we take the vector space $\mathbb{R}^3$ and redefine vector addition by adding 4 to each part of the answer. For example, $(3, 9, 20) + (8, -2, 4)$ would equal $(15, 11, 28)$. We will leave scalar multiplication unchanged.

       Is $\mathbb{R}^3$ with this new definition of addition a vector space?

2. Did we ever prove that every vector is a *unique* linear combination of a given basis?

3. Show that $\{(2, 2, 4), (7, 7, 15), (-3, -2, 3)\}$ is a basis for $\mathbb{R}^3$.

4. Suppose that $\mathcal{U}$ and $\mathcal{W}$ are subspaces of vector space $\mathcal{V}$. Is the intersection of $\mathcal{U}$ and $\mathcal{W}$ ($\mathcal{U} \cap \mathcal{W}$) a subspace of $\mathcal{V}$? If it is, explain why it is. If it is not, then find a counterexample.

5. Suppose that $\mathcal{U}$ and $\mathcal{W}$ are subspaces of vector space $\mathcal{V}$. Is the union of $\mathcal{U}$ and $\mathcal{W}$ ($\mathcal{U} \cup \mathcal{W}$) a subspace of $\mathcal{V}$? If it is, explain why it is. If it is not, then find a counterexample.

6. Can the intersection of two subspaces be disjoint? (Restatement: Can the intersection of two subspaces equal the empty set?)

*answers*

1. It is closed under vector addition—add any two vectors, and you get a vector in $\mathbb{R}^3$. Vector addition is commutative and associative. $(-4, -4, -4)$ is the identity for vector addition.

       Scalar multiplication is closed. It is associative.

       All that is left are the two distributive laws . . . and they both fail. For example, $(5 + 6)( (1, 2, 3) )$ should equal $5(1, 2, 3) + 6(1, 2, 3)$.
The left side works out to $(11, 22, 33)$.
The right side works out to $(5, 10, 15) + (6, 12, 18)$ which is $(15, 26, 37)$.

2. Yes. That was the Lemma on the page before the Nice theorem. Because any vector $\mathsf{u}$ is expressible uniquely as a linear combination of a given basis $\mathsf{v}_1, \mathsf{v}_2, \ldots, \mathsf{v}_n$, we could define the coordinates of $\mathsf{u}$ with respect to $\mathsf{v}_1, \mathsf{v}_2, \ldots, \mathsf{v}_n$.

3. First, to show that the vectors are linearly independent. Using the *New Quicker Summary* on page 162, we begin with

$$\begin{pmatrix} 2 & 7 & -3 \\ 2 & 7 & -2 \\ 4 & 15 & 3 \end{pmatrix} \xrightarrow[-2r_1 + r_3]{-r_1 + r_2} \begin{pmatrix} 2 & 7 & -3 \\ 0 & 0 & 1 \\ 0 & 1 & 9 \end{pmatrix} \xrightarrow{1\,r_2,\,r_3} \begin{pmatrix} 2 & 7 & -3 \\ 0 & 1 & 9 \\ 0 & 0 & 1 \end{pmatrix}$$

We are now in echelon form with no zeros on the diagonal. Hence by the New Quicker Summary the three vectors are linearly independent.

Secondly, in order to show that the vectors are a basis, we must establish that they span $\mathbb{R}^3$. But three independent vectors will always span $\mathbb{R}^3$. The short argument would be to say, "See #3 on page 164." The more complete argument would consist of two steps: First, if those three independent vectors didn't span $\mathbb{R}^3$, use the Fill 'er Up theorem and find one or more vectors to add to the list to create a basis with four or more vectors. Second, this would contradict the Nice theorem which states that all bases have the same number of vectors, and we know that $\mathbf{i}$, $\mathbf{j}$, and $\mathbf{k}$ are a basis for $\mathbb{R}^3$. Thus, the three given vectors are a basis.

4. If $\mathcal{U}$ and $\mathcal{W}$ are subspaces of $\mathcal{V}$, then $\mathcal{U} \cap \mathcal{W}$ will also be a subspace. Proof: Let $\mathsf{v}_1$ and $\mathsf{v}_2 \in \mathcal{U} \cap \mathcal{W}$. Then, by definition of intersection, $\mathsf{v}_1$ and $\mathsf{v}_2 \in \mathcal{U}$. Since $\mathcal{U}$ is a subspace, any linear combination of $\mathsf{v}_1$ and $\mathsf{v}_2$ are in $\mathcal{U}$. The same is true for $\mathcal{W}$. Since any linear combination of $\mathsf{v}_1$ and $\mathsf{v}_2$ are in both $\mathcal{U}$ and $\mathcal{W}$, they are in $\mathcal{U} \cap \mathcal{W}$. Thus $\mathcal{U} \cap \mathcal{W}$ is a subspace since every linear combination of vectors in $\mathcal{U} \cap \mathcal{W}$ is also in $\mathcal{U} \cap \mathcal{W}$.

5. Counterexample #1: Span$\{ (1, 0, 0) \}$ and span$\{ (0, 1, 0) \}$ are both subspaces of $\mathbb{R}^3$. (These subspaces could also have been written as span$\{\mathbf{i}\}$ and span$\{\mathbf{j}\}$.) Note that $(51, 0, 0) \in$ span$\{\mathbf{i}\}$ and $(0, 89, 0) \in$ span$\{\mathbf{j}\}$, but $(51, 89, 0)$, which is a linear combination of $(51, 0, 0)$ and $(0, 89, 0)$, is not in span$\{\mathbf{i}\} \cup$ span$\{\mathbf{j}\}$.

Counterexample #2: The set of all vectors whose heads are on $y = 2x$ and whose tails are at the origin is a subspace of $\mathbb{R}^2$. (This was Fred's sixth try to find a subspace of $\mathbb{R}^2$ on page 173.) The set of all vectors whose heads are on $y = 5x$ and whose tails are at the origin is also a subspace of $\mathbb{R}^2$. The vector whose head is at $(20, 40)$ is in the first subspace. The vector $(20, 100)$ is in the second subspace. But their sum, $(40, 140)$, is in neither subspace.

6. The intersection of two subspaces is always a subspace. A subspace is a vector space and by definition, a vector space is nonempty.

## Tightwad

1. Suppose we have a set with exactly one vector in it, say, $\{ \boxed{} \}$. Define vector addition by $\boxed{} + \boxed{} = \boxed{}$ and multiplication by any real number $r$ by $r\,\boxed{} = \boxed{}$. Is this a vector space?

2. We define span{ } = 0 where 0 is the zero vector. Is the span of any set of vectors in a vector space a subspace?

3. In question 2 in the *Your Turn to Play* on page 137, we showed that for any scalar v, 0v = 0, where 0 is the real number zero, and 0 is the zero vector. This *feels* so obviously true. Note that if you were writing this by hand, it would look like: $o\vec{v} = \vec{0}$.

Maybe we could call it micro-theorem #1: 0v = 0. (There is no such word as micro-theorem. English majors are fond of saying that *micro-theorem* is a neologism.)

The challenge and the fun was to find a way of proving micro-theorem #1: 0v = 0. Small story: When my daughter Margaret was about four years old, I pulled out a chess set to teach her how to play. The first step was to teach her the names of the pieces. "This is my king. I'm going to move it here. Now it's your turn. Where are you going to move your king." We moved all the pieces around for a while, and then Margaret asked, "How do you win this game?" I said, "By taking the other person's king."

She grinned and reached over the board and removed my king.

The fun of proving micro-theorem #1: 0v = 0, or of winning at chess, is to do it within the rules of the game. In geometry everyone knows that the base angles of an isosceles triangle are congruent. The trick is to find a proof using the axioms, postulates, definitions, and previously proven theorems as a starting point.

We proved micro-theorem #2 also on page 137: For any vector v, 1v = v.

Prove: micro-theorem #3: For any scalar *r*, *r*0 = 0, where 0 is the zero vector. If you were writing this by hand, it would look like: $r\vec{0} = \vec{0}$.

For your convenience, here is a list of the properties of a vector space.

→   →   →   →   →   →   →   →

This is a par-15 minute question.

4. Prove micro-theorem #4: Suppose *r*v = 0. Show that either *r* = 0 or v = 0. (Notice that the two zeros are different.

---

## Properties of a vector space

A nonempty set with a binary operation of vector addition that is an abelian group . . .
For every u, v, and w in $\mathcal{V}$:
① commutative      u + v = v + u
② associative    (u + v) + w = u + (v + w)
③ identity      there is a vector 0 in $\mathcal{V}$ such that v + 0 = 0 + v = v
④ inverses    for every v in $\mathcal{V}$ there is a –v such that v + –v = 0

. . . together with a binary operation called scalar multiplication which multiplies a real number *s* times a vector v to obtain a vector *s*v with four properties:

For any real numbers *s* and *t* and any vectors v and w:
|1| associative        $(st)v = s(tv)$
|2| identity          $1v = v$
|3| distributive      $s(v + w) = sv + sw$
|4| distributive      $(s + t)v = sv + tv$

If you were writing this by hand, it might look like: $r = o$ *or* $\vec{v} = \vec{o}$.)

5. Prove micro-theorem #5: $-1v = -v$. (Recall, $-v$ was defined in ④.)

6. Prove micro-theorem #6: $-rv = r(-v)$

***answers***

1. It's closed under vector addition. Add a Fred to a Fred, and you get a Fred. It's closed under scalar multiplication. Multiply a Fred by any scalar and you get a Fred. Vector addition is commutative. All the parts of the definition of a vector space hold.

This is the world's smallest vector space. It is often called the **zero vector space**.

2. The span of any set of vectors $\mathcal{W}$ in a vector space $\mathcal{V}$ is a subspace. First we note that the span of any set of vectors is nonempty. (That's why we define span$\{ \ \} = 0$.) By one of the 𝕴𝖓𝖘𝖎𝖌𝖍𝖙𝖘 in the chapter, a nonempty set $\mathcal{W}$ will be a subspace if every linear combination of vectors in $\mathcal{W}$ is also in $\mathcal{W}$. Linear combinations of a vector in $\mathcal{W}$ is, by definition of span, a linear combination of a linear combination of the original set of vectors. We are done when we note that linear combinations of linear combinations of a set are always expressible as linear combinations of the set.

3. 

| | |
|---|---|
| $0 + 0 = 0$ | ③ (see box on previous page) |
| $r(0 + 0) = r0$ | Scalar multiplication is a function (or, more specifically, a binary operation). |
| $r0 + r0 = r0$ | ③ |
| $r0 + r0 = r0 + 0$ | ③ |
| $r0 = 0$ | Cancellation (see the fourth line from the bottom on page 136.) |

4. 

| | |
|---|---|
| $rv = 0$ | given. If $r = 0$, we are done. If not, then $\frac{1}{r}$ exists. |
| $\frac{1}{r}rv = \frac{1}{r}0$ | Multiply both sides of $rv = 0$ *by* $\frac{1}{r}$. Scalar multiplication is a function (or, more specifically, a binary operation). |
| $1v = \frac{1}{r}0$ | Property of any nonzero real number. |
| $v = \frac{1}{r}0$ | ② *Identity*: $1v = v$ |
| $v = 0$ | Micro-theorem #3 ∎ |

5.    $\text{v} + -1\text{v} = 1\text{v} + -1\text{v}$          ☑2

      $\text{v} + -1\text{v} = (1 + (-1))\text{v}$          ☑4

      $\text{v} + -1\text{v} = 0\text{v}$                      property of real numbers

      $\text{v} + -1\text{v} = 0$                        Micro-theorem #1

      $\text{v} + -1\text{v} = \text{v} + -\text{v}$              ④

          $-1\text{v} = -\text{v}$                  Cancellation (see the fourth line from the bottom on page 136.)

6.              $\text{rv} + -\text{rv} = (\text{r} + (-\text{r}))\text{v}$          ☑4

You might ask how I knew that this was a good place to start. I didn't. I tried a bunch of different starting places until I found that this one worked.

          $\text{rv} + -\text{rv} = 0\text{v}$                Property of real numbers.

          $\text{rv} + -\text{rv} = 0$                  Micro-theorem #1

          $\text{rv} + -\text{rv} = \text{r}0$                Micro-theorem #3

          $\text{rv} + -\text{rv} = \text{r}(\text{v} + -\text{v})$            ④

          $\text{rv} + -\text{rv} = \text{rv} + \text{r}(-\text{v})$            ☑3

              $-\text{rv} =\quad \text{r}(-\text{v})$          Cancellation.  ∎

---

## Greasy Corner

1. The set of all vectors of the form $\text{r}(\mathbf{e}_1 + 2\mathbf{e}_2 - 5\mathbf{e}_3)$—where r is a real number and $\mathbf{e}_1 = (1, 0, 0)$, $\mathbf{e}_2 = (0, 1, 0)$, and $\mathbf{e}_3 = (0, 0, 1)$—form a vector space. (It is a subspace of $\mathbb{R}^3$.) Find a basis for this vector space.

2. $\{\mathbf{e}_1, \mathbf{e}_2, \mathbf{e}_3\}$ is a basis for $\mathbb{R}^3$. Are there other possible bases for $\mathbb{R}^3$. If so, give 50 examples. If not, explain why $\{\mathbf{e}_1, \mathbf{e}_2, \mathbf{e}_3\}$ is the only basis for $\mathbb{R}^3$.

3. $\mathbb{C}$ is the set of all complex numbers. ($\mathbb{C}$ is the set of all numbers that can be expressed as $a + bi$ where $a, b \in \mathbb{R}$ and $i = \sqrt{-1}$.) Let the vectors of space $\mathbb{C}^1$ be elements of $\mathbb{C}$ with scalar multiplication by real numbers. For example, a linear combination of $5 + 4i$ and $8 + 3i$ might be $44(5 + 4i) + 17(8 + 3i)$.

        Find a basis for $\mathbb{C}$.

4. Show that $1, x, x^2, x^3$, and $x^4$ form a linearly independent set of vectors in the vector space of all polynomials of degree 4 or less.

5. What can we say about the rank of $\mathsf{A}$, if $\mathsf{A}$ is a $44 \times 79$ matrix?

6. Show that n linearly independent vectors in an n-dimensional space must be a basis.

***answers***

1. The easiest way to find a basis is to start writing out some examples of vectors in that space. For r = 1, we have $(e_1 + 2e_2 - 5e_3)$. For r = 2, we have $2(e_1 + 2e_2 - 5e_3)$. Other vectors in that space are $\pi(e_1 + 2e_2 - 5e_3)$, and $-87(e_1 + 2e_2 - 5e_3)$, and $0(e_1 + 2e_2 - 5e_3)$.

Every vector in that space is a scalar multiple of $(e_1 + 2e_2 - 5e_3)$. So $\{(e_1 + 2e_2 - 5e_3)\}$ would be a nice basis.

2. Every vector space (except the zero vector space) has an infinite number of bases. Here are 50 examples of bases for $\mathbb{R}^3$: $\{101e_1, e_2, e_3\}$, $\{102e_1, e_2, e_3\}$, $\{103e_1, e_2, e_3\}$, $\{104e_1, e_2, e_3\}$, $\{105e_1, e_2, e_3\}$, . . . , $\{150e_1, e_2, e_3\}$.

3. $\{1\}$ wouldn't work as a basis. No linear combination could give us $6309 + 73i$.

$\{i\}$ wouldn't work as a basis. No linear combination could give us $6309 + 73i$.

But the pair $\{1, i\}$ work as a basis. The set is linearly independent, and span$\{1, i\} = \mathbb{C}$.

4. To show that 1, x, $x^2$, $x^3$, and $x^4$ form a linearly independent set of vectors in the vector space of all polynomials of degree 4 or less, we first note that this is equivalent, using the solution to Hot Coffee, Mississippi #3, to showing that (1, 0, 0, 0, 0), (0, 1, 0, 0, 0), . . . , (0, 0, 0, 0, 1) are linearly independent. Placing these row vectors as columns in a matrix A, we note that A is the 5×5 identity matrix. Using the *New Quicker Summary* approach of Yell, Tennessee #3, (there are no zeros on the diagonal of the echelon matrix), we declare that 1, x, $x^2$, $x^3$, and $x^4$ form a linearly independent set of vectors in the vector space of all polynomials of degree 4 or less. As my Granddad used to say, "A long memory makes for short work."

5. Since the rank of A = the row-rank of A = the column-rank of A, we can say that the rank of A is less than or equal to both 44 and 79. In other words, the rank of A ≤ 44.

6. By the Fill 'er Up theorem, I can add vectors to those n linearly independent vectors and create a basis. By the Nice theorem, I don't have

to add any vectors because I already have n vectors. Hence, the n linearly independent vectors are a basis already.

---

## Knockemstiff

---

1. A **positive matrix** is a square matrix with positive real number entries. Does the set of all 5×5 positive matrices form a vector space?

2. We know that the set of all real-valued functions form a vector space. Is the set of all differentiable functions a subspace? If yes, tell why. If no, find a counterexample.

3. Let A be a 17×43 matrix with rank equal to 6. Suppose B is equal to the transpose of A. (*Transposing* matrices was defined on page 96.)

   Question one: What are the dimensions of B?

   Question two: From the given information, can we determine the rank of B?

   Question three: Could the rank of B equal 12?

4. Suppose we reduce matrix C to D which is in echelon form. The number of nonzero rows in D is the rank of C. The question is whether the nonzero rows of D form a basis for the row space of C.

5. In addition to the row space and the column space of a matrix A, there is another space hiding in A. First, a little bit of review . . .

—·•·——start of review——·•·—

When we were solving $Ax = 0$, we formed the augmented matrix and used Gaussian elimination to obtain a matrix in echelon form. It might look like:

$$\begin{pmatrix} 3 & 7 & 8 & 3 & 0 & 9 & 0 \\ 0 & 0 & 1 & 0 & 7 & 9 & 0 \\ 0 & 0 & 0 & 4 & 5 & 3 & 0 \\ 0 & 0 & 0 & 0 & 0 & 2 & 0 \\ 0 & 0 & 0 & 0 & 0 & 0 & 0 \end{pmatrix}$$

The last column corresponds to the b column vector (after the elementary row operations). We'll concentrate on the first six columns, which correspond to the A matrix.

The distinguished elements (the first nonzero elements in each row) are 3, 1, 4, and 2. So the pivot variables (a.k.a. the leading variables) are $x_1$, $x_3$, $x_4$, and $x_6$. As we indicated on page 105, the rank of a matrix is the number of nonzero rows it has after it has been put into echelon form. *The number of pivot variables equals the rank of a matrix.*

The number of pivot variables equals the dimension of the column space of $A$. (Dimension of the column space = column rank = rank.)

The free variables are $x_2$ and $x_5$. They are the loosey-goosey variables that can take on any value. We called them parameters on page 101. Name a value for each of the free variables, then the values of pivot variables are determined, and we have a particular solution to $Ax = 0$.

I'm a free variable

———— ·•· —— end of review —— ·•· ————

The wiggle room is in the free variables.

If we set $x_2 = 1$ and $x_5 = 0$,
we will obtain a solution that looks like
$$\begin{pmatrix} \star \\ 1 \\ \star \\ \star \\ 0 \\ \star \end{pmatrix}$$

On the other hand,
if we set $x_2 = 0$ and $x_5 = 1$,
we will obtain a solution that looks like
$$\begin{pmatrix} \ast \\ 0 \\ \ast \\ \ast \\ 1 \\ \ast \end{pmatrix}$$

We don't care what the $\star$ and the $\ast$ are. They can be any numbers. What is important is that these two column vectors are linearly independent.

Question one: How can we tell they are linearly independent?

If we had more free variables, each of them would yield a column vector that is a solution to $Ax = 0$. All of these solution vectors are linearly independent.

Question two: If $y$ and $z$ are solution vectors for $Ax = 0$, show that any linear combination of $y$ and $z$ is also a solution.

There are 6 columns in A. The rank of A is 4. There are 2 free variables, and hence 2 linearly independent solution vectors for Ax = 0.

Question three: How do we know those solution vectors form a subspace?

𝕯𝖊𝖋𝖎𝖓𝖎𝖙𝖎𝖔𝖓: We call the subspace of all solutions to Ax = 0, the **null space** of A. We call the dimension of the null space the **nullity** of A.

The number of free variables equals the nullity of A.

Question four: If some matrix B is in echelon form, and its rank is 9, can it have 6 pivot variables?

Question five: If some matrix C is in echelon form, and it has 23 pivot variables, and the nullity of C is 5, what can you say about the dimensions of C?

*answers*

1. It is closed under vector addition—if you add two 5×5 positive matrices together, you get a 5×5 positive matrix.
    Vector addition is commutative and associative. What would be the identity vector? It would have to be the 5×5 matrix with every entry equal to zero. That's not a positive matrix.
    Also 5×5 positive matrices are not closed under scalar multiplication. Multiply a positive matrix by –3978 and the answer is no longer a positive matrix.
2. The set of all differentiable functions is a subspace of all real-valued functions. By the Re-re-restatement of 𝕴𝖓𝖘𝖎𝖌𝖍𝖙: Nonempty set $\mathcal{W}$ is a subspace of $\mathcal{V}$ if every linear combination $(r_1 w_1 + r_2 w_2)$ of vectors in $\mathcal{W}$ is also in $\mathcal{W}$. From calculus we know that if you add together two differential functions, or if you multiply a differentiable function by a real number, the result is a differentiable function.
3. Question one: The dimensions of B are 43×17.

Question two: At the end of this chapter, just before the Cities, was: rank A = row-rank A = column-rank A. Interchanging rows with columns won't affect the rank of A. So the rank of the transpose of A is also equal to 6.

Question three: This is just to find out if you are awake. In question two, we found that the rank of B is equal to 6.

4. The nonzero rows of D are linear combinations of the rows of C and hence are in the row space of C. They are linearly independent since D is in echelon form. So the nonzero rows of D are both independent and there are as many of them as the dimension of C. These two facts are enough to ensure that they form a basis for C.

5. Question one: If they were linearly dependent, one of them would be a multiple of the other, but no multiple of the first vector can yield a 1 in the fifth row of the second vector.

Question two: y is a solution to Ax = 0 means Ay = 0. Similarly, Az = 0. Let ry + sz be any linear combination of y and z.

A(ry + sz) = rAy + sAz = r0 + s0 = 0 + 0 = 0.

Question three: We know they form a subspace by the re-re-restatement of 𝕴𝖓𝖘𝖎𝖌𝖍𝖙: Nonempty set $\mathcal{W}$ is a subspace of $\mathcal{V}$ if every linear combination $(r_1w_1 + r_2w_2)$ of vectors in $\mathcal{W}$ is also in $\mathcal{W}$.

Question four: If the rank of B is 9, then there must be 9 nonzero rows. Then there must be 9 pivot variables.

Question five: The columns of C = the number of pivot variables plus the number of free variables = the number of pivot variables plus the nullity of C = 23 + 5 = 28. The dimensions of C are ?×28.

# Chapter Two and Three-quarters
## Inner Product Spaces

Alexander finished his combo pizza. Five minutes later, Betty took her last bite of her whole-wheat bagel sandwich. . . .

**Wait! Stop! I, your reader, am expecting Chapter 3 at this point. We did our trip to the heights in Chapter 2½ generalizing**
    **matrices,**
      **geometric vectors,**
        **polynomials of degree n or less,**
          **real-valued functions and other stuff.**

a trip to the heights

**Now I'm expecting to get back on the ground again with systems of linear equations. You know, " Chapter 3: Systems of Linear Equations with No Solution—Ax = b ☹."**

In a moment we'll be at Chapter 3. Right now, we have to go just a bit higher in the theory. We won't be too long up here. Just one

← new height

← former height

topic—Inner Product Spaces—and then we will descend to the valley floor and show you how to solve unsolvable equations in Chapter 3.

**It's a deal. I hope I don't get a nosebleed being up this high.**

"Let's take a walk around the Great Lake," Betty suggested as she finished her sandwich. They gathered up their picnic supplies. Alexander removed a couple of rabbits from the picnic basket before he picked it up.

The walk wouldn't be that far. The lake is only 600 yards at its widest point. It was with a touch of irony that the students had named it the Great Lake.

"You know what I miss . . . " Fred began.

"I bet it's your mother," Betty interjected. "You told us once that you were only six months old when she died, and you ran away from home. [*Life of Fred: Calculus, Chapters 9 and 10.*] You said that you used to play in the flour on the kitchen floor while your mom baked cookies all day long. That's when you discovered the Mean Value Theorem. [If f is continuous on the closed interval [a, b] and differentiable on the open interval (a, b), then there has to be a point w so that a < w < b

and $f'(w) = \dfrac{f(b) - f(a)}{b - a}$ ]

You told us about the dump truck that would bring her weekly delivery of cookie flour, and how she once stood too close and was buried up to her earrings in cookie flour. It took you four hours to dig her out."

"That's not what I was thinking about," said Fred. "I was thinking about when vectors were simply arrows. [*Life of Fred: Calculus, Chapter 20.*] You could just draw it."

In the Great Lake beach sand Fred drew:

"If it went from the origin to the point (2, 7), you could draw it, you could call it 2**i** + 7**j**, or you could just name it as (2, 7)."

Alexander said, "But you haven't lost any of those things when you abstracted those arrows into vector spaces. The geometric arrows are still vector spaces. The set of all ordered pairs like (2, 7) is $\mathbb{R}^2$, and that's a vector space. You didn't lose vector addition when you moved from geometric vectors to vector spaces.

In fact, vector addition* was a big part of the definition of a vector space."

"And," Betty added, "in moving to vector spaces in general, you didn't lose scalar multiplication** either."

She drew in the sand:

"Besides," she added, "you have a lot of other vector spaces to play with. There's the vector space of all polynomials of degree 6 or less. The vector space of all 3×5 matrices. The vector space of all real-valued functions on the interval [a, b]. You can easily do scalar multiplication with any of these."

---

★ Chant with me: *Vector addition is commutative, associative, identity, and inverses.*

★★ Chant with me: *Scalar multiplication is associative, identity, and double distributive.*

"But, but, but," Fred exclaimed as he objected to those three examples, "What's the *length* of $3 + 8x + 90x^2 + 4x^3 - 6x^4 + \pi x^5 + \sqrt{7}\, x^6$?

Or the angle between $\begin{pmatrix} 8 & 9 & -2 & 6 & 9 \\ 2 & 1 & 29 & 8 & 7 \\ 3 & 0 & \ln 7 & 0 & 1 \end{pmatrix}$ and $\begin{pmatrix} -4 & 5 & 10 & 3 & 0 \\ -7 & 3 & 5 & 2 & 5 \\ \pi & 0 & 4 & 9 & 3 \end{pmatrix}$ ?

Or can you tell me whether the functions $f(x) = \sin \pi x$ and $g(x) = \sin 2\pi x$ are perpendicular to each on the interval $[-1, 1]$?"

Betty and Alexander were stunned by this six-year-old's questions. Length, angles, and perpendicular applied to geometric vectors, but they didn't make any sense in the concept of a generalized vector space.

> ### Intermission
>
> In high school algebra, most of the students were happy when you defined $x^3$ as x times x times x.
> But when you wrote on the board $x^{1/2}$ they screamed, "How can you raise x to the one-half power!"
> Then you went on to raise x to the $-4$ power.
> And take the square root of $-1$.
> And count the number of natural numbers $\{1, 2, 3, 4, \dots\}$ that there are. (*Life of Fred: Calculus,* Chapter 1.)
>
> Mathematicians keep doing the "impossible." That's what makes it fun.

"Hey Fred," said Alexander, "you could find length and angles and perpendicularity with geometric vectors because you had the **dot product**. You defined u·v as the length of u times the length of v times the cosine of the angle between them. $\boxed{\text{u·v} = |\text{u}||\text{v}| \cos \theta}$ And that all made sense because you could measure the length of those geometric vectors, and you could determine the angle between them. You can't find the length of a polynomial. You can't find the angle between two 3×5 matrices."

Alexander should have known better.  Anybody who knows Fred should realize that you don't say *you can't* to him about anything mathematical.

Fred went into deep think.

0.000038 seconds later he had the whole thing resolved.  Smiling, he could rejoin the conversation.  Most people would never realize that he had left it.

With more than five years of university teaching under his belt, he instinctively went into lecture mode: "Just as we looked at geometric vectors and generalized them into the axioms of a vector space[*], we have to look at the dot product for geometric vectors and generalize it."

As the three of them walked around the Great Lake, Fred explained how he could generalize the dot product.

"First notice that we have a binary operation that takes two vectors and gives a vector answer.  That is vector addition.  $u + v = w$ where $u, v, w \in \mathcal{V}$.

"Second, we have a binary operation that takes a scalar and a vector and gives a vector answer.  That is scalar multiplication.  $ru = v$ where $r \in \mathbb{R}$ and $u, v \in \mathcal{V}$."

"And, of course," Betty added, "we have a binary operation that takes two scalars and gives a scalar answer.  $2 \times 3 = 6$ which we learned in the third grade."

Alexander joined in, "So it's obvious.  The only thing left is a binary operation that takes two vectors and gives a scalar answer."

Fred felt so good.  Lay the right groundwork and the students will leap ahead on their own into the new material.

"This new binary operation that takes two vectors and gives a scalar answer needs a new name" Fred continued.  "The definition of the dot product $u \cdot v$ required that we know the length of $u$ and the length of $v$

---

[*] A vector space is a nonempty set with two binary operations—vector addition which is commutative and associative and has an identity and inverses, and scalar multiplication which is associative and has an identity and obeys the two distributive laws.

and the angle between them. We don't know any of those things in most vector spaces.

"I'm going to call the new binary operation the **inner product**. The inner product of $u$ and $v$ will be written as $< u, v >$.

"Back in calculus, when you took the dot product of geometric vectors whose heads were at $(3, 5)$ and $(4, 8)$ and got an answer of $12 + 40 = 52^*$. The first coordinates, 3 and 4, were multiplied together, and the second coordinates, 5 and 8, were multiplied together. That's why I'm going to call this the inner product.

"Now what properties shall $< u, v >$ have? If I name too few properties, the concept of inner product won't be very useful. If the only thing I say is that $< u, v > = < v, u >$ for all $u, v \in \mathcal{V}$, that won't be saying very much.

"If I give a list of 20 properties for $< u, v >$, then the concept of inner product will be so specific, that there won't be many applications for it. You remember the story of *Goldilocks and the Three Bears*. She liked her mush not too hot and not too cold."

Alexander thought about pizzas. He liked his pizzas not too heavy (so that he couldn't lift them) and not too light. (He always wondered who might be satisfied with a "personal-sized pizza."). An eight-pound pizza was just right for him.

Fred listed the three axioms for an inner product:

1. Symmetry          $< u, v > = < v, u >$ for all $u, v \in \mathcal{V}$
2. Linearity          $< ru + sv, w > = r< u, w > + s< v, w >$
                  for all $r, s \in \mathbb{R}$ and for all $u, v \in \mathcal{V}$
3. Positive-definiteness      $< u, u > > 0$ except for $< 0, 0 > = 0$.

---

✷ Fred is skipping a bunch of steps that were done in calculus. The geometric vector with its head at $(3, 5)$ and tail at the origin is $3\mathbf{i} + 5\mathbf{j}$. Similarly, the other vector would be written as $4\mathbf{i} + 8\mathbf{j}$.

We want $(3\mathbf{i} + 5\mathbf{j})\cdot(4\mathbf{i} + 8\mathbf{j})$. In calculus, we just multiplied this out algebraically and got $12\mathbf{i}\cdot\mathbf{i} + 24\mathbf{i}\cdot\mathbf{j} + 20\mathbf{j}\cdot\mathbf{i} + 40\mathbf{j}\cdot\mathbf{j}$.

We note that $\mathbf{i}\cdot\mathbf{i} = 1$ since $\mathbf{i}$ has unit length and the angle between $\mathbf{i}$ and $\mathbf{i}$ is $0°$, so by definition of dot product $\mathbf{i}\cdot\mathbf{i} = |\mathbf{i}||\mathbf{i}|\cos 0° = 1$.

And $\mathbf{i}\cdot\mathbf{j} = 0$ since $\mathbf{i}$ and $\mathbf{j}$ have unit length and the angle between $\mathbf{i}$ and $\mathbf{j}$ is $90°$, so by definition of dot product $\mathbf{i}\cdot\mathbf{j} = |\mathbf{i}||\mathbf{j}|\cos 90° = (1)(1)(0) = 0$.

So $(3\mathbf{i} + 5\mathbf{j})\cdot(4\mathbf{i} + 8\mathbf{j}) = 12 + 40 = 52$.

It's time to unpack a little of what Fred presented to Betty and Alexander.

1. Fred is asserting that the inner product has the symmetry property. Show that the dot product for geometric vectors also has the symmetry property. Start with two vectors whose heads are at (a, b) and (c, d).

2. Fred's linearity rule ($< ru + sv, w > = r< u, w > + s< v, w >$) is really two rules packed into one.

Linearity #2A would state: $< ru, w > = r< u, w >$. This shows that the inner product is linear with respect to scalar multiplication.

Figure out what Linearity #2B would state. (As usual, please don't peek at the answer below before you have done some thinking on your own.)

3. Continuing question 2, show that Linearity #2A and Linearity #2B imply Fred's Linearity rule: $< ru + sv, w > = r< u, w > + s< v, w >$.

4. Given Fred's three axioms for inner product, is it ever possible for an inner product $< u, v >$ to be negative?

### . . . . . . . COMPLETE SOLUTIONS . . . . . . .

1. (a, b)·(c, d) = (ac, bd) = (ca, db) = (c, d)·(a, b)

Another way to show that the dot product for geometric vectors has the symmetry property would be to go back to the definition. Let (a, b) = u and let (c, d) = v. Then u·v = |u||v|cos θ = |v||u|cos (–θ) = |v||u|cos θ = v·u.

(In the third step, we introduced –θ, since if θ is the angle between u and v, then –θ is the angle between v and u. In the fourth step, we used the fact that cosine is an even function.*

2. One hint was that Linearity #2A showed that the inner product *is linear with respect to scalar multiplication*. So we might expect that Linearity #2B would show that the inner product is linear with respect to vector addition. This expectation is correct:

Linearity #2B: $< u + v, w > = < u, w > + < v, w >$

---

\* A function f(x) is an **even function** if, for all x in the domain of f, f(–x) = f(x).

3. $< ru + sv, w > = < ru, w > + < sv, w >$         by #2B

$= r< u, w > + s< v, w >$         by #2A

4. We know from the positive-definiteness property that $< u, u >$ will never be negative, but this question asks if $< u, v >$ can ever be negative. The answer is yes. Suppose that for some vectors $v$ and $w$ we knew that $< v, w > = 6$. Then $< -v, w > = < (-1)v, w > = -1< v, w > = (-1)(6) = -6$.

A vector space in which an inner product is defined (according to Fred's three axioms) is called an **inner product space**.

"Wait a minute!" Alexander said. He stopped walking and set down the picnic basket. Two rabbits jumped into the basket. "All this inner product space stuff still doesn't answer the question that you, yourself, posed: 'What's the *length* of $3 + 8x + 90x^2 + 4x^3 - 6x^4 + \pi x^5 + \sqrt{7} x^6$?' We know that the set of all polynomials of degree 6 or less is a vector space. Now you've turned it into an inner product space, and we still we don't know its length."

"I did not," Fred said as he grinned.

"Did not *what*?" Alexander asked.

"I did not turn it into an inner product space. I said that an inner product space is a space in which an inner product is defined. We haven't yet defined the inner product of two polynomials."

*Your Turn to Play*

1. Let's start with something easier. We know what the inner product in $\mathbb{R}^2$ is. Two pages ago Fred did $< (3, 5), (4, 8) > = 12 + 40 = 52$.

In $\mathbb{R}^5$, what would $< (1, 2, 7, -3, 2), (4, -5, 0, -3, 6) >$ equal?

2. The vector space of all polynomials of degree 2 or less is a perfect match up with $\mathbb{R}^3$

$$\begin{cases} 3 + 8x + 60x^2 & \text{matches up with} & (3, 8, 60) \\ 5 \quad\;\; + 9x^2 & \text{matches up with} & (5, 0, 9) \end{cases}$$

$$\begin{cases} \text{Add } 3 + 8x + 60x^2 \text{ and } 5 + 9x^2 \text{ and get } 8 + 8x + 69x^2. \\ \text{Add } (3, 8, 60) \quad \text{and } (5, 0, 9) \text{ and get } (8, 8, 69). \end{cases}$$

$\left\{ \begin{array}{l} \text{Multiply } 3 + 8x + 60x^2 \text{ by the scalar 5 and get } 15 + 40x + 300x^2. \\ \text{Multiply } (3, 8, 60) \quad \text{by the scalar 5 and get} \quad (15, 40, 300). \end{array} \right._3$

In $\mathbb{R}^3$, we define the inner product of $(r_1, r_2, r_3)$ and $(s_1, s_2, s_3)$ as $\sum_{i=1}^{3} r_i s_i$.

Your turn: In the vector space of all polynomials of degree 2 or less, how would you define $< r_1 + r_2 x + r_3 x^2, s_1 + s_2 x + s_3 x^2 >$?

3. Just to show you how easy this is, find $< 3 + 8x + 60x^2, 5 + 9x^2 >$. The symbolism of the previous question is much harder than actually doing an inner product.

# .......COMPLETE SOLUTIONS.......

1. $< (1, 2, 7, -3, 2), (4, -5, 0, -3, 6) > = (1)(4) + (2)(-5) + (7)(0) + (-3)(-3) + (2)(6) = 15$.

2. $< r_1 + r_2 x + r_3 x^2, s_1 + s_2 x + s_3 x^2 > = \sum_{i=1}^{3} r_i s_i$. This is awfully, really, completely close to the definition of the inner product in $\mathbb{R}^3$. Wouldn't you agree?

3. $< 3 + 8x + 60x^2, 5 + 9x^2 > = (3)(5) + (8)(0) + (60)(9) = 555$.

"Good," announced Alexander. "I'm glad we've got that cleared up." He picked up the picnic basket. The two rabbits decided to go along for the ride.

Betty giggled. Alexander looked down at the basket and removed the rabbits.

"That's not what I was laughing about," she said. "I think Fred has put one over on you again. You still can't tell what the length of $3 + 8x + 90x^2 + 4x^3 - 6x^4 + \pi x^5 + \sqrt{7} x^6$ is."

"It's . . . it's . . . it's," Alexander sputtered. "Fred! I'm going to get you!"

At this point Alexander would often chase Fred around till he caught him and then would tickle him until he got the silly-giggles. Fred was too fast for him this time. During the previous *Your Turn to Play*, he had stripped down to the bathing suit that he had been wearing under his pants. He yelled, "Can't catch me!" as he ran and jumped into the Great Lake.

204

Betty and Alexander stood there on the shore watching Fred paddle around in the lake.

Actually, defining an inner product is the major step. Finding ① the *length* of a polynomial vector or ② the *angle* between two matrices or ③ whether two functions are *perpendicular*—that stuff involves no creativity. Back in calculus, we did all those things with the dot product of two geometric vectors. Once you have got the inner product defined for a particular vector space, *length*, *angles*, and *perpendicularity* follow instantly.

**Um. I, your reader, would appreciate a slight refresher. How did we define length, angles, and perpendicularity for dot products? Wait! Please don't do another** *Your Turn to Play* **. Just spit it out—one, two, three—and I'll remember it.**

Okay.

① Back in calculus, the length of any vector u was defined by u·u = |u||u|. Or, doing the algebra, |u| = √u·u .

The reason that was true is seen by looking at an example. Suppose we want the length of the

vector from the origin to the point (12, 5). That's the length of 12**i** + 5**j**.

By the Pythagorean theorem, its length is √144 + 25.

On the other hand, if we take the dot product of (12, 5) with itself, we have (12, 5)·(12, 5) = 144 + 25.

So | (12, 5) | = √(12, 5)·(12, 5)

② Once we know how to find the length |u| of a vector u, we can find the angle between two vectors u and v by using the definition of the dot product between two geometric vectors:  u·v = |u||v| cos θ.

By algebra, cos θ = $\frac{u·v}{|u||v|}$     By trig, θ = arccos $\frac{u·v}{|u||v|}$

③ Once we know how to find the angle between two vectors u and v, then we can establish that u is perpendicular to v iff < u, v > = 0.

For geometric vectors < u, v > means the same as u·v.

If u ⊥ v, then θ = 90° (or θ = π/2, if you prefer radians).

cos 90° = 0

u·v = |u||v| cos θ  becomes  u·v = 0  or  < u, v > = 0.

So once you have the inner product defined, length, angles, and perpendicularity are all just mechanical applications. It's a lot like applying $x = \dfrac{-b \pm \sqrt{b^2 - 4ac}}{2a}$ to $ax^2 + bx + c = 0$.

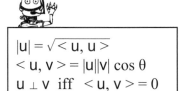

$$|u| = \sqrt{<u, u>}$$
$$<u, v> = |u||v| \cos \theta$$
$$u \perp v \text{ iff } <u, v> = 0$$

Strictly mechanical at this point

In inner product spaces (vector spaces in which an inner product is defined), the length of u has too much of a geometrical sound to it. Instead, |u| is called the **norm** of u. In order to emphasize that this is no longer just a length and that we are in an inner space, the norm of u is written ‖u‖. That makes it look a lot more sophisticated.

The norm of the vector (4, 5) in $\mathbb{R}^2$ is
$$\| (4,5) \| = \sqrt{<(4, 5), (4, 5)>} = \sqrt{16 + 25} = \sqrt{41} .$$

The norm of the vector $4 + 5x$ in the vector space of all polynomials of degree less than 39089 is
$$\| 4 + 5x \| = \sqrt{<4 + 5x, 4 + 5x>} = \sqrt{16 + 25} = \sqrt{41} .$$

The set of all continuous functions on the interval $0 \le x \le 1$ is a vector space. On page 199 Fred asked whether the functions $f(x) = \sin \pi x$ and $g(x) = \sin 2\pi x$ are perpendicular to each on the interval $[-1, 1]$.

*Your Turn to Play*

1. Are they?

**Beats me. You forgot something, Mr. Author. As you said several times, the major step is defining the inner product. Once that is done, then the rest is mechanical.**

Oops. You're right. We need a definition of the inner product of two continuous real-valued functions on the interval [a, b].

First attempt: How about $< f(x), g(x) > = f(x)g(x)$?

**That stinks. An inner product maps pairs of vectors into scalars, and f(x)g(x) isn't a number.**

Second attempt: How about $< f(x), g(x) > = f(½)g(½)$?

**That certainly maps to a scalar answer. It does have symmetry. But it's not positive-definite. < f(x), f(x) > will certainly never be negative, since it is equal to f(½)f(½). But < f(x), f(x) > = 0 only tells me that f(½) = 0. That doesn't make f the zero function—equal to zero for all x in the interval [–1, 1]. Nice try, Mr. Author.**

Third attempt: How about $< f(x), g(x) > = \int_{x=a}^{b} f(x)g(x)\,dx$ ?

**Let me check. It certainly has a scalar answer. It is obviously symmetrical. It has linearity. The last thing to check is positive-definiteness. Certainly, < f, f > could never be negative since**

$$< f, f > = \int_{x=a}^{b} f(x)f(x)\,dx.$$ **The only way that < f, f > could be zero would be if f(x) = 0 for all x in [a, b].*  Okay, Mr. Author. This third try is an inner product.**

Now we can answer Fred's question.

*Your Turn to Play*

1. Are they? (Are the functions f(x) = sin πx and g(x) = sin 2πx perpendicular to each on the interval [–1, 1]?)

. . . . . . .**COMPLETE SOLUTIONS**. . . . . . .

1. That translates into the question $< \sin πx, \sin 2πx > \overset{?}{=} 0$.

*(continued next page)*

---

★ I need to make a technical point. Since no one reads footnotes, this is a good place to do that. We are finding an inner product on the space of all *continuous* functions on [a, b].

The set of all real-valued functions on [a, b] is also a vector space, but my third attempt for a definition of an inner product wouldn't work. You might not be able to integrate f(x)g(x) when f or g are very discontinuous (jumps around too much).

The set of all real-valued functions that can be integrated is also a vector space. This space would include functions like f(x) = 0 for all x, except for x = ⅜ where f(⅜) = 71. Then < f, f > = 0 since $\int_{x=a}^{b} f(x)f(x)\,dx = 0$ (the area under the curve y = f(x) is zero). But f is not the zero function, so this definition of inner product also won't work for the vector space of all real-valued functions that can be integrated.

That translates into $\int_{x=-1}^{1} (\sin \pi x)(\sin 2\pi x)dx \overset{?}{=} 0$

Integration of trig functions is usually encountered in the second semester of calculus. For some readers, this may have been a long time ago. (I took my second semester of calculus at City College of San Francisco in the spring of 1962.) I can never remember the trig formulas, but a good trig book gave me two formulas I could use.

*Approach #1*: Use sin 2A = 2 sin A cos A.

Then the integral at the top of this page becomes

$$\int_{x=-1}^{1} (\sin \pi x)(2 \sin \pi x \cos \pi x) \, dx \; = \; \int_{x=-1}^{1} 2 \sin^2 \pi x \cos \pi x \, dx \; =$$

$$\frac{2}{3\pi} \sin^3 \pi x \, \bigg]_{-1}^{1} \; = \; \frac{2}{3\pi} 0 \; - \; \frac{2}{3\pi} 0 \; = \; 0$$

*Approach #2*: Use sin A sin B = ½ cos (A − B) − ½ cos (A + B).

Then the integral at the top of this page becomes

$$\int_{x=-1}^{1} (½ \cos (-\pi x) - ½ \cos 3\pi x) \, dx \; = \; \int_{x=-1}^{1} (½ \cos \pi x - ½ \cos 3\pi x) \, dx \; =$$

$$\frac{1}{2\pi} \sin \pi x \; - \; \frac{1}{6\pi} \sin 3\pi x \, \bigg]_{-1}^{1} \; = \; 0$$

In any event, < sin πx, sin 2πx > equals zero, and the functions are perpendicular.

---

### Intermission

Almost everything that's fun has its tedious parts. The calculus we just did could have an eyes-glaze-over effect on many people (myself included).

For me, the most dreary part of linear algebra is doing elementary row operations. I always make arithmetic errors.

But let's contrast linear algebra with . . . skiing.

*Intermission (continued)*

Linear algebra beats skiing in virtually every category:

☞ It's less tedious.  Compare those four-hour trips in the car to get to the ski slopes with the time it takes to do a *Your Turn to Play*.

☞ It's a whole lot cheaper.  Add up the cost of this book plus pencils and paper.  It may be less than the cost of the gas for that four-hour trip—not to mention the cost of the boots, skis, lift tickets, etc.

☞ Both can be done outdoors.  Both can build up a sweat and cause you to breathe hard.

☞ Unless you are extraordinarily clumsy with a sharpened pencil, skiing is slightly more likely to put you in the hospital than doing linear algebra.

We changed the length of geometric vector $v$, $|v|$, to the norm of $v$, $\|v\|$, when we moved from $\mathbb{R}^2$ or $\mathbb{R}^3$ to inner product spaces.  Another language change in moving from the geometric approach to the more generalized world of inner product spaces, is the word *perpendicularity*.  It really feels a bit funny to ask if the vector $\sin \pi x$ is perpendicular to $\sin 2\pi x$.  We now say that if $< u, v > = 0$, then $u$ and $v$ are **orthogonal**.

We say a whole set of vectors $\{v_1, v_2, \ldots, v_n\}$ is orthogonal if every pair of vectors in that set is orthogonal.

What is really surprising to many people is that orthogonal sets of vectors are always linearly independent.

Suppose we have a set of six orthogonal vectors $\{v_1, v_2, \ldots, v_6\}$ and suppose we know that $\sum_{i=1}^{6} r_i v_i = 0$.  We would need to show that all the $r_i$ are equal to zero.  That would show that $v_1, v_2, \ldots, v_6$ are linearly independent.

Now we will show that $r_4 = 0$.

Take the inner product of both sides of $\sum\limits_{i=1}^{6} r_i v_i = 0$ with $v_4$.
(Technical point: Since *inner product* is a function, by definition of *function*, $< \sum\limits_{i=1}^{6} r_i v_i \,, v_4 >$ must yield the same result as $< 0, v_4 >$.)

$$< \sum_{i=1}^{6} r_i v_i \,, v_4 > \; = \; < 0, v_4 >$$

$$< \sum_{i=1}^{6} r_i v_i \,, v_4 > \; = \; 0 \qquad$$ We'll show $< 0, u > = 0$ for all $u$ in one of the Cities.

$$\sum_{i=1}^{6} r_i < v_i \,, v_4 > \; = \; 0 \qquad$$ linearity property of inner product

$$r_4 < v_4, v_4 > \; = \; 0 \qquad$$ All the other $< v_i, v_4 > = 0$ since the set is orthogonal.

$$r_4 \; = \; 0 \qquad$$ by positive-definiteness, $< v_4, v_4 > \; > 0$.

We have shown $r_4 = 0$, but there is nothing special about $r_4$.  Give your kid brother an eraser and a pencil, and he could go back and erase every "4" and replace it with "3" to show that $r_3 = 0$.  *Mutatis mutandis =* doing the whole argument over again making the appropriate changes.

---

*Your Turn to Play*

1. Now comes the fun part.  I asked you to take on faith, "We'll show $< 0, u > = 0$ for all $u$ in one of the Cities."  That part is true.

But "Orthogonal sets of vectors are always linearly independent" is not true.

And, of course, the proof is wrong.  (It would bother a lot of mathematicians if we found a valid proof for something that was false.)

How important is it to learn to think critically?  When you were a kid you may have accepted the "fact" that rabbits laid chocolate Easter eggs. As an adult you might not want to accept the statement that orthogonal sets of vectors are always linearly independent just because it was in a book.

As usual, please don't just read the solution given below.  Instead, look at the "proof " above and see where I led you astray.

---

```
.......COMPLETE SOLUTIONS.......
```

1. Often, if you want to convince someone that something is not true, the best place to hide the error in your reasoning is near the end of your argument. Get the person to say, "Yes, yes, yes, yes, . . ." to each of the steps in your reasoning. Then slip in the zinger when they are mechanically nodding agreement.

On the very last line I divided both sides of $r_4 <v_4, v_4> = 0$ by $<v_4, v_4>$ which is a legal algebraic procedure if $<v_4, v_4>$ isn't equal to zero. I assured you that by positive-definiteness, $<v_4, v_4> > 0$.

On page 201, the axiom was, "Positive-definiteness $<u, u> > 0$ except for $<0, 0> = 0$."

If $v_4$ were the zero vector, the last line of the proof falls apart.

---

**Theorem**: Any orthogonal set of *nonzero* vectors is linearly independent.

One last little thing about inner product spaces, and then we can leave the theoretical heights of Chapter 2¾ and get back to terra firma by starting Chapter 3 where we will again look at solving systems of linear equations. (Was that the longest sentence in this book?)

Suppose we're in $\mathbb{R}^2$. And suppose we have the vectors (3, 9) and (2, 7). They make a nice basis for $\mathbb{R}^2$.* If you pictured them as geometrical vectors, they would look like:

Every vector in $\mathbb{R}^2$ could be written as a linear combination of (3, 9) and (2, 7). The numbers are small. Everybody is happy.

Happy. Happy. Happy. Except . . .

Some of us had unhappy teenage years. Not you or me, of course, but some other people had a really rough time with their parents. In order to present the Gram-Schmidt orthogonalization process properly, I need to

---

✶ They are a basis since: (1) there are two of them, and we know the dimension of $\mathbb{R}^2$ is two; and (2) they are linearly independent since neither is a multiple of the other.

describe one of the lurid moments in some people's teenage years—not yours or mine, of course. If you are reading this linear algebra book to your kid sister, you may need to have her put her hands over her ears for the next paragraph.

[R-rated paragraph: extreme emotional violence] Your mother comes into your bedroom and finds you, a teenager, on the floor with this pair of vectors. She says, "Honey, we have company coming over this evening. Could you please straighten out those vectors? I know you can't line them up, since they would lose their linear independence. But could you make them more right angley so they look neater?" You cringe at the expression *right angley*. Why must she continue to use baby talk when you are a teenager? What's wrong with the adult word *orthogonal*? You know that such imperial demands—*could you please straighten out those vectors?*—and deeply crushing insults to your adulthood—*right angley*—could only drive you to a life of crime or years on the analyst's couch. [end of R-rated paragraph.]

You start with one of the vectors. Surely, she can't complain about that. Now, the task is to move the other vector so that it's perpendicular to the first vector.

You phone your friend to say how bad things are at your house and find out that your friend's mom is demanding that your friend's three vectors—(8, 2, –5), (7, 0, 29), and (1, 4, 4)—have to all be "straightened out." Your friend will start with (8, 2, –5) and then adjust (7, 0, 29) so that it is orthogonal to (8, 2, –5). Finally, take the (1, 4, 4) and twist it until it is perpendicular to the first two vectors.

The Gram-Schmidt orthogonalization process helps us do the twisting of the vectors to get them "straightened out." If this were the 1960s, I would have to mention the dance craze, the Twist, which swept the world. Everyone was doing the Twist. Except Erhardt Schmidt. He died in 1959.*

The Twist had its day. *Sic transit gloria mundi.*** It's a fact: Most linear algebra books rarely mention the Twist.

---

✶ Schmidt outlived Jögen Pederson Gram who died in a bicycle collision in 1916. I bet Gram wasn't wearing a helmet.

✶✶ *So passes the glory of the world.* Said when an important person dies or an era is over.

Here are the dance steps for the Gram-Schmidt. We are going to turn the basis $u_1, u_2, \ldots, u_n$ into an orthogonal basis $v_1, v_2, \ldots, v_n$. (The two bases will have the same number of elements in them by the Nice theorem: *If $V$ has a basis with n elements in it, then every basis of $V$ has n elements.*)

Keep in mind, we're turning $u$'s into $v$'s.

Step one:

$$v_1 = u_1$$

Step two:

$$v_2 = u_2 - \frac{<u_2, v_1>}{<v_1, v_1>} v_1$$

Step three:

$$v_3 = u_3 - \frac{<u_3, v_1>}{<v_1, v_1>} v_1 - \frac{<u_3, v_2>}{<v_2, v_2>} v_2$$

. . . and keep dancing up to $v_n$.

Some notes on these Gram-Schmidt dance steps:

♪#1: Everyone likes step one best.

♪#2: In step two, $v_2$ is a linear combination of $u_2$ and $v_1$.
Noticing where $v_1$ came from, we can say that $v_2$ is a linear combination of $u_2$ and $u_1$.

Similarly, $v_3$ is a linear combination of $u_3$, $u_2$, and $u_1$.

And $v_7$ is a linear combination of $u_7, u_6, u_5, u_4, u_3, u_2$, and $u_1$. We are using the given basis to build a new orthogonal basis.

♪#3: If $\dfrac{<u_3, v_2>}{<v_2, v_2>} \, v_2$ looks
vaguely familiar, you were paying
attention back in your calculus
days when you were studying
geometric vectors. It is the
formula for the projection of $u_3$
onto $v_2$.

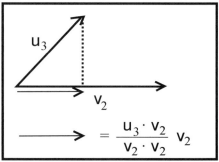

the projection of $u_3$ onto $v_2$

So the third dance step in the
Gram-Schmidt Vector Twist could be written:

$$v_3 \;=\; u_3 \;-\; \text{proj of } u_3 \text{ onto } v_1 \;-\; \text{proj of } u_3 \text{ onto } v_2 \,.$$

♪#4: Since the $v_i$'s are nonzero orthogonal vectors, they are linearly
independent (by the theorem three pages ago). Since there are n of them,
they must span the vector space and be a basis.*

♪#5: I guess, since this is a math book, we should show that the Gram-
Schmidt orthogonalization process works, i.e., that it really twists a basis
$\{u_1, u_2, \ldots, u_n\}$ into an orthogonal set $\{v_1, v_2, \ldots, v_n\}$.

First, we will show that $v_2$ is perpendicular to $v_1$.

$<v_2, v_1> \;=\; <u_2 - \dfrac{<u_2, v_1>}{<v_1, v_1>}\, v_1,\; v_1>$    by definition of $v_2$

$=\; <u_2, v_1> \;-\; <\dfrac{<u_2, v_1>}{<v_1, v_1>}\, v_1,\; v_1>$    by Linearity #2B (p. 202)

$=\; <u_2, v_1> \;-\; \dfrac{<u_2, v_1>}{<v_1, v_1>}\, <v_1,\; v_1>$    by Linearity #2A (p. 202)

$=\; 0$    by high school algebra

---

★ I'm not sure whether we've proved: *Any set of n linearly independent vectors in an
n-dimensional space must be a basis*, but the proof is super-easy. Proof: If they didn't
span the space, then there would be a vector which would not be a linear combination
of the n linearly independent vectors. Those n + 1 vectors would be linearly
independent, which contradicts the 𝕿𝖍𝖊𝖔𝖗𝖊𝖒: *In an n-dimensional vector space, no set
of n + 1 vectors will be linearly independent*, which we did on page 169.

Second, we will do that same four-line procedure to show that if $v_1$, $v_2$, $v_3$, $v_4$, $v_5$, and $v_6$ are orthogonal, then $v_7$ must be orthogonal to each of them. We will show that $v_7$ is perpendicular to $v_3$.

$$< v_7, v_3 > \ = \ < u_7 - \sum_{i=1}^{6} \frac{< u_7, v_i >}{< v_i, v_i >} v_i , v_3 > \qquad \text{by definition of } v_7$$

$$= \ < u_7, v_3 > \ - \ \sum_{i=1}^{6} \frac{< u_7, v_i >}{< v_i, v_i >} < v_i , v_3 > \qquad \text{by linearity}$$

$$= \ < u_7, v_3 > \ - \ \frac{< u_7, v_3 >}{< v_3, v_3 >} < v_3 , v_3 > \qquad \text{since the } v_i \text{ are}$$

$$\text{orthogonal, } < v_j, v_k > = \delta_{jk}$$

$$\text{where } \delta_{jk} \text{ is the Kronecker delta (page 88)}$$

$$= \ 0 \qquad \text{by algebra}$$

♪#6: Reading ♪#5 is dreary work. It is not difficult. It is just long, like adding up a tall column of numbers. Is it a proof? If by the word *proof* we mean an argument that convinces, then I would say that we proved that the Gram-Schmidt process really twists a basis $\{u_1, u_2, \ldots, u_n\}$ into an orthogonal set $\{v_1, v_2, \ldots, v_n\}$.

At the top of this page, I showed that $v_7$ is perpendicular to $v_3$. I didn't show that $v_7$ is perpendicular to $v_2$. My proof lacked what you might call dignity. My purpose was to demonstrate that Gram-Schmidt can do the Twist on a basis—it wasn't to show off. They wouldn't let me[*] write like this in a high brow mathematics journal. In that venue, you are not allowed to be the least bit *silly*.

dignity

Here's how it would have to look:

Theorem: Given any basis $\{u_1, u_2, \ldots, u_n\}$ in an inner product space $\mathcal{V}$, the Gram-Schmidt orthogonalization process will produce a set of n vectors

---

[*]    me

$\{v_1, v_2, \ldots, v_n\}$ with the following properties: First, for any i, $1 \le i \le n$, there will exist scalars $r_j$, $1 \le j \le i$, such that $v_i = \sum_{j=1}^{i} r_j u_j$.

Secondly, span $\{v_1, v_2, \ldots, v_n\} = \mathcal{V}$.

Thirdly, $\{v_1, v_2, \ldots, v_n\}$ is orthogonal.

Proof: We proceed by mathematical induction.

[At this point they stick in the first part of my proof from two pages ago and show that $v_2$ is perpendicular to $v_1$.]

Now proceeding to the induction step, we assume that for some j, $1 \le j \le n-1$, the induction hypothesis holds. We now establish that for any k, $k < j + 1$, that $v_k$ and $v_{j+1}$ are orthogonal.

$$< v_{j+1}, v_k > \; = \; < u_{j+1} - \sum_{i=1}^{j} \frac{< u_{j+1}, v_i >}{< v_i, v_i >} v_i \, , \, v_k > \qquad\qquad \text{etc., etc., etc.}$$

Now look at that line! Three subscripts are running around. Compare it with the equation at the top of the previous page. Sure, this is more precise, more complete, and more dignified, but it takes an Herculean effort to comprehend it.

Rather than put a *Your Turn to Play* here so that you can practice the dance steps of G–S, I'll put one in the first City and include all the steps of the solution in the answer.

---
**Garner**
---

1. Using Gram-Schmidt, transform $\{(2, 2, 1), (-2, 1, 2), (18, 0, 0)\}$ into an orthogonal set.

2. Your father walks into your room when you are a teenager and says, "Look at those vectors!"

"But sir, mother just had me orthogonalize them. Aren't they pretty? Each pair—$(2, 2, 1)$, $(-2, 1, 2)$, and $(2, -4, 4)$—is perpendicular."

"And you call that neat!" he roared. "They are all different lengths. You should be ashamed. Make 'em all unit vectors."

**Unit vectors** are vectors whose lengths are equal to one. In calculus we wrote that $\dfrac{v}{|v|}$ is the unit vector associated with $v$.

Since we are more sophisticated now, we write $\dfrac{v}{\|v\|}$ or $\dfrac{v}{\sqrt{<v, v>}}$

If you obey your father, you will turn this orthogonal set into an **orthonormal** set.

Do it.

*answers*

1. $v_1 = (2, 2, 1)$

$$v_2 = (-2, 1, 2) - \frac{-4 + 2 + 2}{4 + 4 + 1}(2, 2, 1) = (-2, 1, 2)$$

$$v_3 = (18, 0, 0) - \frac{36 + 0 + 0}{4 + 4 + 1}(2, 2, 1) - \frac{-36 + 0 + 0}{4 + 1 + 4}(-2, 1, 2)$$

$$= (2, -4, 4)$$

2. $\dfrac{(2, 2, 1)}{\sqrt{<(2, 2, 1), (2, 2, 1)>}} = \dfrac{(2, 2, 1)}{\sqrt{4 + 4 + 1}} = (\tfrac{2}{3}, \tfrac{2}{3}, \tfrac{1}{3})$

$\dfrac{(-2, 1, 2)}{\sqrt{4 + 1 + 4}} = (-\tfrac{2}{3}, \tfrac{1}{3}, \tfrac{2}{3})$ and $\dfrac{(2, -4, 4)}{\sqrt{4 + 16 + 16}} = (\tfrac{1}{3}, -\tfrac{2}{3}, \tfrac{2}{3})$

---

**Rant Pass**

---

## FOURIER SERIES

1. In the vector space of all continuous functions defined on the interval [−1, 1], why would you *not* use Gram-Schmidt on f(x) = sin πx and g(x) = sin 2πx?

2. For **Fourier series**, the best place to start is with the set {1, cos πx, sin πx, cos 2πx, sin 2πx, cos 3πx, sin 3πx, . . . }. We will represent the n[th] element of that set by $\phi_n$.

   We are first going to make mom happy with that set (by showing it is orthogonal).

   Show that 1 and cos 47πx are orthogonal. We will work in the interval [−1, 1].

3. Showing that cos 6πx and cos 30πx are orthogonal is so similar to showing that sin πx and sin 2πx are orthogonal (which we did in the *Your Turn to Play* on page 207), that we will skip that.

   Show that sin 7πx and cos 55πx are orthogonal. So you don't have to rummage around in your trig book, here is the product-to-sum formula: sin A cos B = ½ sin (A − B) + ½ sin (A + B).

4. We have now done enough calculus that you (correctly) believe that {1, cos πx, sin πx, cos 2πx, sin 2πx, cos 3πx, sin 3πx, . . . } is an orthogonal set. Mom is happy.

   Now to make dad happy (see the previous City) by showing that each of the $\phi_i$ is a unit vector. Show that ‖ cos 7πx ‖ = 1 on the interval [−1, 1].

   We have now shown that {1, cos πx, sin πx, cos 2πx, sin 2πx, cos 3πx, sin 3πx, . . . } is an orthonormal set on [−1, 1].

5. Now we come to the Fourier series part. There are tons of functions—some of them quite complicated—which can be represented as an infinite linear combination of the $\phi_i$.

$$f(x) = \sum_{i=1}^{\infty} r_i \phi_i \text{ where the } r_i \text{ are real numbers.}$$

**Why in the world would you want to do that? Trading plain old *f(x)* for an infinite sum of sines and cosines seems like a bad deal.**

On the contrary, it's a really good deal.  For example, the f(x) function might be messy,* but a sum of sines and cosines is child's play in comparison.

The trick is to find the values of the $r_i$ in $f(x) = \sum\limits_{i=1}^{\infty} r_i \phi_i$.  You know f(x) and you know all the $\phi_i$ to start with.

The $\sum\limits_{i=1}^{\infty} r_i \phi_i$ is the Fourier series for f(x) and the $r_i$ are the Fourier coefficients.

We start with $$f(x) = \sum_{i=1}^{\infty} r_i \phi_i$$

By the symmetric property of equality, $\sum\limits_{i=1}^{\infty} r_i \phi_i = f(x)$.

We're going to find $r_{28}$.  You know the drill by now.  If you want to find $r_{9305}$, just get your kid brother to erase every 28 and replace it with 9305.

Multiply both sides by $\phi_{28}$.  More properly, we should say, "Take the inner product of both sides with respect to $\phi_{28}$.

$$< \sum_{i=1}^{\infty} r_i \phi_i , \phi_{28} > = < f(x), \phi_{28} >$$

$\sum\limits_{i=1}^{\infty} r_i < \phi_i, \phi_{28} > = < f(x), \phi_{28} >$     by linearity of the inner product

$r_{28} < \phi_{28}, \phi_{28} > = < f(x), \phi_{28} >$     because of mom—the $\phi_i$ are orthogonal.  For example,

$$< \phi_{16}, \phi_{28} > = 0$$

---

* The function f might have jump discontinuities— places where its value jumps suddenly.

The function f might have "corners."  These are places on the graph where the slope abruptly changes value—where $\dfrac{d\,f(x)}{dx}$ has a jump discontinuity.

$$r_{28} = \langle f(x), \phi_{28} \rangle$$

because of dad—the $\phi_i$ are unit vectors

$$r_{28} = \int_{x=-1}^{1} f(x)(\phi_{28}) \, dx$$

by definition of the inner product of two functions

We have let $\phi_n$ stand for the $n^{th}$ function in $\{1, \cos \pi x, \sin \pi x, \cos 2\pi x, \sin 2\pi x, \cos 3\pi x, \sin 3\pi x, \dots\}$ defined on $[-1, 1]$. But $\phi_n$ could be the $n^{th}$ function in any orthonormal set.

Suppose you don't like the interval $[-1, 1]$. If you want another interval that is also of length 2—such as $[0, 2]$ or $[1.65, 3.65]$—everything you did in questions 1–4 will still work. You can still use that classic orthonormal set $\{1, \cos \pi x, \sin \pi x, \cos 2\pi x, \sin 2\pi x, \cos 3\pi x, \dots\}$.

If you want a Fourier series on an interval of length 2L—such as $[-L, L]$ or $[1.65, 1.65 + 2L]$—then use this updated orthonormal set:

$$\left\{ \frac{1}{L}, \ \frac{1}{L}\cos\frac{\pi x}{L}, \ \frac{1}{L}\sin\frac{\pi x}{L}, \ \frac{1}{L}\cos\frac{2\pi x}{L}, \ \frac{1}{L}\sin\frac{2\pi x}{L}, \dots \right\}$$

Show that $\dfrac{1}{L}\cos\dfrac{7\pi x}{L}$ has length equal to 1 on $[-L, L]$.

6. Entire books have been written about Fourier series. It is amazing (or distressing, depending on your point of view) what they can do with the relatively simple idea we have outlined in this City: *Take a function f(x) defined on an interval [–1, 1] and take an orthonormal set like {1, cos πx, sin πx, cos 2πx, sin 2πx, cos 3πx, sin 3πx, . . . }. Then we can express f as an infinite trigonometric sum* $f(x) = \sum_{i=1}^{\infty} r_i \phi_i$ *where* $r_i = \int_{x=-1}^{1} f(x)(\phi_i) \, dx.$

The major uses of Fourier series are in the cases in which f is a periodic function—something that repeats itself like sunspots, an alternating electric current, sound waves, etc. The official name for the study of expressing a periodic function f as a sum of sines and cosines is called **harmonic analysis**.

There are two main ways that you can apply Fourier series to periodic phenomena in the world. ① You can use Fourier series to interpret experimental data. You measure the tides on the beach of the

Great Lake and express your results as an infinite trigonometric sum.

② You start with the theory (often it's a differential equation) and solve it using Fourier series. For example, you start with a vibrating string with the ends nailed down, which at time t = 0 looks like:

The height of the string, y, depends on x and it also depends on t. The physicists tell us that the equation of a vibrating string is given by $\frac{\partial^2 y}{\partial t^2} = c \frac{\partial^2 y}{\partial x^2}$ where c is a constant.

We are using partial derivatives, $\partial$, because y is a function of two variables, x and t. A whole page of equations can ensue. (In one engineering-type book, the solution of $\frac{\partial^2 y}{\partial t^2} = c \frac{\partial^2 y}{\partial x^2}$ used Fourier series. And 17 equations!)

Although using Fourier series for periodic phenomena is often the context in which Fourier series is introduced, the definition we have introduced in this City doesn't require periodicity. Take virtually any function f defined on an arbitrary interval [a, b] and use any orthonormal set $\psi_n$ on [a, b], and we have

$$f(x) = \sum_{i=1}^{\infty} r_i \psi_i \ where \ r_i = \int_{x=a}^{b} f(x)(\psi_i) \, dx.$$

One last note on Fourier series. In the vibrating string problem at the top of this page, we had y as a function of two variables, x and t. The solution requires the use of **double Fourier series**.

Your question: Make a guess. Is there such a thing as triple Fourier series? ☐yes ☐no.

*answers*

1. The Gram-Schmidt orthogonalization process is used to make a set of vectors perpendicular to each other. In the *Your Turn to Play* on page 207 we showed that f(x) = sin πx and g(x) = sin 2πx are already perpendicular.

2. $\langle 1, \cos 47\pi x \rangle = \int_{x=-1}^{1} \cos 47\pi x \, dx = \frac{1}{47\pi} \sin 47\pi x \Big]_{-1}^{1} = 0$

3. $\langle \sin 7\pi x, \cos 55\pi x \rangle = \int_{x=-1}^{1} (\sin 7\pi x)(\cos 55\pi x) \, dx =$

$$= \frac{1}{2} \int_{x=-1}^{1} [\, \sin(-48\pi x) + \sin(62\pi x) \,]\, dx$$

$$= \frac{1}{2} \left[ \frac{1}{48\pi} \cos(-48\pi x) + \frac{-1}{62\pi} \cos(62\pi x) \right]_{-1}^{1}$$

$$= \frac{1}{2} \left[ \left( \frac{1}{48\pi} + \frac{-1}{62\pi} \right) - \left( \frac{1}{48\pi} + \frac{-1}{62\pi} \right) \right] = 0$$

4. $\| \cos 7\pi x \| = \displaystyle\int_{x=-1}^{1} \cos^2 7\pi x \; dx$

$$= \int_{x=-1}^{1} (\tfrac{1}{2} + \tfrac{1}{2} \cos 14\pi x) \; dx \qquad \left\{ \begin{array}{l} \text{using the trig formula} \\ \cos^2 A = \tfrac{1}{2} + \tfrac{1}{2} \cos 2A \end{array} \right.$$

$$= \frac{x}{2} + \frac{1}{28\pi} \sin 14\pi x \,\Big]_{-1}^{1} \quad = (\tfrac{1}{2} + 0) - (-\tfrac{1}{2} + 0) = 1$$

5. $\left\| \dfrac{1}{L} \cos \dfrac{7\pi x}{L} \right\| = \dfrac{1}{L} \displaystyle\int_{x=-L}^{L} \cos^2 \dfrac{7\pi x}{L} \; dx$

$$= \frac{1}{L} \int_{x=-L}^{L} (\tfrac{1}{2} + \tfrac{1}{2} \cos \frac{14\pi x}{L}) \; dx$$

$$= \frac{1}{L} \left( \frac{x}{2} + \frac{L}{28\pi} \sin \frac{14\pi x}{L} \right) \Big]_{-L}^{L} = \frac{1}{L} \left( \left( \frac{L}{2} + 0 \right) - \left( \frac{-L}{2} + 0 \right) \right) = 1$$

6. ☒ yes     Triple Fourier series would be encountered when you have a function of three variables. For example, your success in doing Gram-Schmidt orthogonalization problems is a function of (1) the time you put in; (2) your effort; and (3) your native math ability.

---

## Wampum

1. We have found inner products for many common vector spaces.

❀ In $\mathbb{R}^3$, we defined the inner product of $(r_1, r_2, r_3)$ and $(s_1, s_2, s_3)$ as $\displaystyle\sum_{i=1}^{3} r_i s_i$ .

❀ In $\mathbb{R}^n$, we defined the inner product of

     $(r_1, r_2, r_3, \ldots, r_n)$ and $(s_1, s_2, s_3, \ldots, s_n)$ as $\displaystyle\sum_{i=1}^{n} r_i s_i$ .

❀ In the vector space of all polynomials of degree 2 or less, we defined

     $\langle r_1 + r_2 x + r_3 x^2, \; s_1 + s_2 x + s_3 x^2 \rangle$ as $\displaystyle\sum_{i=1}^{3} r_i s_i$ .

✾ In the vector space of all continuous real-valued functions on the interval [a, b], we defined

$$< f(x), g(x) > \text{ as } \int_{x=a}^{b} f(x)g(x) \ dx.$$

But there is one vector space for which we didn't give an inner product definition: the vector space of all m×n matrices. Let's look at the vector space of all 2×3 matrices.

Addition is defined in that space.

$$\begin{pmatrix} 1 & 3 & 5 \\ 0 & 8 & 2 \end{pmatrix} + \begin{pmatrix} 2 & 3 & 2 \\ 3 & 1 & 4 \end{pmatrix} = \begin{pmatrix} 3 & 6 & 7 \\ 3 & 9 & 6 \end{pmatrix} \qquad \text{If } A = (a_{ij}) \text{ and } B = (b_{ij}), \\ \text{then } A + B = (a_{ij} + b_{ij}).$$

Scalar multiplication is defined in that space.

$$7 \begin{pmatrix} 4 & 3 & 9 \\ 9 & 8 & 7 \end{pmatrix} = \begin{pmatrix} 28 & 21 & 63 \\ 63 & 56 & 49 \end{pmatrix} \qquad \text{If } A = (a_{ij}) \text{ and } r \in \mathbb{R}, \\ \text{then } rA = (ra_{ij}).$$

Your challenge is to find a definition of inner product on the vector space of all 2×3 matrices that satisfies the three axioms for inner products: symmetry, linearity, and positive-definiteness.

Remember when I tried to find a definition of two continuous real-valued functions on the interval [a, b] back on page 206? It took me three tries before I found something that fit all three axioms for inner products.

This is a par 20-minute question. That's why there is only one question in this City. Some readers may score a hole-in-one and stumble onto the answer in 2 minutes. Others may get caught in the sand traps and take 40 minutes. It doesn't matter. This is not a race. *Enjoy* the search. If you are to become a mathematician, it is the search for the solution that is the heart of the process—not the memorizing of theorems and definitions.

I have a tape recorder that can easily store every theorem and definition that I read into the microphone. But it will never become a mathematician.

If you jump to the answer before spending time figuring out the answer on your own, you will finish the book more quickly, and you will stay dumber than dirt with respect to linear algebra.

***answer***

1. When I first worked on this problem, I tried using matrix multiplication. If $A$ and $B$ are two 2×3 matrices, then $AB$ doesn't make sense. $\begin{pmatrix} 1 & 3 & 5 \\ 0 & 8 & 2 \end{pmatrix}\begin{pmatrix} 2 & 3 & 2 \\ 3 & 1 & 4 \end{pmatrix}$ don't "fit" together.

Then I tried transposing the second matrix: $AB^T$ so that I could multiply them.

$$\begin{pmatrix} 1 & 3 & 5 \\ 0 & 8 & 2 \end{pmatrix}\begin{pmatrix} 2 & 3 \\ 3 & 1 \\ 2 & 4 \end{pmatrix} = \begin{pmatrix} 21 & 26 \\ 28 & 16 \end{pmatrix}$$

But that wasn't an inner product. An inner product has to give a real number answer.

Then I thought of adding together the four entries in $AB^T$.

$$\begin{pmatrix} 1 & 3 & 5 \\ 0 & 8 & 2 \end{pmatrix}\begin{pmatrix} 2 & 3 \\ 3 & 1 \\ 2 & 4 \end{pmatrix} = \begin{pmatrix} 21 & 26 \\ 28 & 16 \end{pmatrix} \longrightarrow 21 + 26 + 28 + 16$$

How brilliant! It was symmetric. It gave a real number answer. But it wasn't very positive-definite. For example,

$$\begin{pmatrix} 2 & 3 & 5 \\ -2 & -3 & -5 \end{pmatrix}\begin{pmatrix} 2 & -2 \\ 3 & -3 \\ 5 & -5 \end{pmatrix} = \begin{pmatrix} 38 & -38 \\ -38 & 38 \end{pmatrix} \longrightarrow 38 - 38 - 38 + 38 = 0$$

Well, I thought to myself, that's easy to fix. To find the inner product of two 2×3 matrices, $A$ and $B$, first find $AB^T$, then take the absolute value of each row times each column, and then add them up.

How double brilliant! Then the answer will never be negative, and $< A, A >$ can only be zero when $A$ is the zero matrix.

$$\begin{pmatrix} 2 & 3 & 5 \\ -2 & -3 & -5 \end{pmatrix}\begin{pmatrix} 2 & -2 \\ 3 & -3 \\ 5 & -5 \end{pmatrix} = \begin{pmatrix} 38 & -38 \\ -38 & 38 \end{pmatrix} \longrightarrow |38| + |-38| + |-38| + |38| = 152$$

After four minutes of playing with my new double-brilliant definition of inner product of two 2×3 matrices, I was sure that positive-definiteness was okay. Symmetry was fine. But I had lost linearity.

$\langle(-1)A, A\rangle$ should equal $(-1)\langle A, A\rangle$, but it doesn't. If I let A be the same matrix I just worked with, then $\langle(-1)A, A\rangle$ would be

$$\begin{pmatrix} -2 & -3 & -5 \\ 2 & 3 & 5 \end{pmatrix}\begin{pmatrix} 2 & -2 \\ 3 & -3 \\ 5 & -5 \end{pmatrix} = \begin{pmatrix} -38 & 38 \\ 38 & -38 \end{pmatrix}$$

$|-38| + |38| + |38| + |-38| = 152$ which is $\langle A, A\rangle$, and not $(-1)\langle A, A\rangle$.

I guess you could cross off the double brilliant.

But this is the way that it feels when you are creating mathematics. You often go down a bunch of blind alleys before you hit upon the right path.

Little children (Fred excepted) have a low tolerance for going down blind alleys. Teach them how to multiply fractions ("top times top and bottom times bottom") and give them a dozen problems to do. They will be happy and say, "I can do math." But, truthfully, I would never think of asking a little kid to find a definition of an inner product for the vector space of all 2×3 matrices. They couldn't handle the frustration and would whine, "I can't do it! I hate math!"

So, back to the drawing board. Do you remember in the statement of this problem, I showed how similar the inner products of $\mathbb{R}^3$ and the vector space of all polynomials of degree 2 or less were?

$$\langle (r_1, r_2, r_3), (s_1, s_2, s_3) \rangle = \sum_{i=1}^{3} r_i s_i .$$

$$\langle r_1 + r_2 x + r_3 x^2, s_1 + s_2 x + s_3 x^2 \rangle = \sum_{i=1}^{3} r_i s_i$$

In each case, we multiplied corresponding entries and then added them together. This new attempt at finding a definition of an inner product on the vector space of all 2×3 matrices will take that approach:

$$\left\langle \begin{pmatrix} 1 & 3 & 5 \\ 0 & 8 & 2 \end{pmatrix}, \begin{pmatrix} 2 & 3 & 2 \\ 3 & 1 & 4 \end{pmatrix} \right\rangle = 1(2) + 3(3) + 5(2) + 0(3) + 8(1) + 2(4).$$

The axioms for symmetry and for linearity hold for this attempt. Positive-definiteness is where we usually have trouble.

In terms of the symbolism in the statement of the problem, in which $A = (a_{ij})$ and $B = (b_{ij})$, I'm now trying out as a definition:

$$\langle A, B \rangle = \sum_{i=1}^{2} \sum_{j=1}^{3} a_{ij} b_{ij}.$$

To check for positive-definiteness, $<A, A> = \sum_{i=1}^{2} \sum_{j=1}^{3} a_{ij} a_{ij} = \sum_{i=1}^{2} \sum_{j=1}^{3} a_{ij}^{2}$.

Certainly, the sum of squares, $a_{ij}^{2}$, will never be negative.

All that is left to check is that if $<A, A> = 0$, then A must be the zero matrix. But the only way that $\sum_{i=1}^{2} \sum_{j=1}^{3} a_{ij}^{2}$ is going to be equal to zero is if every $a_{ij}$ is equal to zero, which makes A the zero matrix.

We did it.

---

### *Intermission*

Now comes the fun part.

I showed you my first attempt at finding a definition of an inner product on 2×3 matrices to give you the feeling of how one might search for a solution.

My first attempt was to define $<A, B>$ as the sum of the entries in $AB^{T}$.

That didn't work because the sum of the entries in $AA^{T}$ could be negative.

But the weird thing is that we were virtually sitting on the right answer and we didn't know it.

Let me jump out of this box and show you.

---

We'll compare my early attempt with the correct definition for the inner product of $\begin{pmatrix} 1 & 3 & 5 \\ 0 & 8 & 2 \end{pmatrix}$ and $\begin{pmatrix} 2 & 3 & 2 \\ 3 & 1 & 4 \end{pmatrix}$

---

### Early attempt

Using the sum of the entries in $AB^{T}$ approach:

$$\begin{pmatrix} 1 & 3 & 5 \\ 0 & 8 & 2 \end{pmatrix} \begin{pmatrix} 2 & 3 \\ 3 & 1 \\ 2 & 4 \end{pmatrix} = \begin{pmatrix} 1(2) + 3(3) + 5(2) & 1(3) + 3(1) + 5(4) \\ 0(2) + 8(3) + 2(2) & 0(3) + 8(1) + 2(4) \end{pmatrix}$$

and then find the sum of all the entries.

### Correct definition

Multiplying corresponding entries in

$\begin{pmatrix} 1 & 3 & 5 \\ 0 & 8 & 2 \end{pmatrix}$ and $\begin{pmatrix} 2 & 3 & 2 \\ 3 & 1 & 4 \end{pmatrix}$ to get $1(2) + 3(3) + 5(2)$ and $0(3) + 8(1) + 2(4)$

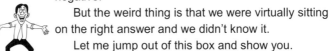

Would you look at that!  In my first attempt, if I had just taken the sum of the diagonal entries instead of all the entries, I would have scored a triple-brilliant.

The sum of the diagonal entries of a matrix is called the **trace** of the matrix.

A second correct way of defining the inner product of two matrices A and B that have the same dimensions is

the trace of $AB^T$

---

**Interlachen**

---

$\mathbb{C}$ instead of $\mathbb{R}$

1.  Near the beginning of this chapter, we said that we were looking for a binary operation that takes two vectors and gives a scalar answer.

But when we gave the three definitions for inner product, we quietly (surreptitiously would be more accurate) used the real numbers as the scalars.

On page 201, we wrote:

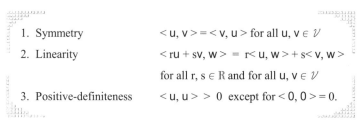

1.  Symmetry             $< u, v > = < v, u >$ for all $u, v \in \mathcal{V}$

2.  Linearity            $< ru + sv, w > = r< u, w > + s< v, w >$

    for all $r, s \in \mathbb{R}$ and for all $u, v \in \mathcal{V}$

3.  Positive-definiteness    $< u, u > > 0$ except for $< 0, 0 > = 0.$

You see that $\mathbb{R}$ in the second axiom?  Suppose we make things a bit more general by substituting $\mathbb{C}$ for $\mathbb{R}$.  $\mathbb{C}$ is the set of all complex numbers—any number that can be written as $a + bi$ where a and b are real numbers and $i = \sqrt{-1}$.

Every real number is a complex number ($5 = 5 + 0i$ and $\pi = \pi + 0i$) so we are just enlarging the possible "answers" for an inner product.

Your question: Which of the three axioms becomes nonsense if we substitute $\mathbb{C}$ for $\mathbb{R}$?

2. What we had defined on page 201 was an inner product on a real vector space. We are now looking for axioms for an inner product on a complex vector space.*

The real trick is to alter the axioms that we have used for this whole chapter so that: ① they will work for complex numbers, and ② they don't mess up all the stuff we have already done.

The key, surprisingly, is changing the symmetry axiom. I say, "surprisingly" since we had trouble in axiom 3 when we changed $\mathbb{R}$ to $\mathbb{C}$ in axiom 2, and we cure that by changing axiom 1.

Here are the new all-purpose inner product axioms:

1.  Symmetry                   $< u, v > = \overline{< v, u >}$   for all $u, v \in \mathcal{V}$
2.  Linearity                   $< ru + sv, w > = r< u, w > + s< v, w >$
                                        for all $r, s \in \mathbb{C}$ and for all $u, v \in \mathcal{V}$
3.  Positive-definiteness       $< u, u > > 0$  except for $< 0, 0 > = 0$.

The bar over the $< v, u >$ in the first axiom is the complex conjugate. The complex conjugate of $5 + 7i$ is $5 - 7i$. The complex conjugate of $-20 - 3i$ is $-20 + 3i$. In general, $\overline{a + bi}$ is $a - bi$.

Show that under these new axioms, $< u, u >$ will always be a real number. (That will prevent nonsense from occurring in axiom 3 as we pointed out in question 1.)

---

✶ In the interests of saving time and effort, mathematicians call what you get when you put an inner product on a complex vector space a **complex inner product space**. That saves about three words of writing. After forty years of using "complex inner product space" instead of "complex vector space with an inner product," that can really add up. It might add two minutes to your life.

3. In the previous question we showed that the "new all-purpose inner product axioms" work for complex numbers. Now we have to show that they don't mess up what we have already done with real numbers.

Show that the new set of axioms applied to real inner product spaces gives the same result as the old axioms we defined on page 201.

*answers*

1. Suppose $< u, u >$ were equal to $-6 + 10i$. Look at axiom 3. It doesn't make much sense to ask if $-6 + 10i > 0$.

2. Let $< u, u >$ equal $a + bi$, where $a, b \in \mathbb{R}$. We want to show that $b = 0$. Since $< u, u > = \overline{< u, u >}$ by our new symmetry axiom, $a + bi = a - bi$. By algebra, $b = 0$.

3. The only thing we changed was the symmetry axiom. If we are using $< u, v > = \overline{< v, u >}$ in the situation where inner products are real numbers, then we can throw away the conjugate bar, since conjugates don't affect real numbers. For $r \in \mathbb{R}$, $\overline{r} = \overline{r + 0i} = r - 0i = r$.

---

**Kelly**

---

1. Is $\{ \, ( \frac{1}{\sqrt{5}} , \frac{2}{\sqrt{5}} ), ( \frac{2}{\sqrt{5}} , \frac{-1}{\sqrt{5}} ) \, \}$ an orthonormal set?

2. There are three axioms for inner products: symmetry, linearity, and positive-definiteness. But linearity is only given for the first "coordinate": $< ru + sv, w > = r< u, w > + s< v, w >$. Prove that for real inner product spaces (as defined on page 201) also holds for the second "coordinate": $< w, ru + sv > = r< w, u > + s< w, v >$.

3. (Continuing the previous question) discover what the corresponding rule for $< w, ru + sv >$ is in complex inner product spaces.

From high school algebra we know:

1. The conjugate of a sum is equal to the sum of the conjugates.
2. The conjugate of a product is equal to the product of the conjugates.
3. The conjugate of the conjugate is equal to the original number.

4. Once more, just for old time's sake, use Gram-Schmidt to transform $\{(1, 1, 1, 1), (1, 0, 0, 1), (0, 2, 1, -1)\}$ into an orthogonal set. And then, to please dad, turn it into an orthonormal set.

***answers***

1. The inner product $< (\frac{1}{\sqrt{5}}, \frac{2}{\sqrt{5}}), (\frac{2}{\sqrt{5}}, \frac{-1}{\sqrt{5}}) >$ equals zero. The vectors are orthogonal.

The length of $(\frac{1}{\sqrt{5}}, \frac{2}{\sqrt{5}})$ is $\sqrt{< (\frac{1}{\sqrt{5}}, \frac{2}{\sqrt{5}}), (\frac{1}{\sqrt{5}}, \frac{2}{\sqrt{5}}) >}$ = $\sqrt{\frac{1}{5} + \frac{4}{5}}$ = 1.

Similarly, the other vector has length equal to one.

2.

$$< w, ru + sv > \; = \; < ru + sv, w > \qquad \text{by symmetry (using the original symmetry axiom)}$$

$$= \; r< u, w > + s< v, w > \qquad \text{by linearity}$$

$$= \; r< w, u > + s< w, v > \qquad \text{by symmetry}$$

3. If we used the new, complex-number inner product axioms introduced in the previous City, the argument would be:

$$< w, ru + sv > \; = \; \overline{< ru + sv, w >} \qquad \text{by symmetry}$$

$$= \; \overline{r < u, w > + s < v, w >} \qquad \text{by linearity}$$

$$= \; \overline{r}\,\overline{< u, w >} + \overline{s}\,\overline{< v, w >} \qquad \text{by } \boxed{1} \text{ and } \boxed{2}$$

$$= \; \overline{r}\,\overline{\overline{< w, u >}} + \overline{s}\,\overline{\overline{< w, v >}} \qquad \text{by symmetry}$$

$$= \; \overline{r}< w, u > + \overline{s}< w, v > \qquad \text{by } \boxed{3}$$

4. $u_1 = (1, 1, 1, 1) \quad u_2 = (1, 0, 0, 1) \quad u_3 = (0, 2, 1, -1)$

$v_1 = (1, 1, 1, 1)$

$v_2 = (1, 0, 0, 1) - \dfrac{1 + 0 + 0 + 1}{1 + 1 + 1 + 1} (1, 1, 1, 1)$ =

$\qquad (1, 0, 0, 1) - (\frac{1}{2}, \frac{1}{2}, \frac{1}{2}, \frac{1}{2}) = (\frac{1}{2}, -\frac{1}{2}, -\frac{1}{2}, \frac{1}{2})$

$v_3 = (0, 2, 1, -1) - \dfrac{0 + 2 + 1 - 1}{1 + 1 + 1 + 1} (1, 1, 1, 1) -$

$\qquad \dfrac{0 - 1 - \frac{1}{2} - \frac{1}{2}}{\frac{1}{4} + \frac{1}{4} + \frac{1}{4} + \frac{1}{4}} (\frac{1}{2}, -\frac{1}{2}, -\frac{1}{2}, \frac{1}{2}) = (\frac{1}{2}, \frac{1}{2}, -\frac{1}{2}, -\frac{1}{2})$

Now to make the $v_i$ into unit vectors: $v_1/\sqrt{1 + 1 + 1 + 1}$ = $(\frac{1}{2}, \frac{1}{2}, \frac{1}{2}, \frac{1}{2})$. $v_2$ and $v_3$ are already unit vectors since their lengths are equal to one.

---

**Lansing**

---

1.  On page 210, we promised, "We'll show $< 0, u > = 0$ for all $u$ in one of the Cities." Please keep that promise for me.

2.  Another easy proof. (The question will be longer than the answer.) Suppose we have a subset S of a vector space $\mathcal{V}$. S is just a subset, it isn't necessarily a subspace. Remember the 𝕴𝖓𝖘𝖎𝖌𝖍𝖙𝖘 back on page 174:

    *Nonempty set $\mathcal{W}$ is a subspace of $\mathcal{V}$ if every linear combination $(r_1w_1 + r_2w_2)$ of vectors in $\mathcal{W}$ is also in $\mathcal{W}$.*

    There are some vectors in $\mathcal{V}$ that are orthogonal to every vector in subset S. One such vector is $0$. I think of this as the vectors in $\mathcal{V}$ that "hate" every member of S. In that mindset, $0$ is a good hater. The set of all vectors in $\mathcal{V}$ that is perpendicular to every element of S is called **S perp**. That set is written $S^{\perp}$. Everyone calls $S^{\perp}$ "S perp." Walk up to people on the street and ask them how to pronounce $S^{\perp}$, and they will say, "S perp."

    Of course, if you were writing some hoity-toity* linear algebra textbook, you would say that $S^{\perp}$ is the **orthogonal complement of S** and give some fancy definition like $S^{\perp} = \{ v \in \mathcal{V}: < v, s > = 0 \text{ for all } s \in S\}$.

    Subset S is not usually a subspace of $\mathcal{V}$, but $S^{\perp}$ always is.

    Show $S^{\perp}$ is always a subspace of $\mathcal{V}$.

*answers*

1.  $< 0, u > = < 00, u >$        since $00 = 0$

    $\phantom{< 0, u > } = 0< 0, u >$        by linearity

    $\phantom{< 0, u > } = 0$        by arithmetic—zero times any number is equal to zero.

2.  If $w_1$ and $w_2 \in S^{\perp}$ and $r_1, r_2 \in \mathbb{R}$ and $v \in \mathcal{V}$,

    $< r_1w_1 + r_2w_2, v > = r_1< w_1, v > + r_2< w_2, v >$        by linearity

    $\phantom{< r_1w_1 + r_2w_2, v >} = r_1(0) + r_2(0) = 0$        since $w_1$ and $w_2 \in S^{\perp}$

    Since every linear combination of elements of $S^{\perp}$ is in $S^{\perp}$, it is a subspace.

---

∗ HOY-tea TOY-tea   Pretentious or haughty.

# Chapter Three
## Systems of Equations with No Solution
Ax = b ☹

We are back on level ground.

In this chapter, we complete the trilogy. The first chapter: Systems of equations with one solution Ax = b ☺. The second chapter: Systems of equations with many solutions Ax = b ☺☺☺.... And now: Systems of equations with no solution Ax = b ☹.

I expect this to be a very peaceful chapter. Fred is paddling around in the Great Lake while Betty and Alexander stand on the shore watching. All we have to do is solve an unsolvable system of equations, and we are done.

**I, your reader, think we are done right now. If you have** $\begin{cases} x + y = 3 \\ x + y = 4 \end{cases}$

**you can't have values of x and y that will satisfy both equations at the same time.**

You're right. It's unsolvable. But why should that stop us? Have you ever had impossible situations in your life? You got through them, didn't you?

So we are going to solve the unsolvable equations . . .

We're going to sing the unsingable song . . .

We're going to Think the Impossible Thought!

**Okay, Mr. Quixote, let's not get carried away with tilting at windmills. Just spit it out. How are you going to do all this wonderworking?**

I already told you.

**What! We just started this chapter, and you haven't said anything yet about HOW to solve unsolvable systems.**

Yes I did.

**No you didn't.**

Yes I did. Back on page 96, when we looked at Nightmare #4, I told you everything you needed to know. My exact words were, "We will turn Ax = b ☹ into $A^T A x = A^T b$ which gives us the best solution in an impossible situation." I even explained what $A^T$ meant. The transpose of

$A$ just flips the rows and columns of $A$. If $A = (a_{ij})$, then $A^T = (a_{ji})$. I even

gave you an example: If $A$ were $\begin{pmatrix} 4 & 5 & 8 & 9 \\ 7 & 2 & 2 & 3 \end{pmatrix}$ then $A^T$ would be $\begin{pmatrix} 4 & 7 \\ 5 & 2 \\ 8 & 2 \\ 9 & 3 \end{pmatrix}$.

Betty and Alexander knew that little boys playing in the water can totally forget about time. They sat down at the edge of the lake to watch and wait.

"I wonder how much this basket weighs," Alexander said. "I guess there's no way to tell until we get home."

"You always ask that question when we go on picnics," Betty responded. "This time we'll find out." She took a large scale out of her backpack and set it on the beach. Alexander put the basket on top.

"See," said Betty. "It's only 10 pounds. That should be nothing for a big guy like you."

"It sure doesn't feel like 10 pounds," said Alexander. He glanced at the scale. "What do you mean—10 pounds? The scale says 22 pounds."

"Silly man. The extra 12 pounds are your passengers." Betty reached into the basket and pulled out a bunny, two mice, and a couple of squirrels.

---

*Your Turn to Play*

1. Let $x_1$ = weight of a bunny. $x_2$ = weight of a mouse. $x_3$ = weight of a squirrel. Let $\mathbf{x}^T = (x_1, x_2, x_3)$. Using the transpose notation is a lot easier

to "set in type" than $\mathbf{x} = \begin{pmatrix} x_1 \\ x_2 \\ x_3 \end{pmatrix}$.

Write the equation showing that the five animals weigh 12 pounds.

2. We now have a "system" consisting of one equation. When you solve that system, are there any free variables?

---

3. Is this system overdetermined (see page 96) or underdetermined (see page 108)?

## .......COMPLETE SOLUTIONS.......

1. $x_1 + 2x_2 + 2x_3 = 12$
2. We are in the $Ax = b$ ☺☺☺. . . case of Chapter 2. The free variables will be $x_2$ and $x_3$.
3. When you have one or more free variables, then $Ax = b$ will have an infinite number of solutions and is called an underdetermined linear system.

"Now the basket weighs 10 pounds," Betty said.

"That's funny," said Alexander. "It sure looks like the scale reads 16 pounds."

"It would read 10 pounds if we didn't have such aggressive wildlife." Betty pulled out a bunny and a mouse that had just hopped in.

In matrix notation, $\begin{pmatrix} 1 & 2 & 2 \\ 1 & 1 & 0 \end{pmatrix} \begin{pmatrix} x_1 \\ x_2 \\ x_3 \end{pmatrix} = \begin{pmatrix} 12 \\ 6 \end{pmatrix}$

The system is still underdetermined, and we still have an infinite number of solutions.

Alexander and Betty were in a playful mood. They would remove the animals from the basket and turn their backs for just a second. When they looked back at the basket, they found it had been repopulated.

Alexander wrote down:

| bunnies | mice | squirrels | total weight |
|---|---|---|---|
| 1 | 2 | 2 | 12 |
| 1 | 1 | 0 | 6 |
| 2 | 2 | 1 | 15 |
| 2 | 4 | 5 | 26 |
| 2 | 2 | 2 | 17 |

When Alexander tried to solve $\begin{pmatrix} 1 & 2 & 2 \\ 1 & 1 & 0 \\ 2 & 2 & 1 \\ 2 & 4 & 5 \\ 2 & 2 & 2 \end{pmatrix} \begin{pmatrix} x_1 \\ x_2 \\ x_3 \end{pmatrix} = \begin{pmatrix} 12 \\ 6 \\ 15 \\ 26 \\ 17 \end{pmatrix}$

he encountered Nightmare #4—a row with all zeros except for the last column. The system was overdetermined. $\mathsf{Ax} = \mathsf{b}$ ☹ It has no solution.

---

### Intermission

This was real life. The bunnies had to weigh something. The math was telling Alexander that he couldn't determine the weight of a bunny. What's wrong?

Actually, nothing is "wrong." When you count bunnies, you get an exact answer. There are either eight bunnies or there are nine bunnies. Counting bunnies involves a discrete variable.

In contrast, weighing a bunny involves a continuous variable. You can't weigh a bunny exactly. Weigh it on Betty's scale and it might weigh 6 pounds. Weigh it on a butcher's scale and it might be 5.7 pounds. On a laboratory scale it might be 5.68 pounds.

Who can build a road exactly one mile long? Who can speak for exactly 46 seconds? Who can build a battery that will deliver exactly 6 volts?

Our world is soft and squishy.

When people in white laboratory coats run their experiments, they invariably get tons of data that yield an overdetermined system. It is built into the way the world works. Errors in measurement of continuous variables give you Nightmare #4. The only way to avoid it is to become what they call a pure mathematician instead of an applied mathematician.

---

At the top of this page we have a matrix equation $\mathsf{Ax} = \mathsf{b}$. There is no value of the vector $\mathsf{x}$ that will make it true.

Instead we solve $\mathsf{A^T Ax} = \mathsf{A^T b}$. This is called the **normal equation**.

Some notes about the normal equation:

♪#1: The normal equation $A^TAx = A^Tb$ will *always* have a solution. You will never run into a ☹ situation with a normal equation. Another way of saying that is $A^TAx = A^Tb$ will always be consistent.

♪#2: In almost all real-life situations there will be a *unique* solution to $A^TAx = A^Tb$. The matrix $A$ is an m×n matrix with m > n. Lots of rows. The only time we might have more than one solution is when the columns of $A$ are linearly dependent. (The fancy way of expressing that is, "The rank of $A < n$.") In the instant case (as lawyers sometimes express it), Betty and Alexander are looking at:

| bunnies | mice | squirrels |
|---------|------|-----------|
| 1 | 2 | 2 |
| 1 | 1 | 0 |
| 2 | 2 | 1 |
| 2 | 4 | 5 |
| 2 | 2 | 2 |

What are the chances that the number of squirrels would be a linear combination of the number of bunnies and mice? Virtually nil. And if you had a hundred observations instead of the five that Alexander made, there's even less of a chance that the columns would be linearly dependent.

And, in any event, if the solution to $A^TAx = A^Tb$ is not unique, who cares?

♪#3: The solution to $A^TAx = A^Tb$ is not usually the solution to $Ax = b$. But it is the best possible answer you could give if you were asked to solve $Ax = b$.

**I, your reader, think that you should define your "best possible answer" phrase. Our world may be "soft and squishy"—as you put it—but this is supposed to be a math book, not some pastoral idyll.**

I don't mean to be critical, but *pastoral idyll* is redundant since idylls typically describe pastoral scenes.

**But not always. Idylls could be some prose or poetry about some tranquil charming scene. I included the word pastoral because of the**

**bucolic aspects of your bunnies, mice, and squirrels. Wait a minute! You are** O.T.S. **(Off The Subject). Define "best possible answer."**

I was about to do that before you interrupted.

**I'm waiting. . . .**

## "BEST POSSIBLE ANSWER"

When Alexander weighed the collections of animals he came up with $b^T = (12, 6, 15, 26, 17)$.

$$Ax = b \qquad \begin{pmatrix} 1 & 2 & 2 \\ 1 & 1 & 0 \\ 2 & 2 & 1 \\ 2 & 4 & 5 \\ 2 & 2 & 2 \end{pmatrix} \begin{pmatrix} x_1 \\ x_2 \\ x_3 \end{pmatrix} = \begin{pmatrix} 12 \\ 6 \\ 15 \\ 26 \\ 17 \end{pmatrix}$$

$$A^TAx = A^Tb \quad \begin{pmatrix} 1 & 1 & 2 & 2 & 2 \\ 2 & 1 & 2 & 4 & 2 \\ 2 & 0 & 1 & 5 & 2 \end{pmatrix} \begin{pmatrix} 1 & 2 & 2 \\ 1 & 1 & 0 \\ 2 & 2 & 1 \\ 2 & 4 & 5 \\ 2 & 2 & 2 \end{pmatrix} \begin{pmatrix} x_1 \\ x_2 \\ x_3 \end{pmatrix} = \begin{pmatrix} 1 & 1 & 2 & 2 & 2 \\ 2 & 1 & 2 & 4 & 2 \\ 2 & 0 & 1 & 5 & 2 \end{pmatrix} \begin{pmatrix} 12 \\ 6 \\ 15 \\ 26 \\ 17 \end{pmatrix}$$

multiplying

$$\begin{pmatrix} 14 & 19 & 18 \\ 19 & 29 & 30 \\ 18 & 30 & 34 \end{pmatrix} \begin{pmatrix} x_1 \\ x_2 \\ x_3 \end{pmatrix} = \begin{pmatrix} 134 \\ 198 \\ 203 \end{pmatrix}$$

solving by the methods of Chapter 1:   $x^T \doteq (5.185, 1.074, 2.278)$.

( $\doteq$ means *equals after rounding*)

So using the normal equation ( $A^TAx = A^Tb$ ), we find that the weight of a bunny is 5.185 pounds. The weight of a mouse, 1.074. The weight of a squirrel, 2.278. These are the best possible answers.

**You still haven't told me why you call them the best possible answers!**

Patience please. I'm almost there.

If I use these values for $x$, then $Ax$ should be pretty close to $b$.

$$Ax = \begin{pmatrix} 1 & 2 & 2 \\ 1 & 1 & 0 \\ 2 & 2 & 1 \\ 2 & 4 & 5 \\ 2 & 2 & 2 \end{pmatrix} \begin{pmatrix} 5.185 \\ 1.074 \\ 2.278 \end{pmatrix} = \begin{pmatrix} 11.889 \\ 6.259 \\ 14.796 \\ 26.056 \\ 17.074 \end{pmatrix} \text{ where b is } \begin{pmatrix} 12 \\ 6 \\ 15 \\ 26 \\ 17 \end{pmatrix}$$

237

Let me make a little chart:

| Scale weights | Weights I computed | Error |
|---|---|---|
| 12.000 | 11.889 | -0.111 |
| 6.000 | 6.259 | 0.259 |
| 15.000 | 14.796 | -0.204 |
| 26.000 | 26.056 | 0.056 |
| 17.000 | 17.074 | 0.074 |

Some of my numbers were too high and some were too low. I can't just add up the errors to see if this normal-equations procedure gives the best answer.*

**I have a suggestion. Why not look at the sum of the absolute values of the errors?**

That almost works. We are looking at the error vector (–0.111, 0.259, –0.204, 0.056, 0.074). In matrix notation, the error vector is b – Ax where x is the computed value of the variables (in this case, animal weights) using the normal-equations procedure.

How big is that vector? (Or, if you are thinking of geometric vectors, you would ask how *long* is that vector.) We want to minimize the norm of that vector.

**Translation please!**

We want to minimize $\| b - Ax \|$.

**More translation!**

We want to minimize $\| (-0.111, 0.259, -0.204, 0.056, 0.074) \|$.

**I think I've forgotten that norm thing. How about expressing yourself in English.**

The norm of a vector is the square root of the sum of the squares of the coordinates of the vector.

If $w = (w_1, w_2, \ldots, w_m)$, then $\| w \| = \sqrt{w_1^2 + w_2^2 + \ldots + w_m^2}$.

If we are going to minimize $\| w \|$, it will be a minimum when $w_1^2 + w_2^2 + \ldots + w_m^2$ is a minimum.

---

\* If I did, then a procedure which gave me errors of +50 and –50 would be as good as one that gave me errors of +0.0003 and –0.0003.

In short, we want the sum of the squares of the errors to be as small as possible. That we will call "our best answer" for x.

It might be clearer in chart form:

| Scale weights | Weights I computed | Error | (Error)² |
|---|---|---|---|
| 12 | 11.889 | -0.111 | 0.012 |
| 6 | 6.259 | 0.259 | 0.067 |
| 15 | 14.796 | -0.204 | 0.042 |
| 26 | 26.056 | 0.056 | 0.003 |
| 17 | 17.074 | 0.074 | 0.005 |
|  | sum of the squares of the errors ➡ | | <u>0.129</u> |

By squaring the errors before we take their sum, we accomplish two things: (1) we make all the entries positive, avoiding the difficulties mentioned in the footnote on the previous page, and (2) we make the procedure more sensitive to any big errors.

Suppose our errors were 1, 1, 1, 1, 10. If we just took the sum of the errors, our answer would be 14. But the sum of the squares of the errors is 104. We want big errors to count more than smaller errors.

This is called the **least squares solution** of $\text{A}\textsf{x} = \textsf{b}$. No other value of x will make the sum of the squares of the errors as small as the least squares solution.

**Okay smarty-pants. Prove it.**
Prove what?
**What you just said. Show that the normal equations procedure gives you the least squares solution. Show that the sum of the squares of the errors made in solving $\text{A}^{\text{T}}\text{A}\textsf{x} = \text{A}^{\text{T}}\textsf{b}$ instead of $\text{A}\textsf{x} = \textsf{b}$ is as small as possible.**
No.
**Yes. Yes. Yes.**
No. No. No. No. No. No.

### What's wrong. Are you chicken?

I am not a coward, but . . .

### Chicken.

.     I don't know of any other linear algebra book that presents an algebraic proof that $A^TAx = A^Tb$ is the least squares solution of $Ax = b$. Some of them draw pretty pictures of geometric vectors and talk about projections onto planes.

### I can draw a pretty picture:

I would have to use partial derivatives from third-semester calculus.

### Chick. Chick. Chick. In fact, $\dfrac{\partial \textbf{\textit{Chick}}}{\partial \textbf{\textit{Stan}}}$

I've had lots of insults in my life, but I've never had one with a partial derivative in it.

### Pretty clever, eh?

Okay. I've got my courage up. Please fasten your seatbelt.

**Proof** of: If we want the sum of the squares of the errors to be a minimum, we should use $A^TAx = A^Tb$ to "solve" $Ax = b$.

If we want the sum of the squares of the errors to be a minimum, we want to minimize $\| b - Ax \|$.

A is given, and b is given. These are the data points. We want to find x. For example, the first row of A is (1 bunny, 2 mice, 2 squirrels), and the first row of b is 12 pounds. The values of x will be our best guess as to x = (weight of one bunny, weight of one mouse, weight of one squirrel).

We are going to minimize $\| b - Ax \|^2$ instead of $\| b - Ax \|$. (A trick from calculus. If f(x) is non-negative, the x that minimizes f(x) is the same x that minimizes $f^2(x)$.)

Since A and b are given, the error $\| b - Ax \|^2$ is just a function of x. Let the error function $E(x_1, x_2, \ldots, x_n)$ equal $\| b - Ax \|^2$.

For each row in A and b, we define the residual:

$$r(1) = b_1 - (a_{11}x_1 + a_{12}x_2 + \ldots + a_{1n}x_n)$$
$$r(2) = b_2 - (a_{21}x_1 + a_{22}x_2 + \ldots + a_{2n}x_n)$$
$$r(3) = b_3 - (a_{31}x_1 + a_{32}x_2 + \ldots + a_{3n}x_n)$$
$$\cdots$$
$$r(m) = b_m - (a_{m1}x_1 + a_{m2}x_2 + \ldots + a_{mn}x_n)$$

Or, in general, we could write $r(i) = b_i - (a_{i1}x_1 + a_{i2}x_2 + \ldots + a_{in}x_n)$,

$$(i = 1, 2, 3, \ldots, m).$$

(line 1) $$r(i) \; = \; b_i - \sum_{k=1}^{n} a_{ik}x_k$$

With this definition of $r(i)$, we have

$$E(x_1, x_2, \ldots, x_n) \; = \; \sum_{i=1}^{m} (r(i))^2 \qquad \text{since this is } \| \, b - Ax \, \|^2.$$

Now is the time to haul out *Life of Fred: Calculus* and turn to the chapter on partial derivatives. Fred and his three friends (the scarecrow, the tin-wood duck, and the lion) were on the world's largest flying carpet. The height, z, of particular points on the carpet was a function of two variables, x and y. Local maximums and minimums occurred when the partial derivatives were equal to zero. Who could forget Duck's Hill?

So we want $\dfrac{\partial E}{\partial x_j} = 0$ for each j.

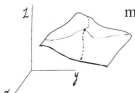

$\dfrac{\partial z}{\partial x} = 0$
$\dfrac{\partial z}{\partial y} = 0$

***Duck's Hill***

$$\sum_{i=1}^{m} 2(r(i))\frac{\partial r(i)}{\partial x_j} \; = \; 0 \qquad \text{using the chain rule* from calculus.}$$

**Wait! You are going to get to $\mathsf{A^{T}Ax = A^{T}b}$ from here! I, your reader, really think you are fooling with my brain.**

Of course I am. That's called *education*.

But my proof is for real. All I have left is a bit of algebra and the definition of partial derivative. The Duck's Hill was the most sophisticated part of the proof.

I divide both sides of the equation by 2.

---

✱ If $y = f(g(x))$, then $\dfrac{dy}{dx} = \dfrac{df}{dg}\dfrac{dg}{dx}$

(line 2)    $$\sum_{i=1}^{m} (r(i)) \frac{\partial r(i)}{\partial x_j} = 0$$

Now I have to see what $\frac{\partial r(i)}{\partial x_j}$ equals.

If $r(i) = b_i - \sum_{k=1}^{n} a_{ik} x_k$, and if $\frac{\partial r(i)}{\partial x_j}$ means take the derivative with respect to $x_j$ while pretending that all the other $x_k$ ($k \neq j$) are constant, then

(line 3)    $$\frac{\partial r(i)}{\partial x_j} = -a_{ij}$$

Putting lines 1 and 3 into line 2    $$\sum_{i=1}^{m} [b_i - \sum_{k=1}^{n} a_{ik} x_k][-a_{ij}] = 0$$

Distributive property    $$\sum_{i=1}^{m} [-a_{ij} b_i - \sum_{k=1}^{n} -a_{ij} a_{ik} x_k] = 0$$

Algebra    $$\sum_{i=1}^{m} \sum_{k=1}^{n} a_{ij} a_{ik} x_k = \sum_{i=1}^{m} a_{ij} b_i$$

And that is    $$A^T A x = A^T b.$$

**Wow. How did you do that? That whole proof seems like magic.**
I guess you could put on my tombstone: Here lies a mathemagician.

In the super simple case, we have one independent variable instead of three. Instead of bunnies, mice, and squirrels, we have someone looking at the night sky. Each night he records the number of hours of observation and the number of new stars discovered. On Monday, one hour of observation yielded one new star found. Tuesday, three hours yielded five stars. Wednesday, six hours yielded six stars. With one independent variable, it is easier to graph.

In a minute I'll show you how to fit a least-squares line through (1, 1), (3, 5), and (6, 6) using a normal equation.

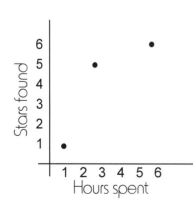

When you fit that line (it's called **data fitting**), it's easy to see the residuals (the dashed lines). It is the sum of the squares of the lengths of those dashed lines that we minimized.

Our data for stargazing look like (1, 1), (3, 5), and (6, 6). For weighing picnic baskets filled with bunnies, mice, and squirrels, it looked like ((1, 2, 2), 12), ((1, 1, 0), 6), (2, 2, 1), 15), etc.

In general, our data look like (a, b), where the first entry is a vector, a, which had one coordinate in the stargazing example and three coordinates in the picnic basket example. The second coordinate of the data, b, is a single number.

What we are looking for is the value of vector x. $x = (x_1, x_2, x_3, x_4, \ldots, x_n)$ is the unknown. For the first data point, $((a_{11}, a_{12}, a_{13}, \ldots, a_{1n}), b_1)$, we want $b_1$ to be a linear combination of the $a_{1i}$. The $x_i$ will be the coefficients.

$$b_1 \approx a_{11}x_1 + a_{12}x_2 + \ldots + a_{1n}x_n$$

" $\approx$ " means approximately equal to.

Since the $x_i$ are the coefficients, it might help to visualize things if we put them in front of the a's.

$$b_1 \approx x_1a_{11} + x_2a_{12} + \ldots + x_na_{1n}$$

The essential point is that everything is nice and linear. A multiple of $a_{11}$ plus a multiple of $a_{12}$, plus . . ., plus a multiple of $a_{1n}$ should roughly equal $b_1$. The weight of one bunny, plus twice the weight of one mouse, plus twice the weight of one squirrel will equal twelve pounds.

We now have to deal with a very delicate topic. Hopefully, no one under the age of 15 is reading this book right now. If you are under 15, please stop reading until we get to the end of this topic.* This sensitive topic goes by many names.

---

* The logical impossibility of not reading until we get to the end of this topic is hereby acknowledged.

We are now going to have to deal with . . . (are you ready?) . . . reality. Accepting reality is not very fashionable among several groups of people in our world. We may experience the disapprobation of such groups, but we will remain intrepid.

> ***What in blazes are you talking about? You, Mr. Author, remind me of the reticence of some parents in explaining the facts of life to their prepubescent daughters. Please get on with it.***

This is a linear algebra book, right?

> ***Yeah.***

But life isn't always . . . linear. There. I've said it. I feel better. We are going to sneak non-linearity into this book. This bumper sticker may be for sale somewhere:

For example, everyone who looks at sunspots notices that their frequency varies on a roughly 11-year cycle. We might want to try to fit a sine curve through those points.

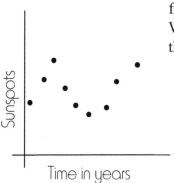

And everyone who has ever driven a car off a sea cliff and measured the height of the car above the ocean knows that the data fit a parabola—a quadratic equation.

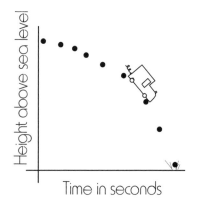

244

In the bunny-mouse-squirrel case, we had for the first data point, $((a_{11}, a_{12}, a_{13}), b_1)$, where

$$b_1 \approx x_1 a_{11} + x_2 a_{12} + x_3 a_{13}.$$

The a's were the number of each animal and the b was the weight of all of the animals. We were looking for the x's, which are the weights of each animal.

In the stargazing, the sunspot, and the car-over-the-cliff examples, things were no longer linear.

For the first data point, $((a_{11}, a_{12}, a_{13}, \ldots, a_{1n}), b_1)$, we will have to represent $b_1$ by

$$b_1 \approx x_1 \phi_1(a_{11}) + x_2 \phi_2(a_{12}) + \ldots + x_n \phi_n(a_{1n}).$$

The $\phi$'s are functions of the a's.

b is no longer linear in the a's.

We can't stick the a's into $\mathsf{Ax = b}$.

But it is linear in the $\phi$(a's).

Back on page 237, with the bunnies, mice, and squirrels, we just put their numbers into $\mathsf{Ax = b}$.

$$\begin{pmatrix} 1 & 2 & 2 \\ 1 & 1 & 0 \\ 2 & 2 & 1 \\ 2 & 4 & 5 \\ 2 & 2 & 2 \end{pmatrix} \begin{pmatrix} x_1 \\ x_2 \\ x_3 \end{pmatrix} = \begin{pmatrix} 12 \\ 6 \\ 15 \\ 26 \\ 17 \end{pmatrix}$$

> We need to fool the $\mathsf{A}$.
> Since $\mathsf{Ax = b}$ can only deal with linear, we
> will stuff $\phi$(a's) into $\mathsf{A}$ instead of a's.

Two steps:   ① Find the $\phi$'s.
              ② Do the stuffing into $\mathsf{A}$.

Think of $\mathsf{A}$ as a turkey

The functions, $\phi_i$ depend, of course, on the real-life example you are looking at. In the stargazing example, we were thinking of finding a straight line that would best fit the data. Why a straight line? Because it seems a good guess that the relationship between the number of hours you spend looking and the number of stars you find is linear. That guess could be wrong. It might be that fatigue would set in* and a logarithmic curve might fit the data better.

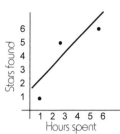

**Wait a minute! This is a math book, and you are asking me to GUESS! You gotta tell me how to find the $\phi_i$ functions.**

I warned you three pages ago that we were about to deal with a very sensitive topic—reality. When you are out there in the real world, virtually everything you do involves guessing. When you pick a spouse, is there a sure-fire formula to use? When you spread butter on your toast, how do you know exactly how much to put on? (My younger daughter, Margaret, really liked butter when she was a kid. She probably would have liked her toast buttered a quarter of an inch deep.)

When it comes to guessing the $\phi_i$ functions, you can always test to see how good your guess was by computing $Ax$ and comparing it with $b$ as we did on the bottom of page 237.

Let's guess that a straight line is the best fit for the stargazing. We have three data points: (1, 1), (3, 5), and (6, 6). Our **model function** will be the equation of a straight line: $y = x_1 + x_2 x$ or
$$b = x_1 \phi_1(a) + x_2 \phi_2(a)$$
$$\text{where } \phi_1(a) = 1 \text{ and } \phi_2(a) = a.$$
The first data point, (1, 1), becomes $((\phi_1(1), \phi_2(1)), 1) = ((1, 1), 1)$.
The second data point, (3, 5), becomes $((\phi_1(3), \phi_2(3)), 5) = ((1, 3), 5)$.
The third data point, (6, 6), becomes $((\phi_1(6), \phi_2(6)), 6) = ((1, 6), 6)$.

In other words, $A = \begin{pmatrix} 1 & 1 \\ 1 & 3 \\ 1 & 6 \end{pmatrix}$     $\begin{array}{l} 1 \to (1, 1) \\ 3 \to (1, 3) \\ 6 \to (1, 6) \end{array}$

---

\* When you are fresh, you find more stars per hour than later in the evening.

$$\text{and } \mathsf{Ax = b} \text{ is} \quad \begin{pmatrix} 1 & 1 \\ 1 & 3 \\ 1 & 6 \end{pmatrix} \begin{pmatrix} x_1 \\ x_2 \end{pmatrix} = \begin{pmatrix} 1 \\ 5 \\ 6 \end{pmatrix}$$

*Your Turn to Play*

1. Solve the above equation.

2. What is the equation of the best-fit straight line through the three stargazing points?

### . . . . . . . C O M P L E T E   S O L U T I O N S . . . . . . .

1. To solve this overdetermined system, we multiply on the left by $\mathsf{A^T}$.

$$\begin{pmatrix} 1 & 1 & 1 \\ 1 & 3 & 6 \end{pmatrix} \begin{pmatrix} 1 & 1 \\ 1 & 3 \\ 1 & 6 \end{pmatrix} \begin{pmatrix} x_1 \\ x_2 \end{pmatrix} = \begin{pmatrix} 1 & 1 & 1 \\ 1 & 3 & 6 \end{pmatrix} \begin{pmatrix} 1 \\ 5 \\ 6 \end{pmatrix}$$

$$\begin{pmatrix} 3 & 10 \\ 10 & 46 \end{pmatrix} \begin{pmatrix} x_1 \\ x_2 \end{pmatrix} = \begin{pmatrix} 12 \\ 52 \end{pmatrix}$$

Solving by the methods of Chapter 1: $x^T \doteq (0.842, 0.947)$.

2. Our model function was $y = x_1 + x_2 x$. In question one, we found the values of $x_1$ and $x_2$.

    Our best-fit straight line through the data is $y = 0.842 + 0.947x$.

      I'll make a chart, like I did back on page 239, to see how close my best-fit straight line, $y = 0.842 + 0.947x$, fits the data points.

| Stars found $y$ | Stars predicted by $y = 0.842 + 0.947x$ | Error | (Error)$^2$ |
|---|---|---|---|
| 1 | $0.842 + 0.947(1) = 1.789$ | 0.789 | 0.62252 |
| 5 | $0.842 + 0.947(3) = 3.683$ | −1.317 | 1.73449 |
| 6 | $0.842 + 0.947(6) = 6.524$ | 0.524 | 0.27458 |
| | sum of squares of the errors ➡ | | 2.63159 |

I want to compare the best-fit straight line equation, y = 0.842 + 0.947x, with an equation that is close to it, say, y = 0.84 + 0.95x. I'll redo the chart and see how that affects the sum of the squares of the error number.

| Stars found y | Stars predicted by y = 0.84 + 0.95x | Error | (Error)² |
|---|---|---|---|
| 1 | 0.84 + 0.95(1) = 1.79 | 0.79 | 0.6241 |
| 5 | 0.84 + 0.95(3) = 3.69 | −1.31 | 1.7161 |
| 6 | 0.84 + 0.95(6) = 6.54 | 0.54 | 0.2916 |

sum of squares of the errors ➡ 2.6318

Changing the equation slightly made the sum of the squares of the errors go up (2.63159 moved up to 2.6318). That makes me feel good that I may have done the arithmetic correctly for the best straight-line fit.

Is y = $x_1$ + $x_2$x the best model function? First, let's add some more data points to make the situation more realistic: (1, 1), (2, 3), (3, 5), (4, 5), and (6, 6). After doing the normal equation stuff, the new straight-line equation will be y = 0.973 + 0.946x.

Would some curved line better fit those points? Trying out a quadratic model function might be a good place to start. y = $x_1$ + $x_2$x + $x_3$x². Data point (a, b) would become ((1, a, a²), b).

$$Ax = b \text{ would become } \begin{pmatrix} 1 & 1 & 1 \\ 1 & 2 & 4 \\ 1 & 3 & 9 \\ 1 & 4 & 16 \\ 1 & 6 & 36 \end{pmatrix} \begin{pmatrix} x_1 \\ x_2 \\ x_3 \end{pmatrix} = \begin{pmatrix} 1 \\ 3 \\ 5 \\ 5 \\ 6 \end{pmatrix}$$

You do the old $A^TAx = A^Tb$ trick and then solve for $x_1$, $x_2$ and $x_3$.

On scratch paper I did the matrix multiplication and the Gaussian elimination. $x_1 \doteq -1.38$, $x_2 \doteq 2.71$, and $x_3 \doteq -0.249$. The best-fit for a quadratic model function: y = −1.38 + 2.71x − 0.249x².

How about a model function involving logarithms? The regular y = ln x curve passes through (1, 0). We'll raise it up a bit by adding a constant. Model function y = $x_1$ + $x_2$ln x.

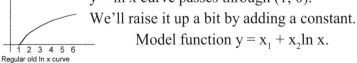
Regular old ln x curve

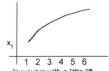
ln x curve with a little lift

Comparing $y = x_1 + x_2 \ln x$ with $y = x_1 \phi_1(a) + x_2 \phi_2(a)$, we see that $\phi_1(a) = 1$ and $\phi_2(a) = \ln a$.

Data point $(a, b)$ would become $((1, \ln a), b)$.

$Ax = b$ would become $\begin{pmatrix} 1 & \ln 1 \\ 1 & \ln 2 \\ 1 & \ln 3 \\ 1 & \ln 4 \\ 1 & \ln 6 \end{pmatrix} \begin{pmatrix} x_1 \\ x_2 \end{pmatrix} = \begin{pmatrix} 1 \\ 3 \\ 5 \\ 5 \\ 6 \end{pmatrix}$

Then you would replace each of the logs by their decimal approximations. E.g., $\ln 4 \doteq 1.3862944$. Then multiply both sides on the left by $A^T$. Then do Gaussian elimination. Then take a vacation for two weeks to recover from all the arithmetic.

After I came back from my favorite vacation spot (Yosemite), the answers were on my doorstep: $x_1 \doteq 1.16$ and $x_2 \doteq 2.86$. So the best-fit for the $y = x_1 + x_2 \ln x$ model would be $y = 1.16 + 2.86 \ln x$.

Just for fun, I'll make a giant chart that compares all three best-fit functions for the stargazing.

| hours spent and stars found | | stars predicted by linear model $y = 0.973 + 0.946x$ | error | (error)$^2$ | stars predicted by quadratic model $y = -1.38 + 2.71x - 0.249x^2$ | error | (error)$^2$ | stars predicted by logarithmic model $y = 1.16 + 2.86 \ln x$ | error | (error)$^2$ |
|---|---|---|---|---|---|---|---|---|---|---|
| x | y | | | | | | | | | |
| 1 | 1 | 1.919 | 0.919 | 0.845 | 1.081 | 0.081 | 0.007 | 1.160 | 0.160 | 0.026 |
| 2 | 3 | 2.865 | -0.135 | 0.018 | 3.044 | 0.044 | 0.002 | 3.142 | 0.142 | 0.020 |
| 3 | 5 | 3.811 | -1.189 | 1.414 | 4.509 | -0.491 | 0.241 | 4.302 | -0.698 | 0.487 |
| 4 | 5 | 4.757 | -0.243 | 0.059 | 5.476 | 0.476 | 0.227 | 5.125 | 0.125 | 0.016 |
| 6 | 6 | 6.649 | 0.649 | 0.421 | 5.916 | -0.084 | 0.007 | 6.284 | 0.284 | 0.081 |
| | | sum of squares of errors | | 2.757 | sum of squares of errors | | 0.483 | sum of squares of errors | | 0.630 |

And the winner (in this case) is the     quadratic model.

---

**El Toro**

---

1. On the previous page we awarded first place to the quadratic model. Does this mean that the quadratic model is the best model for the stargazing data?

2. [A do-you-remember-your-trig question] In the sunspot example, we thought of fitting a sine curve through the data points. Sunspots fluctuate on a roughly 11-year cycle. If we used $y = x_1 \sin x$, (where x is measured in years) as a model function, we wouldn't have the right periodicity. The function $y = x_1 \sin x$ has a period of $2\pi$ (radians). What would you replace the "x" with in order to change the model function into one with a period of eleven?

3. A sine function would be a good model function if the data looked like this:

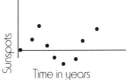

But why couldn't the data look like this graph?

4. Looking at the results of questions 2 and 3, what would be a good model function for sunspots?

*answers*

1. The giant chart told us that the quadratic model fit the data better than the straight-line model or the logarithmic model. But there are approximately 3,282,002,887,194 different models we might have looked

at. For example, $y = x_1 \sinh x + x_2 \sqrt{2 + \log_3 x} + x_3 \int_{t=0}^{x} e^{t^2} \tan \pi t \, dt$.

2. $y = \sin x$ has a period of $2\pi$. (Translation: $\sin(x + 2\pi) = \sin x$ for all x.) We have to do a little experimenting to find the answer.

    If we look at $y = \sin \pi x$, this has a period of 2. (When x goes from 0 to 2, then $\pi x$ will go from 0 to $2\pi$.) So multiplying by $\pi$ divided the period by $\pi$.

If we look at $y = \sin 2\pi x$, this has a period of 1. Multiplying by 2, divide the period in half.

So at this point, how do we get a period of 11? Dividing by 11, will multiply the period by 11. $y = \sin \dfrac{2\pi x}{11}$ will do the trick.

3. The graph is showing the number of sunspots. It would be really tough to have a negative number of sunspots.

4. Three pages ago (at the bottom of the page) we showed how to move a curve upward. Our model function for sunspots: $y = x_1 + x_2 \sin \dfrac{2\pi x}{11}$

---

## Tecumseh

1. At the end of the choir season Pat held a choir party at her house. It was even better than her beginning-of-the-season parties. Everyone skipped the dinner part and went straight for the desserts. Pat had her famous cream puffs, her famous banana splits, and her famous cheesecake.

The butterfat in each of these creations made them all very special. The joke that ran around was that a cow died in order to make one of her cream puffs.

A soprano took one of the cream puffs, two banana splits, and a piece of cheesecake. "Do you think I'm taking too much?" she asked Pat.

Pat smiled and said, "Heavens no. Those things have a total of only 18 calories in them."

An alto took 3 cream puffs and a piece of cheesecake. Pat said there were only 16 calories in that assortment.

A tenor: 2 cream puffs, 1 banana split, 2 pieces of cheesecake, 17 calories.

A bass: 1 cream puff, 1 banana split, 3 pieces of cheesecake, 26 calories.

Entre nous,* Pat had left off three zeros when she reported the number of calories to the choir members. The soprano was actually about to ingest 18,000 calories.

Let $\mathsf{x}^T = (x_1, x_2, x_3) =$ (number of cream puffs, number of banana splits, number of slices of cheesecake).

Express the selections of these four singers as $\mathsf{Ax = b}$.

Then compute $\mathsf{A}^T\mathsf{Ax} = \mathsf{A}^T\mathsf{b}$.

2. For old time's sake, use Gaussian elimination on $\mathsf{A}^T\mathsf{Ax} = \mathsf{A}^T\mathsf{b}$ and find the best fit approximation for the number of calories in the three desserts. As usual, please don't just look at the answer and say to yourself, "I can do that." It is an important skill for linear algebra students to become good at doing Gaussian elimination. You may, of course, use your hand calculator to help with the arithmetic.

*answers*

1. $\mathsf{Ax = b}$ would be
$$\begin{pmatrix} 1 & 2 & 1 \\ 3 & 0 & 1 \\ 2 & 1 & 2 \\ 1 & 1 & 3 \end{pmatrix} \begin{pmatrix} x_1 \\ x_2 \\ x_3 \end{pmatrix} = \begin{pmatrix} 18 \\ 16 \\ 17 \\ 26 \end{pmatrix}$$

$\mathsf{A}^T\mathsf{Ax} = \mathsf{A}^T\mathsf{b}$
$$\begin{pmatrix} 15 & 5 & 11 \\ 5 & 6 & 7 \\ 11 & 7 & 15 \end{pmatrix} \begin{pmatrix} x_1 \\ x_2 \\ x_3 \end{pmatrix} = \begin{pmatrix} 126 \\ 79 \\ 146 \end{pmatrix}$$

2. $\begin{pmatrix} 15 & 5 & 11 & 126 \\ 5 & 6 & 7 & 79 \\ 11 & 7 & 15 & 146 \end{pmatrix} \xrightarrow[\text{rows}]{\text{Interchange}} \begin{pmatrix} 5 & 6 & 7 & 79 \\ 11 & 7 & 15 & 146 \\ 15 & 5 & 11 & 126 \end{pmatrix} \xrightarrow{5r_2} \begin{pmatrix} 5 & 6 & 7 & 79 \\ 55 & 35 & 75 & 730 \\ 15 & 5 & 11 & 126 \end{pmatrix} \xrightarrow[-3r_1 + r_3]{-11r_1 + r_2}$

$\begin{pmatrix} 5 & 6 & 7 & 79 \\ 0 & -31 & -2 & -139 \\ 0 & -13 & -10 & -111 \end{pmatrix} \xrightarrow{31r_3} \begin{pmatrix} 5 & 6 & 7 & 79 \\ 0 & -31 & -2 & -139 \\ 0 & 403 & 310 & 3441 \end{pmatrix} \xrightarrow{13r_2 + r_3} \begin{pmatrix} 5 & 6 & 7 & 79 \\ 0 & -31 & -2 & -139 \\ 0 & 0 & 284 & 1634 \end{pmatrix}$

---

✶ The phrase *entre nous* (AHN-treh noo) used to be French, but we have stolen it. It's in your English dictionary. It means just between you and me.

Back-substituting in

$$5x_1 + 6x_2 + 7x_3 = 79$$
$$-31x_2 - 2x_3 = -139$$
$$284x_3 = 1634$$

yields   $x_3 \doteq 5.75$ (thousand calories per slice of cheesecake)

$x_2 \doteq 4.11$ (thousand calories per banana split)

$x_1 \doteq 2.81$ (thousand calories per cream puff)

---

**Velva**

---

1. You are a door-to-door salesman selling **Fred's Fountain Pens**. The slogan for the company is *Nothing Flows Like a Fred.* "A nice little piece of alliteration," you think to yourself. You are supposed to say that slogan at least twice to every customer.

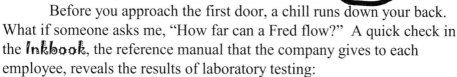

    Before you approach the first door, a chill runs down your back. What if someone asks me, "How far can a Fred flow?" A quick check in the **Inkbook**, the reference manual that the company gives to each employee, reveals the results of laboratory testing:

> test 1: 82 fine point pens + 32 broad point pens flowed a total of 38 miles.
> test 2: 23 fine point pens + 66 broad point pens flowed a total of 24 miles.
> test 3: 55 fine point pens + 17 broad point pens flowed a total of 40 miles.
> test 4: 63 fine point pens + 34 broad point pens flowed a total of 55 miles.
> test 5: 99 fine point pens + 77 broad point pens flowed a total of 112 miles.
> test 6: 28 fine point pens + 73 broad point pens flowed a total of 30 miles.
> test 7: 44 fine point pens + 51 broad point pens flowed a total of 29 miles.

Setting up $Ax = b$ would be long and easy.* Your short and easy question: State the dimensions of $A$, $x$, and $b$.

2. State the dimensions of $A^T A$.

3. State the dimensions of $A^T b$.

---

&#42; "Long and easy" = it would take a long time to do it, but it would be very routine. That would be like asking someone to find the sum of the first hundred digits of $\pi$. When you do those kinds of math problems, you get older, but not much smarter.

    The best kinds of math problems are those that are short and hard. They can be easy to state, but take some head scratching to figure out the answer. An example of a short-and-hard question would be, "Prove the Goldbach conjecture: *Every even number greater than 2 can be written as the sum of two primes.*" $4 = 2 + 2$, $6 = 3 + 3$, $8 = 3 + 5$, $10 = 5 + 5$, $12 = 5 + 7$, $14 = 7 + 7$, $16 = 5 + 11$, $18 = 7 + 11$, $20 = 3 + 17$, $22 = 11 + 11$, etc.

4. The matrix equation $\text{My} = \text{c}$ has a solution iff $M^{-1}$ exists. (Its solution is $\text{y} = M^{-1}\text{c}$.) If $M^{-1}$ exists, then M is called invertible. (We learned that on page 121.) Your question: Will the matrix $A^T A$ always be invertible, regardless of what numbers are reported in the **Inkbook**?

***answers***

1. A is 7×2. x is 2×1. b is 7×1. Recall that when you multiply an m×n matrix times an n×p matrix, you get an m×p matrix. It is as if the n's canceled. (We did that in problem 5 on page 56.)

2. $A^T$ is 2×7. A is 7×2. $A^T A$ is 2×2.

3. Since A is 7×2, $A^T$ is 2×7. Since b is 7×1, $A^T \text{b}$ is 2×1.

4. On the fifth page of this chapter we stated: The normal equation $A^T A \text{x} = A^T \text{b}$ will *always* have a solution. In the problem we said: The matrix equation $\text{My} = \text{c}$ has a solution iff $M^{-1}$ exists, and if $M^{-1}$ exists, then M is called invertible. Putting those three facts together, we have that $A^T A$ will always be invertible.

---

### Darien

---

Robin was still looking for employment. After failing the employment test at *Harry's Hamburgers*, she headed to the hot dog place next door to see if they were hiring. She walked into *Frank's Franks*. There was one customer in the place, and one weenie cooking on the grill. No one was watching it cook. She waited a moment. The customer got up and took the hot dog off the grill, put it on a bun, and went back to his seat.

She saw an office door marked, "F. Frank, owner," knocked on it, and entered. At a desk sat Frances, who was eating her lunch. (It was a *Harry Hamburger*.)

"I'm looking for a job," Robin began.

"Things are kinda slow

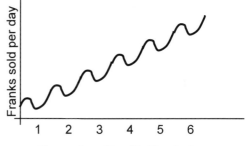

Years since Frank's Franks began

right now," Frances answered. "The hot dog business is seasonal. We sell a lot more in the summer than in the winter." She made a wavy motion with her hand. "But every year things seem to be getting better." She pointed to the sales chart on the wall.

1. We want to find the model function for this curve. Because Frank's business is seasonal, we have the feeling that the sine curve may be a part of the model function. Since the chart is plotted in years along the x-axis, we want a sine function with a period of one year. We did this three Cities ago. Fill in the question mark: $y = \sin\, ?x$.

2. But the chart isn't a sine curve because her business is increasing year-over-year. For example, each July total sales are more than the previous July. If you were to eliminate the seasonal wiggle and straighten out the curve, what would be the model function?

3. Have you ever added two curves together? You don't need one of those $80 graphing calculators to do that. You just add corresponding ordinates* together. Look:

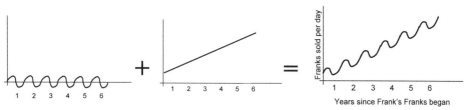

You know the model function for the sine function with a period of 1. You know the model function for a line. What is the model function for the sales at *Frank's Franks*?

4. Many businesses do not grow linearly over the years, but parabolically. (In general, parabolas look like $y = a + bx + cx^2$, where a, b, and c are real numbers.) Draw a graph of Frank's seasonal business if it were growing parabolically and give the model function.

5. Frances had divided the year into ten equal parts. She thought that would be the metric way of doing things. So at the half way point of the first year she was selling hot dogs at the rate of 500 per day. So one data point is (0.5, 500). Frances had 60 records (= 6 years' of sales recorded

---

\* *Ordinates* is another word for y-coordinates. The x-coordinates are called the abscissas.

every tenth of a year).

We'll just look at four data points: (0.5, 500), (0.7, 250), (1.1, 800), (1.6, 540). Use the model function from question 3 of a sine curve ascending linearly: $y = x_1 \sin 2\pi x + x_2 + x_3 x$.

Look at the clock. In the next 20 minutes, go as far as you can in finding $x_1$, $x_2$, and $x_3$.

***answers***

1. $y = \sin 2\pi x$
2. Eliminate the seasonal fluctuations and the chart would look like:

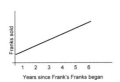

Her business was growing linearly. The model function without the seasonal wiggles would be $y = x_1 + x_2 x$.
3. $y = x_1 \sin 2\pi x + x_2 + x_3 x$.
4.

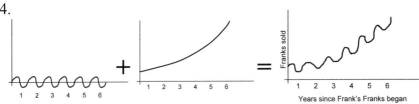

The model function for a sine curve ascending parabolically is $y = x_1 \sin 2\pi x + x_2 + x_3 x + x_4 x^2$.

5. $Ax = b$ would be

$$\begin{pmatrix} 0 & 1 & 0.5 \\ -0.9511 & 1 & 0.7 \\ 0.5878 & 1 & 1.1 \\ -0.5878 & 1 & 1.6 \end{pmatrix} \begin{pmatrix} x_1 \\ x_2 \\ x_3 \end{pmatrix} = \begin{pmatrix} 500 \\ 250 \\ 800 \\ 540 \end{pmatrix}$$

$A^T Ax = A^T b$

$$\begin{pmatrix} 1.60 & -0.951 & -0.960 \\ -0.951 & 4 & 3.9 \\ -0.960 & 3.9 & 4.51 \end{pmatrix} \begin{pmatrix} x_1 \\ x_2 \\ x_3 \end{pmatrix} = \begin{pmatrix} -84.96 \\ 2090. \\ 2169. \end{pmatrix}$$

Then Gaussian elimination on the augmented matrix:

$$\begin{pmatrix} 1.60 & -0.951 & -0.960 & -84.96 \\ -0.951 & 4 & 3.9 & 2090. \\ -0.960 & 3.9 & 4.51 & 2169. \end{pmatrix}$$

and back-substituting yields $x_1 \doteq 306$, $x_2 \doteq 401$, and $x_3 \doteq 199$.

The best-fit equation for the four data points is

$$y = 306 \sin 2\pi x + 401 + 199x.$$

---

## Iona

1. Being a den mother for Cub Scouts isn't always easy. (The previous sentence is an example of litotes.) After each den meeting, Madison's apartment always suffered some damage. Let's take a peek at Madison's journal:

*Diary of Damage*

*April 2ⁿᵈ meeting: 1 dish broken, 1 wall damaged, 3 rug stains. Total $24.*

*April 9ᵗʰ: 3 dishes broken, 2 walls damaged, 2 rug stains. Total $40.*

*April 16ᵗʰ: 1 dish broken, 3 walls damaged, 4 rug stains. Total $45.*

*April 23ʳᵈ: 2 dishes broken, 2 walls damaged, 1 rug stain. Total $30.*

*April 30ᵗʰ: 2 dishes broken, 1 wall damaged, 4 rug stains. Total $33.*

In May, Madison bought some paper plates, sold her rugs, and had her walls plated with steel. Cub meetings were a pleasure after those small changes.

Set up the matrices expressing $A^TAx = A^Tb$. Look at the clock. In the next 13 minutes, go as far as you can in finding out the cost of dishes, wall repair, and rug stain removal.

*answer*

1.
$$\begin{pmatrix} 19 & 16 & 23 \\ 16 & 19 & 25 \\ 23 & 25 & 46 \end{pmatrix} \begin{pmatrix} x_1 \\ x_2 \\ x_3 \end{pmatrix} = \begin{pmatrix} 315 \\ 332 \\ 494 \end{pmatrix}$$

Solving by Gaussian elimination,
$x_1 \doteq \$5.14$/dish    $x_2 \doteq \$8.41$/wall
$x_3 \doteq \$3.60$/rug stain

---

## Keene

1. If you were given the results of some experiment in the form of 16,397 data points, for example, (435.35, 18.46), (477.54, 19.01), (1039.56, 23.68), (1188.63, 24.48), (469.86, 18.91), (990.43, 23.39), (159.43, 12.43), (1493.35, 25.85),

(1218.86, 24.63), (380.40, 17.65), (369.38, 17.47), (1380.17, 25.38), (618.10, 20.56), (1485.49, 25.82), (145.95, 11.90), (1163.49, 24.36), (512.61, 19.44), (1696.97, 26.62), (108.83, 10.14), (397.34, 17.91), (1541.89, 26.04), (467.16, 18.88), (1581.46, 26.20), (1688.46, 26.59), (668.51, 21.03), (543.40, 19.79), (1508.40, 25.91), (330.98, 16.81), (1318.65, 25.11), (206.06, 13.97), (1577.10, 26.18), (1029.33, 23.62), (1359.61, 25.29), (885.51, 22.72), (1399.53, 25.46), (169.47, 12.80), (100.22, 9.64), (621.40, 20.59), (1306.27, 25.05), (1338.77, 25.20), (271.10, 15.61), (1621.01, 26.34), (539.76, 19.75), (1538.30, 26.03), (1106.38, 24.05), (108.17, 10.10), (588.95, 20.27), (370.40, 17.49), (638.81, 20.76), (1052.41, 23.75), (1077.55, 23.89), (1029.20, 23.62), (1680.14, 26.56), (334.00, 16.87), (1259.21, 24.83), (155.22, 12.27), (1584.51, 26.21), (282.03, 15.85), (808.92, 22.17), (160.02, 12.45), (1672.74, 26.53), (1050.36, 23.74), (779.44, 21.95), (1074.59, 23.88), (1698.39, 26.62), (101.38, 9.71), (1673.86, 26.54), (344.63, 17.05), (250.25, 15.13), (1091.42, 23.97), (1550.52, 26.08), (656.56, 20.92), (1150.35, 24.29), (101.76, 9.74), (101.41, 9.72), (1698.37, 26.62), (101.78, 9.74), (116.47, 10.55), (204.94, 13.94), (617.05, 20.55), (264.52, 15.47), (405.87, 18.04), (1.36, 2.9), (127.52, 11.09), (213.03, 14.17), (1537.38, 26.03), (1519.81, 25.96), (743.13, 21.67), (1133.99, 24.20), (1609.69, 26.30), (132.68, 11.33), (1174.24, 24.41), (739.08, 21.63),

(240.31, 14.89), (1683.53, 26.57), (1009.33, 23.50), (102.14, 9.76), (1644.28, 26.43), (1699.58, 26.63), . . ., (1065.36, 23.83), and you were setting up Ax = b, what dimensions of A, x, and/or b would you know?

2. If some scientist (with his rat) came to you with those 16,397 data points and offered to pay you a lot of money to find the equation that best fit those points, how would you respond?

*answers*

1. What you haven't been given is the model function for this experiment. That will give you the number of rows in x. (We know that x has one column.) We know that both A and b have as many rows as data points (which is 16,397). The number of columns of A will match the number of rows of x.

2. If it is a lot of money, you first tell him what a lovely rat he has. Second, you would ask how he got that data. Is there a physical model for his experiment? He might already know the model function, and all you would have to do is set up the Ax = b, and do the A^T trick, and then solve the resulting equation using Gaussian elimination.

If he tells you that the stuff he is working on is TOP SECRET and all you get are the 16,397 data points, ask for more money. The first step, after getting more money, is to plot those points on a graph. Just looking at the graph may suggest what model functions to try out.

For each model function, you set up Ax = b, then A^TAx = A^Tb, then Gaussian elimination, then find x, then compute Ax as we did in the chart on the page before the Cities of this chapter. Fill in the rest of the chart to find the sum of the squares of the errors for each model function. Pick the model function that gives the smallest value for the sum of the squares of the errors.

If the scientist with the rat is paying you by the hour ($350/hour), try lots of model functions.

# *Chapter Three and a Half*
## *Linear Transformations*

**B**etty and Alexander had played the bunnies-mice-and-squirrels game five times. That was enough. Alexander threw his coat over the basket to keep the animals out. They lay* back on the blanket and looked out over the Great Lake—a peaceful sight of the sailboats silently gliding . . .

"Where's Fred!" Betty shouted as she sat bolt upright. "Where's Fred?" she repeated. "He was out there swimming just a minute ago." She stood up and yelled, "Fred! Fred!"

Alexander took a little longer to stand up—perhaps the effect of consuming his 8000-calorie linear combo pizza. Together they stared out over the lake.

"Did a sailboat run over him? Did he. . . ?" Betty couldn't say the word *drown*. "Keep looking!" she urged Alexander. She ran up and down the shoreline. She pulled out her cell phone and called information.** She shouted into the phone, "What's the number of the Kansas Coast Guard?" The operator responded with a giggle and hung up.

Alexander shouted to the nearest sailboat, "Hey. Have you seen a little kid swimming around?" He was going to add, ". . . in the water," but realized that that would be redundant. The other two sailboats were too far away to hear Alexander. Alexander started stripping down to his underwear. He was going to plunge in and find Fred.

---

✶ Not *laid* back! You don't want to sound unedukated. There are two verbs, *lay* and *lie*. The confusion comes from the fact that the past tense of *lie* is the same as the present tense of *lay*.
**Lay:** When you set something down, you *lay* it down. "Please go and lay an egg." In the past tense: "Yesterday you laid three eggs."
**Lie:** When you take a rest, you *lie* down. "Please lie on this recliner for your blood donation." In the past tense: "Yesterday I lay on my couch for two hours."

"Yesterday as I lay on the street, I laid three eggs."

✶✶ Calling 9-1-1 would have been a better choice in a life-or-death situation.

A fourth sailboat came into view. "Wait! Alexander!" Betty shouted. "That fourth thing isn't a sailboat. It looks like a shark."

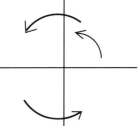

Alexander ignored Betty's warning and dove into the water. If it were a shark, he knew that it might have gobbled up Fred. Irrationally, he thought that he'd just drag the shark ashore, open it up, and get Fred out. Fred is so small, Alexander thought, that the shark wouldn't have to chew him—it could just swallow him whole.

As Alexander swam toward the shark, he realized that it wasn't a shark, but someone with a very pointy nose who was floating on his back in the water.

Alexander didn't know what to say. He blurted out, "What are you doing?"

Fred answered as any six-year-old might, "I'm playing. I was thinking about linear transformations." Fred played with ideas the way most kids play with toys. "Imagine that the surface of this Great Lake were a plane, maybe like $\mathbb{R}^2$, with a whole bunch of little points floating on it."

Alexander looked at the lake. There was a whole bunch of cigarette butts floating on the surface.

"Now suppose I were to spin around in the water and stir the lake. All the cigarette butts, I mean all the points, would rotate around me. It would be a linear transformation of the points in $\mathbb{R}^2$ to $\mathbb{R}^2$.

"Or I could say some magic words* and all the cigarette butts, I mean points, would double their distance from me. That would be a linear transformation.

Points on the lake revolving around Fred

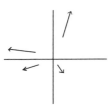

Points on the lake fleeing from Fred

---

★ "Surgeon General"

"Or with a flip of my nose, I could turn over the whole lake so that all the points would be reflected through an east-west line. That would be a great linear transformation."

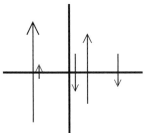

Points on the lake
reflected about
the x-axis

**Wait! Stop! I, your reader, have got to take a little more control of this book. Besides nearly drowning the little guy . . .**

Fred was never in any danger. Only Betty and Alexander thought he was.

**Not true. I was also worried. But my point—before you interrupted me again—was that in all this talk about "linear transformations" you have never bothered to define what a linear transformation is. And don't you dare mention cigarette butts in your definition.**

I think you want me to define linear transformation.

**Duh. Either that or go play hopscotch in Holland.**

I think I'll define it.

> A **linear transformation** T is a function from one vector space to another.
> $$T: \mathcal{V} \to \mathcal{W}$$
> It has two properties.
> If $u, v \in \mathcal{V}$, then
> (1) $T(u + v) = T(u) + T(v)$
> (2) $T(ru) = rT(u)$    for any $r \in \mathbb{R}$.

*Your Turn to Play*

1. I could have made the definition of linear transformation shorter by saying it has one property: If $u, v \in \mathcal{V}$ and r and s are real numbers, then $T(ru + sv) = rT(u) + sT(v)$ is an alternate definition of linear transformation. Show that (1) and (2) in the original definition imply this.

2. [Fill in the blank] To show that $T(ru + sv) = rT(u) + sT(v)$ implies (1) in the definition, just let r and s equal _____.

3. [Fill in the blank] To show that T( r$u$ + s$v$ ) = $\dot{r}$T($u$) + sT($v$) implies (2) in the definition, just let s equal _____.

4. Let $0$ be the zero vector in $\mathcal{V}$. Show that T($0$) always equals the zero vector in $\mathcal{W}$. In other words, T takes the zero vector in $\mathcal{V}$ into the zero vector in $\mathcal{W}$.

## . . . . . . . C O M P L E T E   S O L U T I O N S . . . . . . .

1.  T( r$u$ + s$v$ ) = T(r$u$) + T(s$v$)          by (1) on the previous page
              = $\dot{r}$T($u$) + sT($v$)          by (2) on the previous page

2.  one

3.  zero

4.  T($0$) = T($00$)          by micro-theorem #1: $0v = 0$ for every $v$. This theorem can be found at the top of page 189.

     = $0$T($0$)          by (2)

     = $0$          by micro-theorem #1 again. T($0$) is a vector and plays the role of $v$ in the theorem.

Linear transformations are also called **linear mappings** or **vector space homomorphisms**. As usual, the English is at least as tough as the math. I like the word *homomorphism*. Where else could you find a word with *omomo* in it?

Fred is playing in the lake with linear transformations from $\mathbb{R}^2$ to $\mathbb{R}^2$. This is the special case of a vector space homomorphism in which a vector space is mapped into itself. T: $\mathcal{V} \rightarrow \mathcal{V}$. This is called a **linear operator**, or a **linear transformation on** $\mathcal{V}$.

Let's look at the three linear operators that Fred told Alexander about.

He said that with a flip of his nose, he would send each point to its reflection through the x-axis.

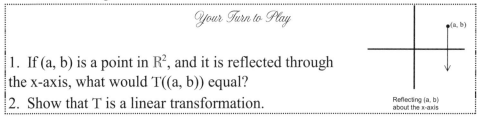

*Your Turn to Play*

1. If (a, b) is a point in $\mathbb{R}^2$, and it is reflected through the x-axis, what would T((a, b)) equal?

2. Show that T is a linear transformation.

Reflecting (a, b) about the x-axis

> . . . . . . . **COMPLETE SOLUTIONS** . . . . . . .

1.  T((a, b)) = (a, –b).
So, for example, T((5, 7)) = (5, –7) and T((–4, –6)) = (–4, 6)
2.  Let (a, b) and (c, d) be points in $\mathbb{R}^2$.  Let r and s $\in \mathbb{R}$.
(We are going to use the alternate definition of a linear transformation as described in questions 1–3 of the previous *Your Turn to Play*.  Namely, we will show T( r**u** + s**v** ) = rT(**u**) + sT(**v**).)

T( r(a, b) + s(c, d) ) = T( (ra, rb) + (sc, sd) )  vectors multiplied by scalars
= T( (ra + sc, rb + sd) )   addition of vectors
= (ra + sc, –rb – sd)   definition of T, see answer 1, above
= (ra, –rb) + (sc, –sd)   "unaddition" of vectors
= r(a, –b) + s(c, –d)   vectors "unmultiplied" by scalars
= rT((a, b)) + sT((c, d))   definition of T, see answer 1, above

    Fred claimed that taking every point and doubling its distance from the origin would be a vector space homomorphism (a.k.a. a linear transformation).

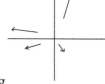

*Your Turn to Play*

1.  If (a, b) is a point in $\mathbb{R}^2$, what would be the coordinates of a point twice as far from the origin?
2.  Show that your answer to question 1 is correct by using the distance formula: the distance between $(x_1, y_1)$ and $(x_2, y_2)$ is given by $\sqrt{(x_2 - x_1)^2 + (y_2 - y_1)^2}$

Points on the lake
fleeing from Fred

3.  Let (a, b) and (c, d) be points in $\mathbb{R}^2$.  Let r and s $\in \mathbb{R}$.  Show that T is a linear transformation.

> . . . . . . . **COMPLETE SOLUTIONS** . . . . . . .

1.  (2a, 2b)
2.  The distance between (0, 0) and (2a, 2b) is $\sqrt{(2a - 0)^2 + (2b - 0)^2}$ which equals $2\sqrt{a^2 + b^2}$ which is twice the distance of the origin to (a, b).

3.  $T(\ r(a, b) + s(c, d)\ )\ =\ T(\ (ra, rb) + (sc, sd)\ )$    vectors multiplied by
                                                    scalars

         $=\ T(\ (ra + sc, rb + sd)\ )$          addition of vectors

         $=\ (2(ra + sc), 2(rb + sd))$       definition of T, see answer 1, above

         $=\ (2ra + 2sc, 2rb + 2sd)$        distributive law

         $=\ (2ra, 2rb) + (2sc + 2sd)$       "unaddition" of vectors

         $=\ r(2a, 2b) + s(2c, 2d)$         distributive law

         $=\ rT(a, b) + sT(c, d)$           definition of T, see answer 1, above

If you have been cheating in the previous two *Your Turn to Play* sections by just reading the questions and then the answers, rather than working them out for yourself on a piece of paper, Judgement Day will come when I ask you in one of the Cities to show that Fred's third example is a linear transformation.

Points on the lake
revolving around Fred

"Let's head back to Betty," Alexander suggested. "I'm sure she'll be happy to see you on dry land again."

"Okay. But I want to talk about linear transformations as we swim."

"But all the examples," Alexander said, "that you have given me are linear operators. $T: \mathcal{V} \to \mathcal{V}$. That's kid stuff. How about some healthy-looking linear transformations that take an m-dimensional space into an n-dimensional space?"

All the world must have broken out in a cold sweat when Alexander said that to Fred. You — just — don't — say — to — Fred that what he's talking about is "kid stuff." He can so easily crank it up several notches.

**You are right. I, your reader, am trembling.**

"So you want a linear transformation that's not a linear operator," Fred said as he began at a higher level than kid stuff. "How about something from $\mathbb{R}^{45}$ to $\mathbb{R}$?"

Alexander kept swimming. Then he remembered that the Great Lake wasn't that big—or deep. He stood up and began to walk. His shoulders were out of the water. Fred kept paddling next to him.

"Something from $\mathbb{R}^{45}$ to $\mathbb{R}$?" Alexander asked. "That sounds like when I stomp on aluminum cans and crush them."

"When I step on a can, it doesn't crush," Fred admitted. "But that's a good name for T:$\mathbb{R}^{45} \rightarrow \mathbb{R}$ defined by T( $(x_1, x_2, \ldots, x_{45})$ ) = $x_{17}$"*

"Or," Alexander said, "how about T:$\mathbb{R}^{45} \rightarrow \mathbb{R}^3$ given by T( $(x_1, x_2, \ldots, x_{45})$ ) = $(x_1, x_2, x_3)$?"

"Or," Fred continued, "how about T:$\mathbb{R}^4 \rightarrow \mathbb{R}^{45}$ defined by T( $(x_1, x_2, x_3, x_4)$ ) = $(x_1, x_2, x_3, x_4, 0, 0, \ldots, 0)$?"

The boys could have kept creating linear transformations all day long, except that they reached the shore.

Betty rushed over to Fred and picked him up. "Oh Fred, Fred, Fred, I was so worried," she exclaimed.

Fred didn't have a clue what she was talking about, but he liked being five feet high. He cranked up the discussion a notch: "How about the vector space of all polynomials of degree three or less? The derivative would be a linear transformation. Call it T. It would take the vector space of all polynomials of degree three or less into the vector space of all polynomials of degree two or less."

---

*Your Turn to Play*

1. A polynomial in the vector space of all polynomials of degree three or less looks like $r_0 + r_1x + r_2x^2 + r_3x^3$.    $T(r_0 + r_1x + r_2x^2 + r_3x^3) = ?$

---

* This can-crushing linear transformation is usually called a **projection**. If Fred had begun his linear-transformations-that-are-not-linear-operators discussion at a more elementary level, he might have given the more standard watch-out-there's-a-pigeon-overhead example: T: $\mathbb{R}^3 \rightarrow \mathbb{R}^2$ defined by T( $(x, y, z)$ ) = $(x, y)$.

2. I'm getting tired of writing *the vector space of all polynomials of degree three or less*. Let me abbreviate that as $\mathscr{P}_3$. Give an example of a nice-looking basis for $\mathscr{P}_3$.

3. What is the dimension of $\mathscr{P}_3$?

4. Give another example of a basis for $\mathscr{P}_3$.

5. Show that the derivative transformation

$T(r_0 + r_1 x + r_2 x^2 + r_3 x^3) = r_1 + 2r_2 x + 3r_3 x^2$ is a linear transformation.

6. Recall that the norm of a vector $r_0 + r_1 x + r_2 x^2 + r_3 x^3$ in $\mathscr{P}_3$ is given by $\sqrt{r_0^2 + r_1^2 + r_2^2 + r_3^2}$. Are the vectors in the basis $\{1, x, x^2, x^3\}$ unit vectors?

7. Is the basis $\{1, x, x^2, x^3\}$ orthogonal? (Back on page 209, we said that a whole set of vectors $\{v_1, v_2, \ldots, v_n\}$ is orthogonal if every pair of vectors in that set is orthogonal. Before that, we said that $u$ and $v$ are orthogonal if $<u, v> = 0$. On page 204, in $\mathscr{P}_2$ we defined the inner product of two polynomials as $<r_1 + r_2 x + r_3 x^2, s_1 + s_2 x + s_3 x^2> = \sum_{i=1}^{3} r_i s_i$, and commented that this definition was "awfully, really, completely close to the definition of the inner product in $\mathbb{R}^3$." And just for completeness sake, in $\mathbb{R}^3$, we defined the inner product of $(r_1, r_2, r_3)$ and $(s_1, s_2, s_3)$ as $\sum_{i=1}^{3} r_i s_i$ .)

. . . . . . . **COMPLETE SOLUTIONS** . . . . . . .

1. $T(r_0 + r_1 x^1 + r_2 x^2 + r_3 x^3) = r_1 + 2r_2 x + 3r_3 x^2$.

2. The nicest basis I know of for $\mathscr{P}_3$ is $\{1, x, x^2, x^3\}$. Those elements of that basis are linearly independent—none of them can be expressed as a linear combination of the others. And those elements span $\mathscr{P}_3$ since every polynomial in $\mathscr{P}_3$ can be expressed as a linear combination of them.

3. The dimension of a vector space is the number of elements in any basis for that space. The dimension of $\mathscr{P}_3$ is four.

4. Answers will vary. For example, $\{7, 3298x, \pi x^2, \sqrt{391} x^3\}$.

5. This proof is easier to do in English. From calculus we know that the derivative of the sum of two polynomials is the sum of the derivatives of those polynomials. That establishes (1) of the definition (given on page 261) $T(u + v) = T(u) + T(v)$. From calculus we know that the

derivative of a constant times any function is equal to the constant times the derivative of that function.

That establishes $T(r\mathbf{u}) = rT(\mathbf{u})$ for any $r \in \mathbb{R}$.

You could prove that $T$ is a linear transformation using symbols. It would begin: $T(\ r(r_0 + r_1x + r_2x^2 + r_3x^3) + s(s_0 + s_1x + s_2x^2 + s_3x^3)\ ) = \ldots$ but if you do, I don't want to watch.

6. $\| 1 \| = \sqrt{1^2 + 0^2 + 0^2 + 0^2} = 1$, so 1 is a unit vector.

$\| x^2 \| = \sqrt{0^2 + 0^2 + 1^2 + 0^2} = 1$, so $x^2$ is a unit vector. Same is true for x and $x^3$.

7. Wow. The hints for this problem were seven times longer than the problem. And the solution is really easy. If I want to find out if, say, x and $x^3$ are orthogonal, I need to find out if $< x, x^3 >$ equals zero, which is equivalent to computing $< (0, 1, 0, 0), (0, 0, 0, 1) >$, which equals $(0)(0) + (1)(0) + (0)(0) + (0)(1)$, which is zero. Any two elements of the basis $\{1, x, x^2, x^3\}$ are orthogonal. The set is orthogonal.

Suppose we have a linear transformation $T: \mathcal{V} \rightarrow \mathcal{W}$ from one vector space to another. Let $\{v_1, v_2, \ldots, v_m\}$ be an ordered basis for $\mathcal{V}$, and let $\{w_1, w_2, \ldots, w_n\}$ be an ordered basis* for $\mathcal{W}$.

**Theorem**: The action of $T$ on $\mathcal{V}$ is completely determined by what $T$ does to the basis of $\mathcal{V}$. Once you know the values of $T(v_1)$, $T(v_2)$, $\ldots$, $T(v_m)$, you can predict the value of $T(\mathbf{v})$ for any $\mathbf{v} \in \mathcal{V}$.

**Proof**: Given any $\mathbf{v} \in \mathcal{V}$. Since $\{v_1, v_2, \ldots, v_m\}$ is a basis, it spans $\mathcal{V}$. That means any $\mathbf{v} \in \mathcal{V}$ can be written as a linear combination of $v_1, v_2, \ldots, v_m$.

I.e., $\mathbf{v} = \sum\limits_{i=1}^{m} r_i v_i$ where the $r_i$ are scalars.

Then $T(\mathbf{v}) = T(\sum\limits_{i=1}^{m} r_i v_i)$

$$= \sum\limits_{i=1}^{m} r_i T(v_i) \quad \text{by the definition of linear transformation,}$$

and since we know all the $T(v_i)$ and all the $r_i$, we know the value of $T(\mathbf{v})$.

Q.E.D.

---

✱ An **ordered basis** is a basis in which the order is fixed. In regular set theory, the set { 🐫 , 🐫 } is equal to { 🐫 , 🐫 }. The order doesn't matter.

But when we were graphing points in algebra, the ordered pair (5, 7) was a different point than (7, 5).

Besides rotating, expanding, reflecting, projecting, and taking derivatives,

$$T(4x^7) = 28x^6$$

there is one other major class of linear transformations that you should know about.

$\mathbb{R}^m$ was the set of m-tuples (the set of vectors with m entries). In the previous chapters, when we wrote $Ax = b$, we let $x$ be a column vector. So, if $x \in \mathbb{R}^m$, then for $Ax$ to make sense, $A$ will have to have m columns:

$$\begin{pmatrix} a_{11} & a_{12} & a_{13} & \cdots & a_{1m} \\ a_{21} & a_{22} & a_{23} & \cdots & a_{2m} \\ & & \cdots & & \\ a_{n1} & a_{n2} & a_{n3} & \cdots & a_{nm} \end{pmatrix} \begin{pmatrix} x_1 \\ x_2 \\ \cdots \\ x_m \end{pmatrix}$$

Multiplying these two together yields a column vector with n entries.

Claim: Every n×m matrix will give us a linear transformation from $\mathbb{R}^m$ to $\mathbb{R}^n$ defined by $T(x) = Ax$.

Does $T(x) = Ax$ satisfy (1) and (2) of the definition of linear transformation? Yes.

(1) If $u, v \in \mathcal{V}$, then $T(u + v) = T(u) + T(v)$ translates into $A(B + C) = AB + AC$.
(2) $T(ru) = rT(u)$ for any $r \in \mathbb{R}$ translates into $ArB = rAB$.*

Use the definition of matrix multiplication: $(AB)_{ij} = \sum\limits_{k=1}^{n} a_{ik} b_{kj}$

and matrix addition $(A + B)_{ij} = a_{ij} + b_{ij}$. The proof at this point becomes a meat-grinder proof—five thousand obvious steps involving zillions of subscripts resulting in [check two boxes]: □ blindness, □ insanity.

Instead, I offer you the chart on the next page.

---

✱ Both $A(B + C) = AB + AC$ and $ArB = rAB$ can be established by letting $(A)_{ij} = a_{ij}$ and $(B)_{ij} = b_{ij}$ and $(C)_{ij} = c_{ij}$.

| Types | Picture | Example |
|---|---|---|
| **Rotate** each point of $\mathbb{R}^2$ through an angle of $\theta$ | | when $\theta = 45°$ $T((1, 0))$ $= (\frac{1}{\sqrt{2}}, \frac{1}{\sqrt{2}})$ |
| **Dilate** the graph by doubling the distance from the origin | | $T((3, 4))$ $= (6, 8)$ |
| **Reflect** each point through an axis | | $T((2, 7))$ $= (-2, 7)$ |
| **Project** each point into a lower dimension | | $T((4, 5, 6))$ $= (4, 5)$ |
| **Take** the derivative | | $T(4x^7) = 28x^6$ |
| **Multiply** by a matrix | $T(v) = Ax$ | |

The title of the table above:

**Handy-Dandy Catalog of Linear Transformations**

This chart is true, but it is misleading.

It's not misleading because I left out some linear transformations.

There are approximately 334,901,655 linear transformations that weren't mentioned, such as . . .

The donut-hole transformation which takes a vector space of donuts and turns each vector into a doughnut hole. $T(v) = 0$ for all $v$. Traditional textbooks call this the **zero transformation**.

The leave-it-alone transformation which maps each vector into itself. T(v) = v for every v. Traditional textbooks call this the **identity transformation**.

The integration linear transformation that maps $\mathscr{P}_3$ to $\mathscr{P}_4$ defined by T(v) = $\displaystyle\int_{x=0}^{x}$ v dx where v is a polynomial in $\mathscr{P}_3$.✶

*Okay. I, your reader, give up. What's so terrible about that Handy-Dandy Chart you have on the previous page? You said it was misleading.*

It is misleading. Two pages ago I showed that every n×m matrix could give me a linear transformation. You give me A, and that will define T.

*And you even called that a "major class of linear transformations."* I'm changing my mind.

*Big deal. We'll call it a minor class of linear transformations.*

No. No. No. We need to go in the other direction. Every linear transformation on finite dimensional vector spaces can be represented by a matrix multiplication. There is no other type of linear transformation. The chart should have looked like:

| Handy-Dandy Catalog of All Linear Transformations | | |
| --- | --- | --- |
| Type | Picture | Example |
| Multiply by a matrix | T(v) = Ax | every T is an example |

---

✶ The multiple roles of x in $\int_{x=0}^{x}$ v dx may cause a little consternation. We have to integrate with respect to x (the "dx") because v is a polynomial in x. That x acts as a dummy variable. I would much rather have defined T(v) to equal ∫v dx. The only problem with that is that T wouldn't be a function! The indefinite integral ∫v dx always has an answer with a "+ C" attached. $\int_{x=0}^{x}$ v dx gets rid of the + C problem.

**Wait one silly little minute! Why in blazes do we need this linear transformation stuff if it's all really just the old matrix multiplication that we've been doing for most of this book? Is this Chapter 3½, as you call it, just a waste of time?**

There is a reason for introducing linear transformations. It is the same reason we introduce a lot of mathematics. It makes our work easier. Remember back in algebra when you first learned about absolute value? It was a lot easier to consider $|x|$ than $\sqrt{x^2}$.

You see, there are two worlds. For finite-dimensional vector spaces[*], these two worlds are essentially equivalent. You can sit in either world.

**That looks like twice as much work for me—having two worlds when one world will do just fine. I'm used to matrix multiplication. Is there any real harm if we just forget this linear transformation stuff?**

As you wish. Does this matrix "sing" to you?

$$\begin{pmatrix} 0 & 1 & 0 & 0 \\ 0 & 0 & 2 & 0 \\ 0 & 0 & 0 & 3 \\ 0 & 0 & 0 & 0 \end{pmatrix}$$

**Yeah. It sings like a duck. It's a 4×4 matrix. It has non-negative integer entries. I don't get it. Is it supposed to have some special meaning?**

It should. You used it for two years.

**You must be thinking of someone else.**

How about on page 265? We talked about the linear transformation $T\colon \mathscr{P}_3 \to \mathscr{P}_2$ which mapped the vector space of all polynomials of degree three or less into the vector space of all polynomials

---

[*] For infinite-dimensional vector spaces, $T\colon \mathcal{V} \to \mathcal{W}$ would have to correspond to a matrix with an infinite number of rows and columns. The last time I checked in my matrix closet I couldn't find any of those.

of degree two or less by taking the derivative. (Two years of calculus = two years of taking derivatives.)

It's a lot easier to think of derivatives than to think of that 4×4 matrix. That is why I drew a nice comfortable arm chair 🪑 in the linear transformation world.

**Hey. I'm easy to get along with. Let's eliminate all the matrix stuff and stick with the linear transformation world. We could make this book a lot shorter by crossing out Chapters 1, 1½, 2, 2¾ and 3. We have to leave Chapter 2½, since it deals with vector spaces, which we need for linear transformations.**

As you wish. Now solve
$$\begin{cases} 3x_1 + 5x_2 + 2x_3 = 3.20 \\ 5x_1 + 3x_2 + x_3 = 2.85 \\ x_1 + 4x_2 + 3x_3 = 2.35 \end{cases}$$

using linear transformations to find out that the whole-wheat bagel costs 30¢, a slice of turkey costs 40¢, and a peach costs 15¢. (This was from the second *Your Turn to Play* in this book on page 28.)

**How about our keeping both worlds? But under one condition. Four pages ago you showed me how to go from 🪑 to 🪑 —from Ax to T(x) in that** Claim **thing.**

**Now you have to show me how to go the other direction—from a linear transformation** T:$\mathcal{V} \to \mathcal{W}$ **to** Ax.

Once I do that, you will be able to go from "derivative linear transformation" to
$$\begin{pmatrix} 0 & 1 & 0 & 0 \\ 0 & 0 & 2 & 0 \\ 0 & 0 & 0 & 3 \end{pmatrix}$$

**I like that. Then I could choose which chair to sit in.**

I am starting with a linear transformation T:$\mathcal{V} \to \mathcal{W}$ from one vector space to another, where $\{v_1, v_2, \ldots, v_m\}$ is an ordered basis for $\mathcal{V}$, and $\{w_1, w_2, \ldots, w_n\}$ is an ordered basis for $\mathcal{W}$. Where I will end up at is an n×m matrix A.

Are you ready?

**I have my seatbelt fastened.**

It will take two steps to go from T to A, which I call 𝔖trip a𝔫d 𝔖tuff.

272

The vectors in $\mathcal{V}$ and $\mathcal{W}$ could be almost anything: geometric arrows, polynomials, matrices, or even Freds (see page 188). Our goal is to arrive at a matrix **A** whose entries are scalars. The first step is to ṣtrip away all the arrows and polynomials and matrices and ⟋'s, and turn everything into numbers.

☞ $\mathcal{V}$ and its basis $\{v_1, v_2, \ldots, v_m\}$ will become $\mathbb{R}^m$.

☞ $\mathcal{W}$ and its basis $\{w_1, w_2, \ldots, w_n\}$ will become $\mathbb{R}^n$.

☞ And T will now go from the column vectors of $\mathbb{R}^m$ to the column vectors of $\mathbb{R}^n$.

Underneath every vector space $\mathcal{V}$ with an ordered basis $\{v_1, v_2, \ldots, v_m\}$ is a skeleton of column vectors of $\mathbb{R}^m$.* Let me explain.

Start with any $v \in \mathcal{V}$. It has a unique representation as a linear combination of the basis $\{v_1, v_2, \ldots, v_m\}$. $v = \sum_{i=1}^{m} r_i v_i$ where the $r_i$ are real numbers.

Instead of $v$, use $\begin{pmatrix} r_1 \\ r_2 \\ \ldots \\ r_m \end{pmatrix}$.

The same for $\mathcal{W}$. If $w \in \mathcal{W}$ is equal to $\sum_{i=1}^{n} s_i w_i$, then use $\begin{pmatrix} s_1 \\ s_2 \\ \ldots \\ s_n \end{pmatrix}$.

Now, instead of $T(v) = w$, we have $T(r_1, r_2, \ldots, r_m)^T = (s_1, s_2, \ldots, s_m)^T$. (Recall, that little $^T$ stood for "transpose.")

The explanation is harder than the doing. Suppose we have the derivative transformation $T: \mathscr{P}_3 \to \mathscr{P}_2$.

The standard basis for $\mathscr{P}_3$ is $\{1, x, x^2, x^3\}$.

The standard basis for $\mathscr{P}_2$ is $\{1, x, x^2\}$.

We are going to ṣtrip away the polynomial language.

$T( 5 + 27x + 36x^2 + 4x^3 ) = 27 + 72x + 12x^2$ becomes

$T( (5, 27, 36, 4)^T ) = (27, 72, 12)^T$.

---

✶ Formally: Every m-dimensional vector space is isomorphic to $\mathbb{R}^m$. Isomorphic means that they have essentially the same structure.

The 🛋️ step is just as fast: A is the matrix whose columns are $T(v_1), T(v_2), \ldots, T(v_m)$.

So you don't have to waste your highlighter pen, here it is in a box.

# How to Go from 🛋️ to 🪑

If $T: \mathcal{V} \to \mathcal{W}$ is a linear transformation from one vector space to another, where $\{v_1, v_2, \ldots, v_m\}$ is an ordered basis for $\mathcal{V}$, and $\{w_1, w_2, \ldots, w_n\}$ is an ordered basis for $\mathcal{W}$, then

A is the matrix whose columns are $T(v_1), T(v_2), \ldots, T(v_m)$,

where the $T(v_i)$ are the stripped down versions.

Let's finish up the derivative transformation $T: \mathcal{P}_3 \to \mathcal{P}_2$ example. Step 1: Take T of each basis element of $\mathcal{P}_3$, express the answer as a linear combination of the basis elements of $\mathcal{P}_2$, and 🛋️ away everything except the coefficients.

$$
\begin{aligned}
T(1) &= 0 = 0(1) + 0(x) + 0(x^2) & (0, 0, 0)^T \\
T(x) &= 1 = 1(1) + 0(x) + 0(x^2) & (1, 0, 0)^T \\
T(x^2) &= 2x = 0(1) + 2(x) + 0(x^2) & (0, 2, 0)^T \\
T(x^3) &= 3x^2 = 0(1) + 0(x) + 3(x^2) & (0, 0, 3)^T
\end{aligned}
$$

Step 2: 🛋️ these column vectors into A: $\begin{pmatrix} 0 & 1 & 0 & 0 \\ 0 & 0 & 2 & 0 \\ 0 & 0 & 0 & 3 \end{pmatrix}$

Did you notice that A depends on which basis you chose for $\mathcal{V}$?
Did you notice that A depends on which basis you chose for $\mathcal{W}$?
And, of course, A depends on which linear transformation you are using.

People who do linear algebra are sometimes called linear algebraists. [al-gee-BRAY-ists]. Have you noticed that you have been doing linear algebra?

*Your Turn to Play*

For these four problems, you are given a linear transformation T: $V \to W$ from one vector space to another, where $\{v_1, v_2, \ldots, v_m\}$ is an ordered basis for $V$, and $\{w_1, w_2, \ldots, w_n\}$ is an ordered basis for $W$.

1. From the given information, do we know the dimension of $W$?
2. From the given information, do we know whether m < n, m = n, or m > n?
3. For the matrix A associated with the given information, do we know its dimensions?
4. Suppose $V$ and $W$ are both $\mathbb{R}^2$ with the standard basis $\{e_1, e_2\}$. There is a slight chance that this abbreviation for the standard basis may have slipped your mind.

   A million pages ago (on page 149), we let **i**, **j**, and **k** be the standard basis $\{(1, 0, 0), (0, 1, 0),$ and $(0, 0, 1)\}$ for $\mathbb{R}^3$. Then we noted in $\mathbb{R}^4$ we started to run out of letters for (1, 0, 0, 0), (0, 1, 0, 0), (0, 0, 1, 0), and (0, 0, 0, 1), and decided to use the letters $e_1$, $e_2$, $e_3$, and $e_4$ instead.

   So in $\mathbb{R}^2$, $e_1 = (1, 0)$ and $e_2 = (0, 1)$.

   Suppose T is a dilation defined by T( (a, b) ) = (5a, 7b). Find A.

.......**COMPLETE SOLUTIONS**.......

1. Since we know that one basis for $W$ has n elements, every basis of $W$ must have n elements. In symbols, dim($W$) = n.
2. We don't have a clue. It could be any of these. The derivative and the projection linear transformations sent us into lower dimensional spaces. The integration linear transformation, mentioned on page 270, sent us into a higher dimensional space. The rotation, dilation, and reflection transformations, mentioned on page 269, went from $\mathbb{R}^2$ to $\mathbb{R}^2$.
3. A will consist of m column vectors—one for each $v_i$. Therefore, A will have m columns. Each column vector was a stripped-down version of a linear combination of the $w_i$. Each column vector will have n elements. Thus, A will have n rows.

It's just the way things work out: A linear transformation from m dimensions to n dimensions yields an n×m matrix representation. It would have been a lot nicer if it had been an m×n matrix, but I suspect that you won't get much sympathy if you tell your friends this is the hardest burden you have to bear in life.

4. For each basis element of $\mathcal{V}$, we find T(v), express it as a linear combination of the basis elements of $\mathcal{W}$, and Strip away everything except the coefficients.

$$T((1, 0)) = (5, 0) = 5e_1 + 0e_2 \qquad (5, 0)$$
$$T((0, 1)) = (0, 7) = 0e_1 + 7e_2 \qquad (0, 7)$$

and then Stuff these column vectors into A: $\begin{pmatrix} 5 & 0 \\ 0 & 7 \end{pmatrix}$

Of course, I could have written the above equations as:

$$T(e_1) = (5, 0) = 5(1, 0) + 0(0, 1)$$
$$T(e_2) = (0, 7) = 0(1, 0) + 7(0, 1)$$

or as:

$$T(e_1) = (5, 0) = 5e_1 + 0e_2$$
$$T(e_2) = (0, 7) = 0e_1 + 7e_2$$

Can you deal with this much freedom?

As a memory aid for the process of going from a linear transformation to its associated matrix, one student once wrote in his notes:

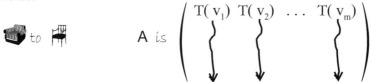

It's time we go on a little safari. We are going to hunt down several vector spaces which have been lurking around but haven't been identified thus far.

One of these vectors spaces is slightly weird.

One is full-scale, up-to-your-eyeballs-in-theory weird.

And one is completely over the top.

Before we start our hunt, let's do a little chanting to get us in the mood for finding vector spaces.

A vector space was some set (of arrows, polynomials, matrices, etc.) cooking over some field ($\mathbb{R}$, $\mathbb{Q}$, or $\mathbb{C}$)

with a vector addition that

is commutative, associative, has an identity, and has inverses.

is commutative, associative, has an identity, and has inverses.

is commutative, associative, has an identity, and has inverses.

and a scalar multiplication that

is associative, has an identity, and is double distributive.

is associative, has an identity, and is double distributive.

is associative, has an identity, and is double distributive.

We will start in the middle with the full-scale, up-to-your-eyeballs-in-theory weird vector space. From about page 260 when Fred was stirring cigarette butts around in the Great Lake until now, you have seen a bunch linear transformations T: $\mathcal{V} \rightarrow \mathcal{W}$ defined from one vector space to another.

**You haven't lost me so far.**

Good, since I haven't said anything new yet.

**Okay. Keep going.**

Now, for the time being, pick any vector space you like for $\mathcal{V}$ and pick any vector space you like for $\mathcal{W}$.

**Okay. I've got them in my mind. I'm choosing $\mathbb{R}^3$ for $\mathcal{V}$ and . . .**

You don't have to tell me. Just keep them fixed in your mind and don't change them.

**Is it okay if I write them down?**

Puleez, give me a break. You know, in a regular linear algebra book, all they say is "Fix any two vector spaces."

**Yeah, I know. But they are sooo unclear when they talk about**

**"fix."** $\begin{cases} \text{WHEN I FIX A PICTURE TO THE WALL—IT'S STATIONARY.} \\ \text{WHEN I FIX A DRINK—I STIR IT.} \end{cases}$

$\begin{cases} \text{WHEN I FIX A CLOCK—I MAKE IT WORK.} \\ \text{WHEN I FIX MY CAT—I MAKE IT NON-WORKING.} \end{cases}$

Now think of a linear transformation between those two vector spaces you picked.  $T: \mathcal{V} \to \mathcal{W}$.

**I'm so glad I picked $\mathbb{R}^3$ for $\mathcal{V}$ and $\mathbb{R}^2$ for $\mathcal{W}$. If I had selected the set of all geometric arrows from the origin to any point in three-dimensional space and the set of all polynomials of degree 49 or less, I would have had more difficulty finding an example for T.**

Can you keep a secret?

**Sure.**

Then don't tell me what you have selected for $\mathcal{V}$ and $\mathcal{W}$.

Next think of another linear transformation between those two vector spaces.  We'll call it $T'$.  So $T': \mathcal{V} \to \mathcal{W}$.

And another, $T''$.  And another, $T'''$.  And another, $T''''$.

**My brain is full.**

Now think of every possible linear transformation from $\mathcal{V}$ to $\mathcal{W}$.

**Brain overflowing.**

Now for any homomorphism T that you thought of, I can think of a dozen more: 2T, 3T, 4T, 5T, 6T, 7T, 8T, 9T, 10T, 11T, and 12T.

**That's eleven.**

Oops.  Sorry.  And 13T.

**I may be brain dead at this point, but I don't think you ever defined what 5T means.**

You're right.  We are given $T: \mathcal{V} \to \mathcal{W}$ and I should define 5T.

Definition: 5T is the linear transformation from $\mathcal{V}$ to $\mathcal{W}$ obtained by first computing $T(v)$ and then multiplying by 5.  In short, $(5T)(v)$ is defined as $5T(v)$.

Also, for any two homomorphisms, T and $T'$, you thought of, I can think of a new linear transformation $T + T'$.

Definition: $(T + T')(v)$ is defined as $T(v) + T'(v)$.

And now for your full-scale, up-to-your-eyeballs-in-theory weird vector space, I offer you the set of all linear transformations from $\mathcal{V}$ to $\mathcal{W}$ with scalar multiplication and vector addition as just defined.

The vectors in this new vector space are homomorphisms from $\mathcal{V}$ to $\mathcal{W}$.  This new vector space is called Hom($\mathcal{V}$, $\mathcal{W}$).  That's what everyone calls it.

When the concept of *function* is first introduced in high school math, it usually takes students somewhere between two days and a blue moon[*] to start to feel comfortable with the idea.

Hom($V$, $W$) usually generates that same feeling among linear algebra students.

**Hey. I was just starting to feel comfortable with what a vector space is, and now you are telling me that we take any two vector spaces and create a new vector space consisting of all possible linear transformations between them.**

Worse yet, I am about to prove that Hom($V$, $W$) is a vector space.

**No. No. No. Mercy! That means that you are going to have to show that this vector addition in** Hom($V$, $W$) **is** *commutative, associative, has an identity, and has inverses* **and that scalar multiplication is** *associative, has an identity, and is double distributive.* **That's eight separate little proofs. A real meat-grinder proof as you call it.**

Cool it. The whole proof is approximately two steps long.

**Did I ever tell you I like this book?**

My address is on the PolkaDotPublishing.com Web site.

---

[*] The common expression is *once in a blue moon*. "Everybody" knows that the definition of a blue moon is the second of two full moons occurring in the same calendar month. A whole year can go by without a blue moon.

In a year with thirteen full moons, one month must have a blue moon in it. In mathematics, this is an example of the **pigeonhole principle**. (That's its real name.) If you have 100 pigeons and only 99 holes for them, some hole is going to contain at least two pigeons.

Last year there were two people in Lima, Peru who took exactly the same number of breaths during the year. Did you know that? Ignoring the less than 0.001% of the people who took more than 8 million breaths in the year, we have more than 8,200,000 people in Lima each breathing less than 8 million breaths. Two of them, by the pigeonhole principle, must have breathed exactly the same number of breaths.

For those who know about functions: the pigeonhole principle is the theorem that states that if we have a function on finite sets in which the domain is larger than the codomain, then that function can never be one-to-one.

"Everybody's" definition of a full moon as the second of two full months in a calendar month is, of course, wrong. This silly definition came from an error in the March 1946 issue of *Sky & Telescope* magazine. It was only in 1999 that that error was discovered. But by then, everyone (including the game Trivial Pursuit) adopted this definition. I forget what the original definition (pre-1946) was.

Proof that Hom($\mathcal{V}$, $\mathcal{W}$) is a vector space:

Step 1: Every vector in Hom($\mathcal{V}$, $\mathcal{W}$), which is a linear transformation from m-dimensional vector space $\mathcal{V}$ to n-dimensional vector space $\mathcal{W}$, can be completely described by an n×m matrix **A** multiplication.

Step 2: Every set of matrices of a given dimension is a vector space.

We've found the full-scale, up-to-your-eyeballs-in-theory weird vector space, Hom($\mathcal{V}$, $\mathcal{W}$). If this were the Goldilocks story, this would be the mama bear.

Two bears to go: the baby, slightly weird bear, and the completely over-the-top papa bear. We'll discover two more vector spaces that have been hiding.

Baby bear first. We defined a vector space as any set "cooking" over some field like the real numbers, the rational numbers, or the complex numbers. Vectors over scalars.

Let's suppose, for a moment, that the scalars are the real numbers. What is perhaps the weirdest set we might choose for the vectors? In the past, we have used:

     geometric arrows from the origin,

     polynomials of degree less than three, $\mathscr{P}_3$,

     5×1 matrices, 5-tuples, "skinny" matrices,

     Freds,

     all 3×4 matrices, and

     all linear transformations between $\mathcal{V}$ and $\mathcal{W}$.

What we have never used is the set right in front of our noses. What if we use $\mathbb{R}$ as both the scalars and the vectors?

When I'm thinking of $\mathbb{R}$ as vectors, I'll use a dark font like this: **3.83**, **−77**, **0**, $\mathbf{\sqrt{4.4}}$ . As scalars, I'll write: 3.83, −77, 0, $\sqrt{4.4}$ . In reality, they are the same old numbers, just thought of in two different ways.

To show that the set of vectors $\mathbb{R}$ over the field of $\mathbb{R}$ is a vector space, I have to check the eight properties.

Vector addition is commutative: $\mathbf{5} + \mathbf{7} = \mathbf{7} + \mathbf{5}$. Yup.

Vector addition is associative: $\mathbf{5} + (\mathbf{7} + \mathbf{8}) = (\mathbf{5} + \mathbf{7}) + \mathbf{8}$. Yup.

Vector addition has an identity: $\mathbf{5} + \mathbf{0} = \mathbf{5}$. Yup.

Vector addition has inverses: $\mathbf{5} + (\mathbf{-5}) = \mathbf{0}$. Yup.

Scalar multiplication is associative: $[(5)(7)]\mathbf{8} = (5)[(7)(\mathbf{8})]$. Yup.

Scalar multiplication has an identity: $(1)(\mathbf{5}) = \mathbf{5}$. Yup.

Scalar multiplication distributes over vector addition:
$$5(\mathbf{7} + \mathbf{8}) = (5)(\mathbf{7}) + (5)(\mathbf{8}). \text{ Yup.}$$

Scalar multiplication distributes over scalar addition:
$$(5 + 7)\mathbf{8} = (5)(\mathbf{8}) + (7)(\mathbf{8}). \text{ Yup.}$$

**Theorem**: Every field (like, for example, $\mathbb{R}$, $\mathbb{Q}$, or $\mathbb{C}$) can be thought of as a vector space over itself.

One last vector space to find. Mr. Papa Bear—the mind-melting, over-the-top, final vector space. See how far you can walk with me.

First step: For any two vector spaces, $\mathcal{V}$ and $\mathcal{W}$, the set of all linear transformations from $\mathcal{V}$ to $\mathcal{W}$, $\text{Hom}(\mathcal{V}, \mathcal{W})$ is a vector space.

Second step: For any field F, such as $\mathbb{R}$, $\mathbb{Q}$, or $\mathbb{C}$, it can be considered as a vector space over itself.

Third step: Take any vector space $\mathcal{V}$ over its scalar field F. Think of scalar field F as a vector space. Consider a linear transformation from vector space $\mathcal{V}$ to vector space F.

⏸ pause . . . It used to take two vector spaces in order to get a homomorphism. Now, any old vector space that you find on the street has a linear transformation built right inside of itself. ▶

Fourth step: A linear transformation from a vector space to its scalars (when considered as a vector space) is called a **linear functional**. So given $\mathcal{V}$ over F, we have the linear functional $\phi: \mathcal{V} \rightarrow F$. Every linear functional is a homomorphism.

Fifth step: For a given vector space $\mathcal{V}$, consider all linear functionals, $\phi : \mathcal{V} \to F$. We could write that as $\text{Hom}(\mathcal{V}, F)$.

Sixth step: So, starting with any vector space $\mathcal{V}$, we have a vector space $\text{Hom}(\mathcal{V}, F)$. This is called the **dual space** of $\mathcal{V}$. The dual space of $\mathcal{V}$ is usually abbreviated by $\mathcal{V}^*$.

⊞ pause . . . The vectors of $\mathcal{V}^*$ are linear functionals from $\mathcal{V}$ to its underlying scalar field. ▶

Seventh step: If $\{v_1, v_2, \ldots, v_m\}$ is a basis for $\mathcal{V}$, we are going to find a basis for $\mathcal{V}^*$.

In order to define the basis $\{T_1, T_2, \ldots, T_m\}$ for $\mathcal{V}^*$, all we have to do is tell what each $T_i$ does to each basis element of $\mathcal{V}$. (This is true by the theorem we proved on page 267: "The action of $T$ on $\mathcal{V}$ is completely determined by what $T$ does to the basis of $\mathcal{V}$.")

We define $T_1$ by          $T_1(v_1) = 1, T_1(v_2) = 0, \ldots, T_1(v_m) = 0$
We define $T_2$ by          $T_2(v_1) = 0, T_2(v_2) = 1, \ldots, T_2(v_m) = 0$

. . .

We define $T_m$ by         $T_m(v_1) = 0, T_m(v_2) = 0, \ldots, T_m(v_m) = 1$

In other words, $T_i(v_j) = \delta_{ij}$
                                       where $\delta_{ij}$ is the Kronecker delta (page 88).
I will show that this definition really is a basis for $\mathcal{V}^*$, but first. . . .

⊞ pause . . . A short overview of the properties of the sigma notation.

First property:          $5 \sum\limits_{i=1}^{3} a_i = \sum\limits_{i=1}^{3} 5a_i$

means $5(a_1 + a_2 + a_3) = 5a_1 + 5a_2 + 5a_3$
So the first property is the distributive law in disguise.

Second property:     $\sum\limits_{i=1}^{n} \sum\limits_{j=1}^{p} b_{ij} = \sum\limits_{j=1}^{p} \sum\limits_{i=1}^{n} b_{ij}$

We are just rearranging the order in which we do the addition, so this is just the commutative property of addition in disguise. ▶

Now to show that the $T_i$, as defined, do form a basis for $\mathcal{V}^*$.

$\{T_1, T_2, \ldots, T_m\}$ is a basis iff the span of $\{T_1, T_2, \ldots, T_m\} = \mathcal{V}^*$ and the $T_i$ are linearly independent.

First, to show that the span of $\{T_1, T_2, \ldots, T_m\} = \mathcal{V}^*$, which means that every $T \in \mathcal{V}^*$ could be expressed as a linear combination of the $T_i$. There are six obvious reasons I could use in the proof.

    (I)  the definition of $T_i$        $T_i(v_j) = \delta_{ij}$

    (II)  since the $v_i$ are a basis, we can say, for any $v \in \mathcal{V}$, $v = \sum_{i=1}^{m} r_i v_i$ for some scalars $r_i$

    (III)  Since for any $T$ in $\mathcal{V}^*$, $T$ being a linear transformation from $\mathcal{V}$ to $\mathbb{R}$, for each i we have $T(v_i) = s_i$ where $s_i \in \mathbb{R}$

    (IV)  the first property of sigma notation     $5\sum_{i=1}^{3} a_i = \sum_{i=1}^{3} 5a_i$

    (V)  the second property of sigma notation   $\sum_{i=1}^{n}\sum_{j=1}^{p} b_{ij} = \sum_{j=1}^{p}\sum_{i=1}^{n} b_{ij}$

    (VI)  the definition of a linear transformation   If $u, v \in \mathcal{V}$, then

        (1) $T(u + v) = T(u) + T(v)$

        (2) $T(ru) = rT(u)$   for any $r \in \mathbb{R}$

---

*Your Turn to Play*

For each step in this proof that the span of $\{T_1, T_2, \ldots, T_m\} = \mathcal{V}^*$, supply a reason from the list above.

Line 1     For any $v \in \mathcal{V}$, $T(v) = T\left(\sum_{i=1}^{m} r_i v_i\right)$

Line 2              $= \sum_{i=1}^{m} r_i\, T(v_i)$

Line 3              $= \sum_{i=1}^{m} r_i s_i$

Line 4              $= \sum_{i=1}^{m} s_i\left(\sum_{j=1}^{m} r_j \delta_{ij}\right)$   This line is true because
$\sum_{j=1}^{m} r_j \delta_{ij} = 0r_1 + 0r_2 + \ldots + 0r_{i-1} + 1r_i + 0r_{i+1} + \ldots + 0r_m$ which equals $r_i$.

Line 5          $= \sum\limits_{i=1}^{m} s_i \sum\limits_{j=1}^{m} r_j T_i(v_i)$

Line 6          $= \sum\limits_{i=1}^{m} s_i \sum\limits_{j=1}^{m} T_i(r_j v_j)$

Line 7          $= \sum\limits_{i=1}^{m} s_i T_i( \sum\limits_{j=1}^{m} (r_j v_j) )$

Line 8          $= \sum\limits_{i=1}^{m} s_i T_i(v)$

which shows that every T can be expressed as a linear combination of the $T_i$.

## ......COMPLETE SOLUTIONS.......

Reason for Line 1          (II) since the $v_i$ are a basis, we can say, for any $v \in \mathcal{V}$, $v = \sum\limits_{i=1}^{m} r_i v_i$ for some scalars $r_i$

Reason for Line 2          (VI) the definition of a linear transformation   If $u, v \in \mathcal{V}$, then

      (1)  $T(u+v) = T(u) + T(v)$

      (2)  $T(ru) = rT(u)$   for any $r \in \mathbb{R}$

      Some people might think the reason for Line 2 might be (IV) the first property of sigma notation, but it isn't. $T ( \sum\limits_{i=1}^{m} r_i v_i )$ doesn't mean T *times* $\sum\limits_{i=1}^{m} r_i v_i$.

Reason for Line 3          (III)  Since for any T in $\mathcal{V}^*$, T being a linear transformation from $\mathcal{V}$ to $\mathbb{R}$, for each i we have $T(v_i) = s_i$ where $s_i \in \mathbb{R}$

Reason for Line 4 was given in the proof itself

Reason for Line 5          (I)  the definition of $T_i$          $T_i(v_j) = \delta_{ij}$

Reason for Line 6          (VI)(2)  the definition of a linear transformation

Reason for Line 7          (VI)(1)  the definition of a linear transformation

Reason for Line 8          (II)  since the $v_i$ are a basis. . . .

Some notes about the proof:

♪#1: It took me a little longer than two hours to create those eight lines. There were a lot of false starts—a lot of sheets of paper were tossed into the trash. I was starting with the standard definition of a basis for $\mathcal{V}^*$, namely, $T_i(v_j) = \delta_{ij}$. The first three lines of the proof came easily:

$$T(v) \;=\; T(\sum_{i=1}^{m} r_i v_i) \;=\; \sum_{i=1}^{m} r_i T(v_i) \;=\; \sum_{i=1}^{m} r_i s_i$$

but then I was stuck.

How did I get Line 4:     $= \displaystyle\sum_{i=1}^{m} s_i (\sum_{j=1}^{m} r_j \delta_{ij})$   ?

Here's the trick I used. I started at the other end of the proof. I knew I wanted to show that $T(v)$ was some linear combination of the $T_i$, so I wrote the last line: $= \displaystyle\sum_{i=1}^{m} s_i\, T_i(v)$ and then began to work ꙅbɿɒwʞɔɒd.

Going from Line 5 to Line 4 was easy and natural. It was just using the definition of the $T_i$, viz., $T_i(v_j) = \delta_{ij}$.

♪#2: In elementary school, in high school, in calculus, the students who seemed to make the best grades were the good little tape recorder students.

In elementary school, they memorized the names of all the birds that the teacher was talking about.

In high school, they memorized $\tan 2\theta = \dfrac{2 \tan \theta}{1 - \tan^2 \theta}$

In calculus, they committed to memory the formula for the derivative of the arc hyperbolic tangent function: $\dfrac{d}{dx} \tanh^{-1} x \;=\; \dfrac{1}{1 - x^2}$

But memorizing isn't the same as thinking. The stuff you use often enough will automatically memorize itself.* I use paper and pencils to help me remember things. Back on page 279, when I mentioned the population of Lima, that wasn't something I memorized in elementary school. I have books in which that information is stored.

---

* You should have fully committed to memory the name of your spouse, the names of your kids, and which hand you use to hold your toothbrush.

Even this linear algebra stuff . . . when I'm confronted with a system of linear equations $Ax = b$, where the $b$ can change from day to day, I think of "Love You" and then of $LU$ and then of a picture: Then I go digging through the book to find that picture so that I can recall how to do it.

Linear algebra may be the turning point in which the good little tape recorder students take their places behind the individuals for whom cognition takes precedence over regurgitation.*

Finally, if we are to show that the $T_i$ form a basis for $V^*$, we have to show that the $T_i$ are linearly independent.

Here's a perfect time to show the difference between how mathematicians think when no one is watching and how they write when they want to impress the world.

| How I Think | How to Impress and Obscure |
|---|---|
| $T_7$ couldn't be a linear combination of the other $T_i$'s because $T_7(v_7)$ is equal to one, and all of the other $T_i(v_7)$'s are equal to zero. | Let $\{v_1, v_2, \ldots, v_m\}$ be a basis for $V$, and $\{T_1, T_2, \ldots, T_m\}$ be a basis for $V^*$. |

In order to establish that $\{T_1, T_2, \ldots, T_m\}$ is linearly independent, we must show that there must exist an i, where $1 \le i \le m$, for which there do not exist scalars $r_1, r_2, \ldots, r_m$, such that for all $v \in V$,

$$T_i(v) = \sum_{\substack{j=1 \\ j \ne i}}^{m} r_j T_j(v). \qquad \text{(equation 1)}$$

Assume that such scalars exist. We shall seek a contradiction.

Fix i, and set $v = v_i$. Equation 1 becomes

$$T_i(v_i) = \sum_{\substack{j=1 \\ j \ne i}}^{m} r_j T_j(v_i). \qquad \text{(equation 2)}$$

Using the fact that $T_i(v_j) = \delta_{ij}$, we obtain

---

* Please delete this sentence if your instructor requires you to memorize tons of stuff and has closed book exams. Note that when you get out in the real world, your boss won't require that you do linear algebra with your book closed.

$$1 = \sum_{\substack{j=1 \\ j \neq i}}^{m} r_j(0) \qquad \text{(equation 3)}$$

which simplifies to

$$1 = 0 \qquad \text{(equation 4)}$$

which is a contradiction

which implies that our assumption that scalars

$r_1, r_2, \ldots, r_m$ exist is false

which establishes that $\{T_1, T_2, \ldots, T_m\}$ is linearly independent.

Quod Erat Demonstrandum

**Hey! You said that the last vector space would be a "mind-melting, over-the-top" vector space. I, your reader, feel very proud. Except for that junk in the right column, I think I almost understand these seven steps and this dual space $\mathcal{V}^*$ thing.**

**You start with a vector space. You consider its scalars as another vector space. You consider all the linear transformations from the vectors in the vector space to the underlying scalars. And that forms a new vector space which you call the dual space to the original vector space.**

**My brain might have become a little soft, but it isn't melted. HaHaHaHaHa.**

Eighth step: . . .

**Hold it! Stop! What in blazes are you doing!!??**

You told me your mind wasn't melted yet.

**Do you HAVE to do that?**

I try to keep my promises.

Eighth step: $\mathcal{V}^{**}$

**AAAAAaaaaaauuuuuugggggghhhhh!**

Okay. I understand your pain. You have my permission to skip to the Cities at this point. I won't mention $\mathcal{V}^{**}$ in the Cities or in Chapter 4. But on the next page I would like to talk to myself about the **second dual** of $\mathcal{V}$. I trust that you won't be listening.

$\mathcal{V}$* is the set of all linear functionals on $\mathcal{V}$.
$\mathcal{V}$** is the set of all linear functionals on $\mathcal{V}$*.

$\mathcal{V}$* = Hom($\mathcal{V}$, $\mathbb{R}$) is a vector space.
$\mathcal{V}$** = Hom($\mathcal{V}$*, $\mathbb{R}$) is a vector space.

Vectors in $\mathcal{V}$ I think of as nuts.
Linear functionals on $\mathcal{V}$ are like monkeys that eat the nuts and spit out real numbers.

Linear functionals on $\mathcal{V}$* are like lions that eat monkeys and spit out real numbers.

T's are lion food. T:$\mathcal{V}$→$\mathbb{R}$. Somehow, I've got to turn every T into a real number. That's the work of the lion. That's what the linear functionals in Hom($\mathcal{V}$*, $\mathbb{R}$) do.

I'm used to seeing T(v)—monkey eats a nut and gives me a real number. T is the function and v is the thing that the function chews on. Now I need something to chew on the T's.

The nut is going to eat the lion! Define a function v which eats T's and spits out real numbers. Official definition: v(T) = T(v).

So the elements of $\mathcal{V}$** are nuts (vectors) that have transmogrified* themselves into functions.

$\mathcal{V}$** is isomorphic to $\mathcal{V}$.

---

✳ trans-MOG-reh-fied  To strangely change shape.

1. If T and T′ are linear transformations from $\mathcal{V}$ to $\mathcal{V}$, will it always be true that $T(T'(v)) = T'(T(v))$?

2. Way back in Chapter 2¾, we defined a function from pairs of vectors, u and v in $\mathcal{V}$ to the scalar field associated with $\mathcal{V}$. What was that function called?

3. If $T:\mathcal{V} \to \mathcal{W}$ is a linear transformation, then it is a function whose domain is $\mathcal{V}$ and whose codomain is $\mathcal{W}$ with the special property that if u, v $\in \mathcal{V}$ and r and s are real numbers, then $T( ru + sv ) = rT(u) + sT(v)$.

    If, instead, T is a linear operator on $\mathcal{V}$, what would T's domain and codomain be?

4. Suppose $\mathcal{V}$ is an inner product space. Fix a particular vector, say w, in $\mathcal{V}$. Define $T(v) = <v , w>$ for every v $\in \mathcal{V}$. This T is not a linear operator since its codomain is not $\mathcal{V}$. T's codomain is the field of scalars associated with $\mathcal{V}$. T is a special kind of linear transformation called a linear _____.

        [fill in the blank]

5. Show that the T defined in the previous question is a linear transformation. Namely, show that if u, v $\in \mathcal{V}$ and r and s are scalars, then $T( ru + sv ) = rT(u) + sT(v)$.

6. One central part of this chapter was that every linear transformation on finite dimensional vector spaces, $T:\mathcal{V} \to \mathcal{W}$, could be expressed as matrix multiplication. We jumped back and forth between 🪑 and 🪑 .

    Another grand unification ties together linear operators and inner products on finite dimensional vectors spaces: **Theorem**: Every linear operator $T:\mathcal{V} \to \mathbb{R}$ can be expressed as $T(v) = <v , w>$ for some w $\in \mathcal{V}$. Everything all fits together now.

    Here's the proof of this theorem. Supply a reason for each line.

    We are given a linear operator $T:\mathcal{V} \to \mathbb{R}$ defined on a finite dimensional vector space $\mathcal{V}$.

Line 1      We know there exists a finite basis for $\mathcal{V}$.

Line 2      Turn that basis into an orthonormal basis.

           Call it $\{v_1, v_2, \ldots, v_m\}$.

Line 3      Let w $\in \mathcal{V}$ be the vector defined by $\sum_{i=1}^{m} T(v_i)v_i$

Line 4       Let T′ be the linear transformation defined by

$$T'(v) = <v, w> \text{ for every } v \in \mathcal{V}.$$

Line 5       $T'(v_j) = <v_j, w>$

Line 6            $= <v_j, \sum_{i=1}^{m} T(v_i)v_i>$

Line 7            $= \sum_{i=1}^{m} T(v_i) <v_j, v_i>$

Line 8            $= \sum_{i=1}^{m} T(v_i) \, \delta_{ji}$

Line 9            $= T(v_j)$

Line 10       T and T′ are the same linear transformation.

                                                ■

### *answers*

1. There are many ways to show that the answer is no.

     One way: Suppose that $\mathcal{V}$ is $\mathbb{R}^2$, that T is "reflect through the x-axis" and T′ is "rotate counterclockwise by 90°."

     If we drew the letter $\mathcal{R}$ on the graph and computed $T'(\mathcal{R})$ we would get ℞. And $T(T'(\mathcal{R}))$ would be ℞.

     In contrast, $T(\mathcal{R})$ would be ℞, and $T'(T(\mathcal{R}))$ would be ℞.

     A second way to show that T and T′ don't commute is to recall that linear transformations are faithfully mirrored (are isomorphic) to multiplication by matrices. Since matrix multiplication is not commutative (we showed that in problem 3 on page 56), linear transformations are not commutative.

2. It was called the inner product, <u, v >. Any vector space with a scalar product defined was called an inner product space.

3. A linear operator maps a vector space to itself. $T: \mathcal{V} \rightarrow \mathcal{V}$. Its domain and codomain would both be $\mathcal{V}$.

4. A linear transformation from a vector space to its scalars (when considered as a vector space) is called a linear functional.

5. Fix a particular $w \in \mathcal{V}$. Then

    $T( ru + sv ) = <ru + sv, w>$             by definition of T

$$= r< u, w > + s< v, w >$$  by the linearity property of inner product, given on page 201.

$$= rT(u) + sT(v)$$  by definition of T

6.

Reason for Line 1    We are given that $\mathcal{V}$ is finite dimensional.

Reason for Line 2    The dance steps for the Gram-Schmidt orthogonalization will turn any basis into an orthogonal one.

Reason for Line 3    Since $\mathcal{V}$ is a vector space, it is, by definition, closed under scalar multiplication and vector addition.

Reason for Line 4    By question 5, we showed this definition will give us a linear transformation.

Reason for Line 5    The definition of T′ from line 4.

Reason for Line 6    The definition of w from line 3.

Reason for Line 7    By the linearity property of inner product, given on page 201.

Reason for Line 8    If the $v_i$ are orthogonal, $< v_j, v_i > = \delta_{ji}$ where $\delta_{ji}$ is the Kronecker delta (page 88).

Reason for Line 9    Definition of $\delta_{ji}$.

Reason for Line 10    The theorem (from page 267) which stated that the action of T on $\mathcal{V}$ is completely determined by what T does to the basis of $\mathcal{V}$.

---

## Ranger

1. Let's look at the vector space ⌨ of all n×m matrices A. What is the dimension of this space? (This is a question that could have been asked several chapters earlier. We ask it now, because we might (hint! hint!) need it sometime in the near future.)

2. Define $T:\mathbb{R}^2 \to \mathbb{R}^2$ by $T((x, y)) = (x + 6, y + 6)$.

   Is T a function?

   Is T a linear transformation?

   Is T a linear operator?

3. Define $T:\mathbb{R}^3 \to \mathbb{R}^3$ by $T((x, y, z)) = (y, z, x)$. It should be noted that some books abbreviate $T((x, y, z))$ by $T(x, y, z)$.

    i) Show that T is a linear transformation.

    ii) $T(T((5, 6, 7))) = ?$

    iii) If we abbreviate $T(T((5, 6, 7)))$ by $T^2((5, 6, 7))$, what is another really simple name for $T^3$?

    The answer isn't just $T(T(T((x, y, z))))$.

***answers***

1. To find the dimension of the vector space of all n×m matrices A, we can find a basis and then count the number of elements in that basis. One element of that basis could have a 1 in the row 1, column 1 position and zeros everywhere else. A second element could have a 1 in the row 1, column 2 position and zeros everywhere else. Etc. There are nm such matrices. They are linearly independent—no linear combination of any of them could equal another one. They span the space of all n×m matrices. The vector space of all n×m matrices has dimension nm.

2. To ask if T is a function is to ask if every element of $\mathbb{R}^2$ has a unique image in $\mathbb{R}^2$. Can we perform $T((x, y)) = (x + 6, y + 6)$ on every point $(x, y)$ in $\mathbb{R}^2$? Yes. For example, $T((19, -\pi)) = (25, -\pi + 6)$. Will we always get a unique answer? Yes. Adding six to each coordinate of $(x, y)$ will give us just one answer.

    One property all linear transformations possess is that $T(0) = 0$ where the first $0$ is the zero vector in the domain of T, and the second $0$ is the zero vector in the codomain of T (proved in problem 4 on page 262). Since $T((0, 0)) = (6, 6) \neq (0, 0)$, T is not a linear transformation.

    Since all linear operators are linear transformations, *a fortiori*, T is not a linear operator. The Latin phrase *a fortiori* ( The *a* is pronounced like the *a* in st*a*r. four-ti-OH-ree) means "even more obviously true." If no six people can drink all the water in the ocean, then, *a fortiori*, no married couple can do that.

    The more common logical phrase is a priori. (It doesn't have to be italicized like *a fortiori*, since a priori is now considered an English expression rather than a foreign one.) Reasoning is considered a priori if it goes from the general law to a specific example. If we know that no six people can drink all the water in the ocean, then by a priori reasoning, we know that Dave and Jane and their four kids can't do that. (The a in a priori is a "long a." My elementary school teacher told me that long vowels are vowels that say their own name. She had us memorize: Mate, Meet, Mite, Mote, and Mute. There are a half-dozen standard ways to pronounce *priori*. I like pry-OR-ee, but almost anything you try will be okay.)

A less common logical phrase, but still not italicized, is a posteriori reasoning. (The a in a posteriori is a long a. poe-STEER-ee-OR-i, the i being a long i.) This is the kind of reasoning that scientists use in going from experiments and observations to formulating their hypotheses, theories, and laws. The conclusions of a posteriori reasoning are always only probable conclusions, not certain ones. Newton's law of gravitation was supported by centuries of observations. But in the early 1900s, some guy who used to work in a patent office somewhere in Europe showed that Newton's law of gravitation was not true.

In everyday life, we use a posteriori reasoning to go from our known I've-seen-it-with-my-own-eyes facts to their probable causes.

He smoked 292,000 cigarettes (two packs/day for 20 years), and because of that got lung cancer. (a posteriori)

He spent $43,800 ($6/day for 20 years) on cigarettes, and therefore someone (or ones) received $43,800 from him. (a priori)

He died, and therefore, was no longer buying cigarettes. (a fortiori)

3. i) We need to show that if $u$, $v \in \mathbb{R}^3$ and r and s are scalars, then $T(ru + sv) = rT(u) + sT(v)$.

I'm going to show that this is true for (1, 2, 3) and (4, 5, 6) and real numbers 7 and 8. When I'm done, go back and replace (1, 2, 3) by $(x_1, y_1, z_1)$, replace (4, 5, 6) by $(x_2, y_2, z_2)$, replace 7 and 8 by $r_1$ and $r_2$. Then all the fussbudgets* will be happy.

| | |
|---|---|
| $T(7(1, 2, 3) + 8(4, 5, 6)) = T((39, 54, 69))$ | scalar multiplication and vector addition in $\mathbb{R}^3$ |
| $= (54, 69, 39)$ | definition of T |
| $= 7(2, 3, 1) + 8(5, 6, 4)$ | scalar multiplication and vector addition in $\mathbb{R}^3$ |
| $= 7T(1, 2, 3) + 8T(4, 5, 6)$ | definition of T |

---

✶ A fussbudget is a person who loves to find fault. The word came into our language sometime between 1900 and 1905.

3. ii)  This can make you dizzy as it spins around.

T(T((5, 6, 7))) = T((6, 7, 5)) = (7, 5, 6)

3. iii)  $T^3((x, y, z))$ = (x, y, z) for all x, y, z ∈ ℝ, so $T^3$ is the identity linear transformation.

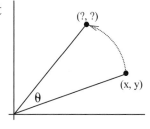

---

**Wanette**

---

1.  The number of elements in a basis of a vector space is the dimension of that vector space.  (The Nice theorem said that if $\mathcal{V}$ has a basis with n elements in it, then every basis of $\mathcal{V}$ has n elements.  We branded each ~~cow~~ vector space with a number.  )

If $\mathcal{V}$ is an m-dimensional vector space, what is the dimension of $\mathcal{V}^*$?

2.  If T: $\mathcal{V} \to \mathcal{W}$ is a linear transformation, is T a 1-1 function?  (The definition of a one-to-one function is given in the Quick Review of Functions on page 300.)

3.  Let T: $\mathbb{R}^2 \to \mathbb{R}^2$ be the linear transformation that rotates each point (x, y) in $\mathbb{R}^2$ counterclockwise around the origin by an angle of θ.

Then T((x, y)) = ?

It has been shown that, in general, students who don't attempt to solve these Cities problems on their own, but just look at the questions and then look at the answers, don't learn as much.  Lazy implies ignorant is clearly an a posteriori argument.*

4.  If T is the linear transformation defined in the previous problem, find the associated matrix A.  This is your chance to hop from 🛋 to 🪑 .

*answers*

1.  In the "Seventh step" on page 282, we started with a basis {$v_1$, . . ., $v_m$} for $\mathcal{V}$ and defined as basis {$T_1$, $T_2$, . . ., $T_m$} for $\mathcal{V}^*$.  So Hom($\mathcal{V}$, ℝ) —which is another name for $\mathcal{V}^*$—also has dimension m.

---

★ Ignorant implies lazy is an argument with a special name.  It's called false.

2.  Linear transformations are not necessarily 1-1.  Rotations in the X-Y plane are linear transformations, and they are 1-1.  Two different points, each rotated by 38° will always end up in different places.  But the derivative transformation will take both $4x^3 + 17$ and $4x^3 + 23$ both into the same vector $12x^2$.

   The projection linear transformation $T((a, b, c)) = (a, b)$ will take both (7, 8, 9) and (7, 8, 10) into the vector (7, 8).

3.  Let r be the distance from the origin to each of the two points.

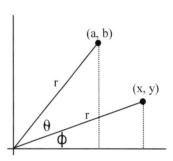

By the definition of cosine: $a/r = \cos(\phi + \theta)$
(*)                                 $x/r = \cos \phi$
By the definition of sine: $b/r = \sin(\phi + \theta)$
(**)                                $y/r = \sin \phi$
Using trig identities:
$$a/r = \cos \phi \cos \theta - \sin \phi \sin \theta$$
$$b/r = \sin \phi \cos \theta + \cos \phi \sin \theta$$

Stuffing (*) and (**) into these equations to eliminate the $\phi$'s:
$$a/r = x/r \cos \theta - y/r \sin \theta \quad \Longrightarrow \quad a = x \cos \theta - y \sin \theta$$
$$b/r = y/r \cos \theta + x/r \sin \theta \quad \Longrightarrow \quad b = x \sin \theta + y \cos \theta$$

$$T((x, y)) = (a, b) = (x \cos \theta - y \sin \theta, \ x \sin \theta + y \cos \theta)$$

4.  The matrix A that we find that is associated with T will depend on the basis chosen for $\mathbb{R}^2$.  If you choose $\{(397.7, \frac{3}{4}), (16\pi, \sqrt{0.31})\}$ as the basis, you might consider writing the scripts for horror movies.  I'm choosing (1, 0) and (0, 1), which are affectionately known as $\mathbf{e}_1$ and $\mathbf{e}_2$.
From the previous question we have:
$$T((x, y)) = (x \cos \theta - y \sin \theta, \ x \sin \theta + y \cos \theta).$$
$$T((1, 0)) = (\cos \theta, \sin \theta) \ \text{ and } \ T((0, 1)) = (-\sin \theta, \cos \theta)$$

Making these the columns of A, we have  $A = \begin{pmatrix} \cos \theta & -\sin \theta \\ \sin \theta & \cos \theta \end{pmatrix}$

## Inverness

1. Hom($\mathcal{V}$, $\mathcal{W}$) is the vector space of all linear transformations from $\mathcal{V}$ to $\mathcal{W}$. (Everybody knows that.) Suppose that the dimension of $\mathcal{V}$ is m, and the dimension of $\mathcal{W}$ is n. What is the dimension of Hom($\mathcal{V}$, $\mathcal{W}$)?

2. Define T: $\mathcal{P}_4 \rightarrow \mathcal{P}_5$ by T(p(x)) = xp(x) where p(x) ∈ $\mathcal{P}_4$. Find T($3 + 5x^2 + 80x^4$).

3. Is the function T as defined in the previous question a linear transformation? If so, then prove it. If not, show that it is not.

*answers*

1. With $\mathfrak{S}$trip and $\mathfrak{S}$tuff we went from 🪑 to 🪑—from linear transformations T in Hom($\mathcal{V}$, $\mathcal{W}$)—to the vector space of n×m matrices **A**.

   So the question boils down to asking what is the dimension of an n×m matrix. We did that two Cities ago (in Ranger #1). It's nm.

2. T($3 + 5x^2 + 80x^4$) = $3x + 5x^3 + 80x^5$

3. Let p(x) and p′(x) be any vectors in $\mathcal{P}_4$ and let r and s be any scalars.

$$T( rp(x) + sp'(x) ) = x[rp(x) + sp'(x)] \quad \text{by definition of T}$$
$$= rxp(x) + sxp'(x) \quad \text{by algebra}$$
$$= r\,T(p(x)) + s\,T(p'(x)) \quad \text{by definition of T}$$

and this establishes that T is a linear transformation.

## John Day

1. Hom($\mathcal{V}$, $\mathcal{W}$) is the vector space of all linear transformations from $\mathcal{V}$ to $\mathcal{W}$. But Hom($\mathcal{V}$, $\mathcal{W}$) has something that regular old vector spaces like $\mathcal{V}$ don't have.

   In $\mathcal{V}$, you could add two vectors: v + v′ and get a vector answer.

   In $\mathcal{V}$, you could multiply a vector by a scalar: rv and get a vector answer.

   In Hom($\mathcal{V}$, $\mathcal{W}$), you could add two vectors: T + T′.

   In Hom($\mathcal{V}$, $\mathcal{W}$), you could multiply a vector by a scalar: rT.

Does the subtraction of linear transformations, $T - T'$, make any sense at all—or is it just nonsense?

2.  Given arbitrary vector spaces $V$ and $W$, will there always be an identity transformation in Hom($V$, $W$)?

3.  Suppose we have Hom($U$, $V$) and Hom($V$, $W$) where $T_1 \in$ Hom($U$, $V$) and $T_2 \in$ Hom($V$, $W$).

We can define the product of $T_1 T_2$ to be $T_1(T_2(v))$ where $T_1(T_2(v))$ is the composition of two functions.

Show that $T_1 T_2$ is a linear transformation.

4.  When you multiply two matrices together, they have to have the correct dimensions in order to make the multiplication possible. If A is 3×5, and we want AB to make sense, B needs to have what dimensions?

5.  All the stuff from matrix multiplication seems to carry over nicely to this multiplication of linear transformations. There are two distributive laws and two associative laws:

    i)  $T(T' + T'') = TT' + TT''$

    ii)  $(T + T')T'' = TT'' + T'T''$

    iii)  $T(T'T'') = (TT')T''$

    iv)  $r(TT') = (rT)T' = T(rT')$

Actually, three associative laws if you count the last one (iv) as two laws.

To prove iii), we must show that $(T(T'T''))(v) = ((TT')T'')(v)$ for every $v$ in $V$. I'll write the proof; you supply the reasons. One possible reason to use is the definition of the product of two linear transformations as given in question 1 above where we defined $(TT')(v)$ as $T(T'(v))$.

$$(T(T'T''))(v) = T((T'T'')(v)) \qquad \underline{\qquad ? \qquad}$$
$$= T(T'(T''(v))) \qquad \underline{\qquad ? \qquad}$$
$$= ((TT')T''(v)) \qquad \underline{\qquad ? \qquad}$$
$$= ((TT')T'')(v) \qquad \underline{\qquad ? \qquad}$$

***answers***

1.  Back in the City of Teller, problem 5 in chapter 1½, we defined the subtraction of matrices as $A - B = A + (-1)B$.

We can do virtually the same thing with linear transformations.

Definition: $T - T' = T + (-1)T'$.

2. Suppose $\mathcal{V}$ is the vector space of polynomials of degree 800 or less. Suppose $\mathcal{W}$ is the vector space of all 2×3 matrices. An identity transformation on $\mathcal{P}_{800}$ would map $2 + 9x^3 - 5x^7 + x^{624}$ to $2 + 9x^3 - 5x^7 + x^{624}$. That's not a 2×3 matrix. An identity transformation on $\mathcal{V}$ maps vectors in $\mathcal{V}$ to vectors in $\mathcal{V}$ and not to any other space.

3. For $T_1T_2$ to be a linear transformation, we need to show $(T_1T_2)(\,r\mathsf{u} + s\mathsf{v}\,) = r(T_1T_2)(\mathsf{u}) + s(T_1T_2)(\mathsf{v})$ for all $\mathsf{u}, \mathsf{v} \in \mathcal{V}$ and $r, s \in \mathbb{R}$. (This was the alternative definition given on the third page of the chapter in problem one of the *Your Turn to Play* .

This is what I've called a meat-grinder proof.

$$
\begin{aligned}
(T_1T_2)(\,r\mathsf{u} + s\mathsf{v}\,) &= T_1(\,T_2(\,r\mathsf{u} + s\mathsf{v}\,)) && \text{by definition of } T_1T_2 \\
&= T_1((\,rT_2(\mathsf{u}) + sT_2(\mathsf{v})) && \text{since } T_2 \text{ is a linear transformation} \\
&= r\,T_1(\,T_2(\mathsf{u})) + s\,T_1(T_2(\mathsf{v})) && \text{since } T_1 \text{ is a linear transformation} \\
&= r(T_1T_2)(\mathsf{u}) + s(T_1T_2)(\mathsf{v}) && \text{by definition of } T_1T_2
\end{aligned}
$$

4. $5 \times ?$

5. All four reasons are the same: the definition of the product of linear transformations. In symbols: $(TT')(\mathsf{v}) = T(T'(\mathsf{v}))$, which in words is: *The product of two linear transformations is defined to equal the composition of the two functions.*

---

## Kelso

Let T, T′ be linear operators on vector space $\mathcal{V}$. (Linear operator was defined on page 262.) We are going to use the definition of TT′ which was given in the previous City (John Day).

We know that, in general, TT′ ≠ T′T (from the first question in the first City of this chapter). But I had the feeling that if TT′ were the identity transformation,[*] that would be enough to ensure that TT′ = T′T.

In other words, for all $\mathsf{v} \in \mathcal{V}$, I wanted to show
$$(TT')(\mathsf{v}) = (T'T)(\mathsf{v}).$$

At first, I thought that the proof would be like ![sketch]. I spent an hour and a half in my office trying to prove it.

---

[*] That means that $(TT')(\mathsf{v}) = \mathsf{v}$ for all $\mathsf{v} \in \mathcal{V}$.

Then at the breakfast table I filled another couple of pages.

After a nap in the afternoon, I reached for a clipboard and "nailed" the proof.

While I was working on the proof, I kept wondering whether the claim was true. Not knowing whether it is true, makes it a lot harder to find a proof. There have been many times in the history of mathematics in which the proof of a conjecture took decades to find, but once someone had established its truth, several other different proofs were quickly found.

Now I ask you, would it be fair to give you this theorem to prove:

**Theorem**: If $T$, $T'$ $\in$ Hom($\mathcal{V}$, $\mathcal{V}$), and if $TT'$ is the identity homomorphism, then $T'T$ is also the identity homomorphism.

☐ *Yes! What's a little effort on my part? As somebody once said, "That which doesn't kill me, will make me stronger."*

Friedrich Nietzsche said something like that. But he was wrong. If you talk with soldiers who have been in heavy, prolonged combat, they often report that they bear psychological scars for the rest of their lives.

But he also wrote, "The doer alone learns." By checking the yes box, you have my admiration. Without the hints given below, go ahead and tackle this proof.

☐ *No! Tremulous is my middle name. I prefer*  *to* .

Okay. We'll chop it up into bite-sized pieces. Someone, not Nietzsche, once said that the best way to eat an elephant is a bite at a time.

1. We are given that $TT'$ is the identity function. $(TT')(v) = v$ for every $v \in \mathcal{V}$. Show that $T'$ is a 1-1 function.

## *Quick Review of Functions*

f:A→B is a function whose name is f, whose domain is A and whose codomain is B.

f is a rule which assigns to each element of A exactly one element of B.

f is 1-1 ("one-to-one") if no two elements of A are assigned to the same element of B.

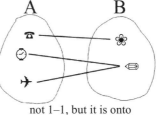

not 1–1, but it is onto

f is onto B if every element of B is the image of at least one element of A.

2. We now know that TT′ is the identity function and that T′ is 1-1. Let $\{v_1, v_2, \ldots, v_m\}$ be a basis for $\mathcal{V}$. In this question we are going to show that $\{T'(v_1), T'(v_2), \ldots, T'(v_m)\}$ is also a basis for $\mathcal{V}$. All we need to show is that $\{T'(v_1), T'(v_2), \ldots, T'(v_m)\}$ is linearly independent.* Do it.

3. Show that T′ is onto $\mathcal{V}$.

4. Wait a minute. If T′ is 1-1 (problem 1) and T′ is onto (problem 3), then the

**Proposition about functions**: If f:A→A and g:A→A and f∘g is the identity function and g is 1-1 and onto A, then g∘f must also be the identity function.

would finish the proof. Very nice. Prove this proposition. (With functions the usual notation for "the composition of f with g" is f∘g which is defined as (f∘g)(x) = f(g(x)) for all x in the domain of g.)

———————————

✶  Since T′ is 1-1, we know that if we started with m vectors in the original basis $\{v_1, v_2, \ldots, v_m\}$, we would get m vectors in the set of images $\{T'(v_1), T'(v_2), \ldots, T'(v_m)\}$. In problem 6 of the City of Greasy Corner in Chapter 2, we showed n linearly independent vectors in an n-dimensional space must be a basis.

5. I wondered if all this linear algebra material—linear transformations, bases, vector spaces, etc.—was really necessary. Maybe this theorem was true for all functions, not just linear operators.

I wondered: *If f and g are any two functions, and if f∘g is the identity function, then must g∘f also be the identity function?*

To test that thought, I let f:A→A and g:A→A and let f(g(a)) = a for all a ∈ A. By the first problem in this City, g is 1-1.

I first considered the case in which A was *finite*. If g:A→A and g were 1-1 and A were finite, then g would have to be onto A,* and the **Proposition about functions** in #4 above establishes that g∘ f must also be the identity function.

I then knew that if the conjecture: *If f and g are any two functions, and if f∘g is the identity function, then must g∘f also be the identity function where f:A→A and g:A→A* was going to turn out to be false, it would have to be when A was infinite.

So I let A be the positive real numbers $\mathbb{R}^+$. I was trying to make the conjecture false. I knew that g would have to be 1-1 by problem 1. But I also knew that g couldn't be onto or the conjecture would be true by the proposition about functions in the previous question.

So I needed a g:$\mathbb{R}^+$→$\mathbb{R}^+$ that is 1-1 but not onto.

Find such a g. (This is the kind of puzzle you can think about while walking in the woods.)

*answers*

1. If T′ were not 1-1, then for some u, v ∈ $\mathcal{V}$, where u ≠ v, T′(u) = w and T′(v) = w. Since TT′ is the identity function, T(w) would have to equal both u and v.    The details:  T T′(u) = u

$$\qquad\qquad T\,T'(u) = T(w)$$
$$\qquad\qquad T\,T'(v) = v$$
$$\qquad\qquad T\,T'(v) = T(w)$$

Then T wouldn't be a function. Contradiction. Therefore T′ is 1-1.

---

* For finite sets, if h:C→D is a 1-1 function, then the number of elements in D must be greater than or equal to the number of elements in C with equality occurring only if h is onto.

In our situation, g:A→A where g is 1-1, A must certainly have the same number of elements as A, so g must be onto.

2. If $\{T'(v_1), T'(v_2), \ldots, T'(v_m)\}$ were linearly dependent, then

$$0 = \sum_{i=1}^{m} r_i T'(v_i) \qquad \text{where not all the } r_i \text{ are equal to zero.}$$

$$0 = T'\left(\sum_{i=1}^{m} r_i v_i\right) \qquad \text{since T' is a linear transformation}$$

$$T(0) = TT'\left(\sum_{i=1}^{m} r_i v_i\right) \qquad \text{taking " T " of both sides. Since T is a function, every element in its domain has a unique image.}$$

$$0 = \sum_{i=1}^{m} r_i v_i \qquad \text{The left side is true because } T(0) = 0, \text{ which was proved in the first *Your Turn to Play* of the chapter in problem 4.}$$

The right side is true because TT' is the identity transformation.

But this last line can't be true, since the $v_i$ are linearly independent.

3. Let w be any vector in $\mathcal{V}$. We must find a $v \in \mathcal{V}$ such that $T'(v) = w$.

Since $\{T'(v_1), T'(v_2), \ldots, T'(v_m)\}$ is a basis (from the previous problem), we know that w is a linear combination of $T'(v_1), T'(v_2), \ldots, T'(v_m)$.

$$w = \sum_{i=1}^{m} r_i T'(v_i) \qquad \text{for some scalars } r_i.$$

$$= T'\left(\sum_{i=1}^{m} r_i v_i\right) \qquad \text{since T' is a linear transformation.}$$

$$\text{Let } \sum_{i=1}^{m} r_i v_i = v \qquad \text{and we're done.}$$

4. Given $f(g(a)) = a$ for all $a \in A$ where $f:A \rightarrow A$ and $g:A \rightarrow A$ and g is 1-1 and onto A.

Let a be any element of A. I want to find $g(f(a))$.

$g(f(a)) = g(f(g(b)))$    where $g(b) = a$.   b exists since g is onto.

$= g(b)$        since $f(g(b)) = b$ by the given

$= a$

It took me a whole bowl of cereal and two pieces of toast (20 minutes) to work out these three steps.

5. A function g:$R^+ \to R^+$ that is 1-1 but not onto.

At first, I tried the sine function. That would compress all of $R$ down to [−1, 1] and then if I added 5 to that I would compress all of $R^+$ to [4, 6]. g(x) = sin x + 5. Oops, that's not 1-1 since sin 90° = sin 450°.

Then I tried g(x) = x + 6. This is 1-1 and not onto $R^+$. Now to define f:$R^+ \to R^+$ so that f∘g(x) would equal x for all x in $R^+$. (If I simply used f(x) = x − 6, then f(5) would be −1, which is not in $R^+$. Instead, I defined f by:

$$f(x) = \begin{cases} x - 6 & \text{if } x > 6 \\ 89 & \text{if } x \le 6 \end{cases}$$

With these definitions, f∘g(x) = f(x + 6) = x  for all x in $R^+$, but g∘f(2) = g(89) = 95.

Therefore, g∘f is *not* the identity function. All of this made me feel good. The theorem was true for linear transformations: If T, T' ∈ Hom($V$, $V$), and if TT' is the identity homomorphism, then T'T is also the identity homomorphism, but it wasn't true for just plain old functions.

Betty set Fred down, and they all packed up their picnic supplies. It was time to head back to KITTENS University. They decided to take the path through the Great Woods rather than go back over the Great Lawn. In the early summer afternoon, a walk through the Great Woods was always a treat. The smell of the flowers, the songs of the birds, the shade of the trees—who could ask for anything more?

Their reverie was broken by Fred: "My tummy hurts."

"What did you have to eat today?" Betty asked him.

Fred grinned, shrugged his little shoulders, and was silent.

Betty had been through this so many times in the last five years. Without parents for most of his life, Fred had really missed some of the basics of childhood. That recurring pain in the abdominal region could be appendicitis, gastroenteritis,* food poisoning (which can only occur if you have eaten), urinary infection, intestinal obstruction (which is tough to get if your intestines are empty), peritonitis,** ruptured duodenal ulcers, ileus,*** or carcinoids.****

---

\* Gastroenteritis = irritated digestive tract

\*\* Peritonitis = inflamation of the membrane that covers the tummy, intestines, and other abdominal organs

\*\*\* Ileus = paralyzed intestines (often occurring after abdominal surgery)

\*\*\*\* Carcinoids = tumor in the wall of the intestines or in the lungs—usually small and yellowish (except in the dark)

In Fred's case, Betty made the correct diagnosis: the kid was hungry. Medical students are always taught: *If you hear hoof beats in the city park, think horses, not zebras.*

She knew that if she offered him some food from the picnic basket, he would take the food and thank her (he was always polite), and then put it in his pocket "for later." But she never gave up. She held out a piece of bagel. He took it and thanked her for her thoughtfulness. And put it in his pocket.

Meanwhile, Alexander was having visions of what he would be enjoying when he got back to his apartment.

Neither Betty nor Alexander had noticed a "slight" change in what lay in front of them. Fred did.

"Oh look! Kitties!" Fred exclaimed as he ran ahead to play with them. Before Betty or Alexander could react, Fred was petting them and offering them a piece of bagel. The lions weren't interested. They weren't what you would call vegetarians.

As everyone knows, the wild lions in Kansas only eat monkeys. Fred was in absolutely no danger. He was as safe as a bagel.

Lions thrive on monkeys. With lots of monkeys around, the population of lions will grow year after year. According to Prof. Eldwood's *Lions in the Great Woods of Kansas*, 1848, the population of lions in any particular year depends on two things: how many lions and how many monkeys there were in the previous year. If you look in the appendix of Eldwood's best-seller, you can find his famous equation: Lions in any year = 70% of the lions in the previous year plus 40% of the monkeys in the previous year.

$$( 0.7 \quad 0.4) \begin{pmatrix} \text{lions in 1841} \\ \text{monkeys in 1841} \end{pmatrix} = ( \text{lions in 1842} ) \qquad \text{equation 1}$$

Of course, monkeys are not "as safe as a bagel." The more lions that are running around, the more monkeys disappear. In Prof. Eldwood's inimitable words: Monkeys in any particular year = 120% of the monkeys in the previous year minus 50% of the population of the lions in the previous year. In matrix notation:

$$( -0.5 \quad 1.2 ) \begin{pmatrix} \text{lions in 1841} \\ \text{monkeys in 1841} \end{pmatrix} = ( \text{monkeys in 1842} ) \qquad \text{equation 2}$$

Notice that if there were no lions in the Great Woods, then the population of monkeys would grow by 20% each year.*

Also notice that if there were no monkeys in the Great Woods, then the population of lions would decrease by 30% each year. Soon there wouldn't be any lions for Fred to play with.

Combining equation 1 and equation 2:

$$\begin{pmatrix} 0.7 & 0.4 \\ -0.5 & 1.2 \end{pmatrix} \begin{pmatrix} \text{lions in 1841} \\ \text{monkeys in 1841} \end{pmatrix} = \begin{pmatrix} \text{lions in 1842} \\ \text{monkeys in 1842} \end{pmatrix}$$

Of course, since lions and monkeys don't have calendars, this was true for any year.

$$\begin{pmatrix} 0.7 & 0.4 \\ -0.5 & 1.2 \end{pmatrix} \begin{pmatrix} \text{lions in 1927} \\ \text{monkeys in 1927} \end{pmatrix} = \begin{pmatrix} \text{lions in 1928} \\ \text{monkeys in 1928} \end{pmatrix}$$

For those readers who love to talk about linear algebra at pizza parties, you can call $\begin{pmatrix} 0.7 & 0.4 \\ -0.5 & 1.2 \end{pmatrix}$ a **transition matrix**.

In this chapter, A will be the transition matrix.

This year in the Great Woods there are 100 lions and 1000 monkeys. We will let x = the initial population. $x = (100, 1000)^T$.

---

* In a hundred years, 50 monkeys would become $50(1.2)^{100}$ monkeys which is 50(82,817,975) or approximately 4.1 billion monkeys. The Great Woods would be solid monkeys twenty-six feet deep.

*Your Turn to Play*

1. What does Ax represent?

2. Compute Ax.

3. We know what A(Ax) represents—the populations of the lions and the monkeys two years from now. Does A(Ax) always equal (AA)x?

· · · · · · · **C O M P L E T E   S O L U T I O N S** · · · · · · ·

1. Ax = $\begin{pmatrix} 0.7 & 0.4 \\ -0.5 & 1.2 \end{pmatrix}\begin{pmatrix} 100 \\ 1000 \end{pmatrix}$ = is the number of lions and monkeys in the Great Woods a year from now.

2. Ax = $\begin{pmatrix} 0.7 & 0.4 \\ -0.5 & 1.2 \end{pmatrix}\begin{pmatrix} 100 \\ 1000 \end{pmatrix} = \begin{pmatrix} 470 \\ 1150 \end{pmatrix}$

3. Asking if A(Ax) always equals (AA)x could be easily answered if we knew whether matrix multiplication was always associative. Two months ago (back in question 5 in the *Your Turn to Play* on page 66) you proved that (AB)C = A(BC) by showing that the i-k[th] entry in (AB)C and the i-k[th] entry in A(BC) are equal. It was a long and yucky proof involving subscripts i, j, k, and α. It is quite understandable if you have repressed the memory of that experience. Lions and monkeys are a lot more fun than quadruple subscripts.

    The whole point of this question is that I want to be able to write $A^2(x)$ instead of A(Ax).

This year we have $\begin{pmatrix} 100 \text{ lions} \\ 1000 \text{ monkeys} \end{pmatrix}$      = x

Next year we will have $\begin{pmatrix} 470 \text{ lions} \\ 1150 \text{ monkeys} \end{pmatrix}$      = Ax

Next year there will be lots more lions and monkeys for Fred to play with.

Two years from now ($A^2x$) the lion population will have increased even more, and the monkey population will have leveled off.

$$A^2x = \begin{pmatrix} 789 \text{ lions} \\ 1145 \text{ monkeys} \end{pmatrix}$$

$$A^3x = \begin{pmatrix} 1010 \\ 980 \end{pmatrix}$$

$$A^4x = \begin{pmatrix} 1099 \\ 670 \end{pmatrix}$$

$$A^5x = \begin{pmatrix} 1037 \\ 255 \end{pmatrix}$$

$$A^6x = \begin{pmatrix} 828 \\ \text{zero!} \end{pmatrix}$$  and we have some mighty unhappy lions.

The lions are doomed to extinction in the Great Woods now that the monkeys are gone.

If the lions had gastric bypass surgery so that their appetites were greatly decreased, the transition matrix might have been $\begin{pmatrix} 0.7 & 0.4 \\ -0.1 & 1.2 \end{pmatrix}$ instead of $\begin{pmatrix} 0.7 & 0.4 \\ -0.5 & 1.2 \end{pmatrix}$

Then $A^6x$ would be $\begin{pmatrix} 1988 \\ 2224 \end{pmatrix}$ and $A^8x$ would be $\begin{pmatrix} 2585 \\ 2736 \end{pmatrix}$ and everyone would be living happily ever after, just like in the fairy tales.*

In fact, with this new transition matrix, it wouldn't be that long before the Great Woods would be solid lions and monkeys twenty-six feet deep.

---

★ But, as you know, *Life of Fred: Linear Algebra* is not a work of fiction. This book, with Fred, Alexander, and Betty walking through the Great Woods, is classified by every librarian that uses the Library of Congress system in the QA section. The QA section (mathematics) is solidly nonfiction.

Things get a little tricky if the appetite of the lions is only partially decreased—somewhere between $\begin{pmatrix} 0.7 & 0.4 \\ -0.1 & 1.2 \end{pmatrix}$ and $\begin{pmatrix} 0.7 & 0.4 \\ -0.5 & 1.2 \end{pmatrix}$ —such as $A = \begin{pmatrix} 0.7 & 0.4 \\ -0.18 & 1.2 \end{pmatrix}$

What's going to happen over the years?  When I did all the matrix multiplications, I found:

$$A^6 x = \begin{pmatrix} 1742 \text{ lions} \\ 1660 \text{ monkeys} \end{pmatrix}$$

$$A^7 x = \begin{pmatrix} 1883 \text{ lions} \\ 1678 \text{ monkeys} \end{pmatrix}$$

$$A^8 x = \begin{pmatrix} 1990 \text{ lions} \\ 1675 \text{ monkeys} \end{pmatrix}$$

It is hard to tell what is going to happen twenty years from now.

Alternative A: The monkeys become extinct, and then the lions all die.

Alternative B: The populations continue to grow.

Alternative C: Things sort of level out and reach an equilibrium.

I am really tired of doing all those matrix multiplications.  In the previous *Your Turn to Play* I asked you to do just one matrix multiplication. What needs to be done now is to motivate you, the reader, by having you compute what will happen a century from now with this in-between transition matrix $\begin{pmatrix} 0.7 & 0.4 \\ -0.18 & 1.2 \end{pmatrix}$

Suppose, just suppose . . .

---

*Your Turn to Play*

1. Compute $A^{100} x$.

---

**I, your reader, ask that you don't suppose.  I've got better things to do with my afternoons, evenings, and weekends, than multiply A times A times A times A times A times A times A times A times A times A times A times A times A times A times A times A times A times A times A times A times A times A times A times A times A times A times A times A times A . . . times A times x.  Besides, I'm getting to know you a little better, Mr. Author.  I'm**

*guessing you have a little trick up your sleeve that will make computing* $A^{100}$ *a snap, a breeze, a walk in the park, a moment's exertion.*

In other words, you would like an easy way to compute $A^{100}$.

**Brilliant! I couldn't have expressed that more succinctly.**

You want "easy"?

**Yes.**

I don't have "easy." It's going to take me nine steps to compute $A^{100}$. It is a lot shorter than doing a hundred matrix multiplications. The good news is that six of the steps are really trivial.

The three nontrivial steps will involve: ① solving an equation; ② solving systems of equations like those in Chapter 2 ( $Ax = b$ ☺☺☺... ); and ③ finding the inverse of a matrix, which we did in Chapter 1½.

**I want a ONE-step procedure.**

Okay. Step one: Hire a linear algebraist.

**Real funny. I want a one-step <u>math</u> procedure.**

Okay. But you're going to be sorry. Here goes. I'm going to put the whole chapter into a one-step math procedure. Here it is with a bow on it.

Step 1: To find $A^{100}$, simply compute $PD^{100}P^{-1}$
where $D$ is the diagonal matrix consisting of
the eigenvalues $\lambda_1, \lambda_2, \ldots, \lambda_n$ of $A$, and
where $P$ is the matrix whose columns are the
eigenvectors $v_1, v_2, \ldots, v_n$ which correspond
to the eigenvalues $\lambda_1, \lambda_2, \ldots, \lambda_n$.

The three nontrivial parts of this procedure involve ① solving an equation in order to get the eigenvalues $\lambda_1, \lambda_2, \ldots, \lambda_n$ of $A$, ② solving systems of equations in order to get the eigenvectors $v_1, v_2, \ldots, v_n$, and ③ finding the inverse of $P$ in order to get $P^{-1}$.

**Um. Could you please go a little more slowly? In fact, please bust this thing into the nine steps. But before you begin, tell me how to pronounce "eigen."**

EYE-gin where the *gin* is pronounced like the second syllable in *begin. Eigenvalue* is a half-German and half-English word.

The first step in the Nine Steps to Find $\mathbf{A}^{100}$ will require that you remember something from your second year of high school algebra. You will need to find a **determinant** of a (square) matrix. In Chapter 5 of *Life of Fred: Advanced Algebra Expanded Edition* we solved systems of linear equations using **Cramer's Rule**, and Cramer's Rule used determinants.

**Hey! We've been studying how to solve systems of linear equations all the way through this book. Why haven't you talked about Cramer's Rule?**

Cramer's Rule is kind of cute, but it is a lousy way to solve systems of equations. It is much slower than Gaussian elimination. It was included in *Life of Fred: Advanced Algebra* because we needed a reason to teach how to find determinants. The real reason you learned determinants in high school was for today—for the first step in the Nine Steps to Find $\mathbf{A}^{100}$.

**Your phrase, "learned determinants," is a bit overreaching. I don't think I could tell a determinant from a dandelion at this point. If you want me to get through the first step in the Nine Steps to Find $\mathbf{A}^{100}$, you might consider doing a little overview of determinants.**

Good thought.

## Overview of Determinants

A determinant takes a square ($n \times n$) matrix and turns it into a number. Start with a matrix like $\begin{pmatrix} 6 & 7 \\ 2 & 9 \end{pmatrix}$ and its determinant is $\begin{vmatrix} 6 & 7 \\ 2 & 9 \end{vmatrix}$ which equals $(6)(9) - (2)(7) = 40$.

For $2 \times 2$ determinants, it is the product of the main diagonal minus the product of the other two numbers: $\begin{vmatrix} 6 & 7 \\ 2 & 9 \end{vmatrix}$

$\begin{vmatrix} 4 & -5 \\ 8 & 11 \end{vmatrix} = (4)(11) - (8)(-5) = 84$

For 3×3 determinants, you have your choice of two methods: (1) what I call the "hairnet" and (2) expansion by minors.

The hairnet: $\begin{vmatrix} 8 & 2 & 3 \\ 2 & 5 & -1 \\ 4 & 9 & 12 \end{vmatrix}$ is the sum of the three "southeast" diagonals minus the sum of the three "northeast" diagonals.

SE diagonals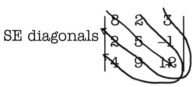

$$(8)(5)(12) + (2)(-1)(4) + (3)(9)(2)$$

NE diagonals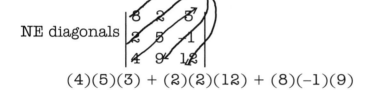

$$(4)(5)(3) + (2)(2)(12) + (8)(-1)(9)$$

Putting these all together

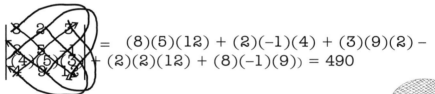 $= (8)(5)(12) + (2)(-1)(4) + (3)(9)(2) - (2)(2)(12) + (8)(-1)(9)) = 490$

To me that looks like a hairnet, or a beehive, or a Los Angeles freeway interchange.

The expansion by minors: First, you need to know what a minor is. The minor of 3 in $\begin{vmatrix} 8 & 2 & 3 \\ 2 & 5 & -1 \\ 4 & 9 & 12 \end{vmatrix}$ is $\begin{vmatrix} 2 & 5 \\ 4 & 9 \end{vmatrix}$ which is obtained by crossing out the row and the column that 3 is in.

312

For a matrix $A = (a_{ij})$, we can now find the minor of $a_{ij}$. Next, define the cofactor of $a_{ij}$ to be the minor of $a_{ij}$ times $(-1)^{i+j}$. Or, more simply, the minor of $a_{ij}$ times its position

on the waffle iron

```
+ – + – + –
– + – + – +
+ – + – + –
– + – + – +
+ – + – + –
```

Finally, to find the determinant of a 3×3 matrix, pick any row. The determinant is equal to the sum of the elements in that row times their cofactors. (You will also get the same answer if you pick any column.)

Selecting the third column,

$$\begin{vmatrix} 8 & 2 & 3 \\ 2 & 5 & -1 \\ 4 & 9 & 12 \end{vmatrix} = +(3)\begin{vmatrix} 2 & 5 \\ 4 & 9 \end{vmatrix} -(-1)\begin{vmatrix} 8 & 2 \\ 4 & 9 \end{vmatrix} +(12)\begin{vmatrix} 8 & 2 \\ 2 & 5 \end{vmatrix}$$

$$= (3)(18 - 20) + 1(72 - 8) + 12(40 - 4) = 490$$

For 4×4 (or higher) determinants, you have to use the expansion by minors approach.

---

*Your Turn to Play*

1. Evaluate $\begin{vmatrix} 1 & 3 & 4 \\ -2 & 1 & 5 \\ -6 & 0 & 2 \end{vmatrix}$ expanding by the third row.

2. [Fill in the blanks] If you were to evaluate a 50×50 determinant, you would have to evaluate __?__ 49×49 determinants. Each of those 49×49 determinants would require that you evaluate __?__ 48×48 determinants. Each of those 48×48 determinants would require that you evaluate __?__ 47×47 determinants.

3. Finding the determinant of a 50×50 matrix could use up several sheets of paper. For fun, estimate how long it would take you to evaluate a 50×50 determinant. As usual, please don't just look at the answer, but do

the figuring on your own.  This kind of estimating is not just "for fun," but can be very valuable in life.  For example, it may be important to estimate the cost of your wedding if you are going to invite 150 people and have a sit-down dinner afterward.

## . . . . . . . COMPLETE SOLUTIONS . . . . . . .

1. $-6(15 - 4) - 0 + 2(1 + 6) = -52$

2. 50, 49, 48

3. Evaluating a 50×50 determinant will involve evaluating approximately 50! 2×2 determinants.  50! means 50×49×48×. . .×3×2×1 (fifty factorial). If you have a "!" on your calculator, you can punch in 50! and get an answer of about $3×10^{64}$ 2×2 determinants to evaluate.

Do you remember conversion factors?  To change 132 inches into feet, you multiply by the conversion factor $\frac{1 \text{ foot}}{12 \text{ inches}}$ .  The numerator and denominator of a conversion factor are always equal to each other.

$\frac{132 \text{ i\cancel{nches}}}{1} \times \frac{1 \text{ foot}}{12 \text{ i\cancel{nches}}} = 11$ feet.  Notice how the units cancel.

Now let's see how long it will take to work out $3×10^{64}$ 2×2 determinants.

$\frac{3×10^{64} \text{ 2×2 determinants}}{1} \times \frac{1 \text{ line on a paper}}{4 \text{ 2×2 determinants}} \times \frac{1 \text{ sheet of paper}}{26 \text{ lines}} \times$

$\frac{7 \text{ minutes of work}}{1 \text{ sheet of paper}} \times \frac{1 \text{ hour}}{60 \text{ minutes}} \times \frac{1 \text{ day}}{24 \text{ hours}} \times \frac{1 \text{ year}}{365 \text{ days}} \times$

$\frac{1 \text{ lifetime}}{80 \text{ years}}$ .  Multiplying this all out, we find that evaluating a 50×50 determinant (assuming no sleep and working 24 hours a day) will take approximately $4.8 × 10^{55}$ lifetimes.  Even having everyone help you (all six billion of your closest friends), won't make much of a dent in this project.

We have to be a little careful talking about 1×1 determinants because | –9 | looks a lot like the absolute value of –9 which is +9, whereas the determinant of the 1×1 matrix (–9) is equal to –9.

## Some Handy Facts about Determinants

H.F.#1: $|A| = |A^T|$

$A^T$ is the transpose of A—the matrix you get by turning every row of A into a column. This handy fact will allow us to take a theorem about the rows of a determinant and turn it into a theorem about the columns.

H.F.#2: If A is upper triangular, then $|A|$ is equal to the product of the entries on the diagonal.

$$\begin{vmatrix} 3 & 7 & 832 \\ 0 & 2 & 937 \\ 0 & 0 & 5 \end{vmatrix} = 30$$

Of course, by H.F.#1, we could also say that the determinant of any lower triangular matrix is equal to the product of the diagonal entries.

H.F.#3: If A has a row of zeros, then $|A| = 0$.

Using expansion by minors along that row would show that $|A| = 0$.

Using H.F.#1, we could also say that if A has a column of zeros, its determinant is equal to zero.

H.F.#4: Multiplying any row of A by a scalar r multiplies the determinant by r.

$$\begin{vmatrix} 1 & 2 & 3 \\ 4 & 7 & 8 \\ 5000 & 6000 & 2000 \end{vmatrix} = 1000 \begin{vmatrix} 1 & 2 & 3 \\ 4 & 7 & 8 \\ 5 & 6 & 2 \end{vmatrix}$$

This handy fact can help make the numbers inside a determinant more manageable. It also works if a row looks like: 0.00005  0.00006  0.00002, and it also works for columns by H.F.#1.

H.F.#5: Interchange any two rows of A and you will change the sign of $|A|$.

Or any two columns . . . by H.F.#1.

H.F.#6: Add a multiple of any row of A to any other row of A and |A| is unchanged.

> Or . . . column to . . . column . . . by H.F.#1.
>
> Handy Facts #4, #5, and #6 may remind you of the elementary row operations we introduced way back on page 22.

H.F.#7: Just like Cramer's Rule is cute but a lousy way to solve systems of linear equations,[*] so expansion by minors is a cute but lousy way to evaluate a determinant.[**]

For decent-sized systems of linear equations, you use Gaussian elimination.

And for decent-sized determinants, you use Handy Facts #4, #5, and #6 to make it upper triangular and then use H.F.#2 or H.F.#3 to polish it off. (If this is your book and you like to use a highlighter, this is the paragraph to highlight.)

> Two pages ago we showed you couldn't evaluate a 50×50 determinant by using expansion by minors. But a determinant that is 50×50 is not unusual in the applications of linear algebra. When you encounter one of them, you use H.F.#7.

H.F.#8: $|AB| = |A||B|$

> Of course, this only makes sense if A and B are square matrices of the same dimension.

H.F.#9: $A^{-1}$ exists if and only if $|A| \neq 0$.

> "$A^{-1}$ exists" is sometimes expressed as "A is invertible."

Now you can tell determinants from dandelions, and we can begin the Nine Steps to Find $\mathbf{A}^{100}$.

But first, a message from our sponsor, Mr. *Your Turn to Play* .

---

[*] except, maybe, little systems with two or three unknowns.

[**] except, maybe, little determinants with two or three rows.

*Your Turn to Play*

1.  If you had to evaluate $\begin{vmatrix} 6 & 0 & 7 & 9 \\ 2 & 8 & 0 & 8 \\ 0 & 0 & 3 & 4 \\ 0 & 0 & 9 & 5 \end{vmatrix}$ using expansion by minors, which row or column would make things easiest for you?

2.  We know that $|AB| = |A||B|$ by Handy Fact #2. What about $|A| + |B| = |A + B|$? Please don't be a dumb dodo bird and just look at the answer. Try out some possible A's and B's to see if it's true.

3.  Evaluate $\begin{vmatrix} 1 & 2 & 3 \\ 4 & 5 & 6 \\ 7 & 8 & 9 \end{vmatrix}$ using the hairnet procedure described on page 312.

4.  Then evaluate it (question 3) by expansion by minors on the second column.

5.  Then evaluate it by using Handy Fact #7.

6.  Prove $|A^{-1}| = \dfrac{1}{|A|}$

........**COMPLETE SOLUTIONS**.......

1.  I would choose column 2 since it has three zeros. Every other choice of row or column will entail evaluating at least two 3×3 determinants.

2.  Almost any pair of matrices will show that $|A| + |B| \neq |A + B|$.

For example, $\begin{vmatrix} 2 & 3 \\ 1 & 4 \end{vmatrix} + \begin{vmatrix} 5 & 1 \\ 2 & 0 \end{vmatrix} \overset{?}{=} \begin{vmatrix} 7 & 4 \\ 3 & 4 \end{vmatrix}$

$$8 - 3 \;+\; 0 - 2 \overset{?}{=} 28 - 12$$
$$3 \overset{?}{=} 16 \qquad \text{not equal}$$

3.  $1 \cdot 5 \cdot 9 + 2 \cdot 6 \cdot 7 + 3 \cdot 8 \cdot 4 - (7 \cdot 5 \cdot 3 + 4 \cdot 2 \cdot 9 + 1 \cdot 6 \cdot 8)$
$= 45 + 84 + 96 - (105 + 72 + 48) = 0$    The raised dot · is one standard way to indicate multiplication. It gets rid of a lot of parentheses, but using it when there are decimal numbers can be slightly dangerous.

4. $-2\begin{vmatrix} 4 & 6 \\ 7 & 9 \end{vmatrix} + 5\begin{vmatrix} 1 & 3 \\ 7 & 9 \end{vmatrix} - 8\begin{vmatrix} 1 & 3 \\ 4 & 6 \end{vmatrix}$

$= -2(4\cdot9 - 7\cdot6) + 5(1\cdot9 - 7\cdot3) - 8(1\cdot6 - 4\cdot3) = 0$

5. $\begin{vmatrix} 1 & 2 & 3 \\ 4 & 5 & 6 \\ 7 & 8 & 9 \end{vmatrix} \underset{\substack{\text{H.F.#6} \\ -4r_1 + r_2 \\ -7r_1 + r_3}}{=} \begin{vmatrix} 1 & 2 & 3 \\ 0 & -3 & -6 \\ 0 & -6 & -12 \end{vmatrix} \underset{\substack{\text{H.F.#6} \\ -2r_2 + r_3}}{=} \begin{vmatrix} 1 & 2 & 3 \\ 0 & -3 & -6 \\ 0 & 0 & 0 \end{vmatrix} \underset{\text{H.F.#3}}{=} 0$

6.      $|I| = 1$          by H.F.#2 ("I" is the identity matrix.)

    $|A\,A^{-1}| = 1$         by the definition of $A^{-1}$

   $|A\,|\,|A^{-1}| = 1$        by H.F.#8

      $|A^{-1}| = \dfrac{1}{|A|}$      by algebra (Determinants are just numbers.)

---

### Intermission

That's a short proof, but many students ask, "How did you ever think of that? I would never have thought of starting with $|I| = 1$."

There is a trick here that only teachers and advanced math students know about. It makes you look really smart.

Confession: I didn't start with $|I| = 1$.
I started with what I wanted to prove.

I started with $|A^{-1}| = \dfrac{1}{|A|}$

and then said that would be true IF $|A\,|\,|A^{-1}| = 1$

which would be true IF $|A\,A^{-1}| = 1$

which would be true IF $|I| = 1$

and then, while you weren't looking, I just reversed the order of all the steps.

Now, beginning the proof with $|I| = 1$ makes it all look like magic.

Finally, after 15 pages, we get to begin the . . .

## *Nine Steps to Compute* $A^{100}$

*Step One*: Evaluate the determinant $|A - \lambda I|$.
Explanation: $\lambda$ is an "unknown" like x was in high school algebra. ($\lambda$ = lambda, a letter in the Greek alphabet)

We couldn't use x since we were using x as the initial condition $\begin{pmatrix} 100 \text{ lions} \\ 1000 \text{ monkeys} \end{pmatrix} = x$ back on page 307.

$|A - \lambda I|$ is called the **characteristic polynomial**. Everyone uses $\lambda$ in writing the characteristic polynomial. It's a tradition in linear algebra.

Step One is easy. Suppose $A = \begin{pmatrix} 1 & 2 \\ 3 & 2 \end{pmatrix}$

$$\lambda I = \lambda \begin{pmatrix} 1 & 0 \\ 0 & 1 \end{pmatrix} = \begin{pmatrix} \lambda & 0 \\ 0 & \lambda \end{pmatrix}$$

so $A - \lambda I = \begin{pmatrix} 1 - \lambda & 2 \\ 3 & 2 - \lambda \end{pmatrix}$     $A - \lambda I$ is just "subtract $\lambda$ from the diagonal entries of $A$."

$$\begin{vmatrix} 1 - \lambda & 2 \\ 3 & 2 - \lambda \end{vmatrix} = (1 - \lambda)(2 - \lambda) - (3)(2)$$

*Step Two*: Set the characteristic polynomial equal to zero.

$$(1 - \lambda)(2 - \lambda) - (3)(2) = 0$$

This is the **characteristic equation**.

*Step Three*: Solve the characteristic equation.

$$2 - \lambda - 2\lambda + \lambda^2 - 6 = 0$$
$$\lambda^2 - 3\lambda - 4 = 0$$

Solving by factoring
$$(\lambda - 4)(\lambda + 1) = 0$$
$$\lambda - 4 = 0 \ \text{ OR } \ \lambda + 1 = 0$$
$$\lambda = 4 \ \text{ OR } \ \lambda = -1$$

319

**Wait! Stop! I, your reader, have all kinds of comments. First of all, Step Two is soooo easy—just set the characteristic polynomial equal to zero. Why did you make it a separate step?**

For three reasons:

> (i) On page 310, you asked me to "go a little slower."
>
> (ii) Step Two introduces new terminology—*characteristic equation*—and I didn't want you to confuse it with *characteristic polynomial.*
>
> (iii) The pretty illustration of a Victorian staircase had nine steps, and I wanted to use it in our book. I write *"our book"* since you seem to like to write almost as much as I do.

**Thank you. Now, back to my second comment. When I learned how to factor trinomials back in high school algebra, I always wondered when I'd ever use that technique. Now, a hundred years later, here in linear algebra with the lions and monkeys and Fred, I'm actually going to have a use for factoring?**

Yup.* We had the same thing happen in trig when we introduced the secant function.** It was utterly useless in solving triangles, since you could always use the cosine function. And, besides, there is no | sec | button on a calculator. You had to wait for months in calculus before you learned the real use for the secant function: the derivative of the tangent: $\dfrac{d \tan \theta}{d\theta} = \sec^2 \theta$. And since that formula was only good in radians, that turned out to be the real reason you learned about radians in trig class.

**My last comment (on this page): Aren't you going to tell me what the $\lambda$'s are called? You have named everything else.**

I already did. On page 310. $\lambda_1 = 4$ and $\lambda_2 = -1$ are called the **eigenvalues** of A.

_____

★ Yup is short for yeah which is short for yes. When you say *yup*, you finish the pronunciation with closed lips creating an unreleased labial stop. *Labial* is the adjective form of *lips*. Fred's labial features are almost as sharp as his nose.

★★  $\sec \theta = \dfrac{1}{\cos \theta}$

The eigenvalues of a matrix can be positive, or zero, or negative, or complex numbers such as $6 + 5i$.

Step Three, solving the characteristic equation can always be done fairly easily when $A$ is a 2×2 matrix. It will always be a quadratic equation, which can be solved either by factoring, or, if necessary, by using the quadratic formula.*

For 3×3 matrices, you will get a cubic equation, and you should hope that some tricky factoring can be done to solve it so that you can find the eigenvalues.

For 50×50 matrices, you may 🕊️🕊️🕊️🕊️🕊️🕊️ 🕊️🕊️🕊️ 🕊️🕊️🕊️🕊️ of finding the exact values of the eigenvalues. On the other hand, there are ways of approximating the solutions to a 50$^{th}$-degree equation that will give you answers as close to the exact answers as you wish.

If $A$ is an n×n matrix, you now have n eigenvalues $\lambda_1, \lambda_2, \ldots, \lambda_n$. Some of these may be repetitions. For example, when you solve the characteristic equation  $\lambda^3 - 14\lambda^2 + 65\lambda - 100 = 0$

$$(\lambda - 4)(\lambda - 5)^2 = 0$$
$$\lambda_1 = 4, \lambda_2 = 5, \lambda_3 = 5$$

you get two equal eigenvalues. We say that $\lambda = 5$ has an **algebraic multiplicity** of two.

***Step Four***: For each eigenvalue, $\lambda_i$, find a corresponding **eigenvector**, $x_i$.

How to do that: Take the eigenvalue, put it into $(A - \lambda I)x = 0$ and solve. There will always be an infinite number of answers. Pick any nonzero answer. Each eigenvalue gives you an eigenvector.

$$\lambda_i \cdots\!\!\blacktriangleright x_i$$

So far, we have done 4 of the 9 steps.

You are given $A$. You want $A^{100}$.

Step One: Evaluate $|A - \lambda I|$.

Steps Two and Three: Solve $|A - \lambda I| = 0$ to obtain $\lambda_1, \lambda_2, \ldots, \lambda_n$.

Step Four: Stuff each $\lambda_i$ into $(A - \lambda I)x = 0$ to obtain a nonzero $v_i$.

---

✱ If $ax^2 + bx + c = 0$, then  $x = \dfrac{-b \pm \sqrt{b^2 - 4ac}}{2a}$

I have to explain where $(A - \lambda I)x = 0$ came from, and then we will do an example of the first four steps.

$$(A - \lambda I)x = 0$$

is equivalent to    $Ax - \lambda Ix = 0$    by the distributive law

$\qquad\qquad\qquad\quad Ax - \lambda x = 0$    since I is the identity matrix

$\qquad\qquad\qquad\qquad Ax = \lambda x$    by algebra

What this last equation says is that for a given square matrix A, we can find a number $\lambda$ and a nonzero vector x so that multiplying x by A can done more simply by multiplying x by a single number $\lambda$.

$\quad$ Ax is the multiplication of an n×n matrix times an n×1 matrix.

$\quad$ $\lambda$x is the multiplication of a number times an n×1 matrix.

$\qquad$ Let's continue the example we started three pages ago.

We supposed $A = \begin{pmatrix} 1 & 2 \\ 3 & 2 \end{pmatrix}$

We found $\lambda_1 = 4$ and $\lambda_2 = -1$.  Now to find the corresponding eigenvectors.

$$\boxed{\boxed{\text{For } \lambda_1 = 4}}$$

$(A - \lambda I)x = 0$, which is $\begin{pmatrix} 1 - \lambda & 2 \\ 3 & 2 - \lambda \end{pmatrix}\begin{pmatrix} x_1 \\ x_2 \end{pmatrix} = \begin{pmatrix} 0 \\ 0 \end{pmatrix}$

becomes $\qquad \begin{pmatrix} -3 & 2 \\ 3 & -2 \end{pmatrix}\begin{pmatrix} x_1 \\ x_2 \end{pmatrix} = \begin{pmatrix} 0 \\ 0 \end{pmatrix}$

Gaussian elimination for $\begin{pmatrix} -3 & 2 & 0 \\ 3 & -2 & 0 \end{pmatrix}$

$r_1 + r_2 \qquad \begin{pmatrix} -3 & 2 & 0 \\ 0 & 0 & 0 \end{pmatrix}$    $x_2$ is a free variable (Nightmare #2)

which corresponds to the system $\begin{cases} -3x_1 + 2x_2 = 0 \\ \quad\ 0 + 0 = 0 \end{cases}$

We transpose the free variable to the right side of the equation.

$$-3x_1 = -2x_2$$

$x_2$ is a parameter. We can name almost any value we wish. All we need is one particular nonzero eigenvector. (So we can't say $x_2 = 0$, because that would make $x_1 = 0$.)

> Eigenvalues, $\lambda$, can be positive, negative, zero, or complex.
> Eigenvectors must be nonzero vectors.

To make things easy, I'll set $x_2$ equal to 3. Then $x_1$ will be 2.*

So an eigenvector corresponding to $\lambda = 4$ is $\begin{pmatrix} x_1 \\ x_2 \end{pmatrix}$ which is $\begin{pmatrix} 2 \\ 3 \end{pmatrix}$

*Your Turn to Play*

1. Continuing the above example, find an eigenvector which corresponds to $\lambda = -1$.

2. What is the algebraic multiplicity of $\lambda = -1$ in question one?

3. Take a big breath. You are going to do the four steps for a 3×3 matrix and find its eigenvalues and corresponding eigenvectors. Work on this as long as you can hold your breath or for 20 minutes (whichever takes longer).

One sticky point will be when it comes time to factor $\lambda^3 + \lambda^2 - 21\lambda - 45$. It factors into $(\lambda - 5)(\lambda + 3)^2$.

$$A = \begin{pmatrix} -2 & 2 & -3 \\ 2 & 1 & -6 \\ -1 & -2 & 0 \end{pmatrix}$$

This particular A was chosen for several reasons:

♦ It won't involve fractions or decimals—just nice integers.

♦ It gives you a chance to play with finding eigenvalues and their corresponding eigenvectors.

♦ It will reinforce what your first grade teacher told you: write neatly and don't make silly arithmetic errors. If your handwriting is a bit

---

* If I had set $x_2$ equal to $3\pi/(\log_5 12)$, then $x_1$ would be $2\pi/(\log_5 12)$.
When someone says to me, "Think of a number between 1 and 20, and I will try to guess it," I often think of $3\pi/(\log_5 12)$.

**TOO SLOPPY** or you make even one little minus sign error, your 20 minutes of work can become not as pleasant as you might wish.*

### . . . . . . . COMPLETE SOLUTIONS . . . . . . .

1. Gaussian elimination for $\begin{pmatrix} 2 & 2 & 0 \\ 3 & 3 & 0 \end{pmatrix}$

$-1.5r_1 + r_2$    $\begin{pmatrix} 2 & 2 & 0 \\ 0 & 0 & 0 \end{pmatrix}$   which corresponds to $\begin{cases} 2x_1 + 2x_2 = 0 \\ 0 + 0 = 0 \end{cases}$

$x_2$ is the free variable. We transpose and simplify $x_1 = -x_2$

$x_2$ is a parameter. All we need is one nonzero eigenvector. We can set $x_2$ to any nonzero number we like. I like $x_2 = 1$.

The eigenvector corresponding to $\lambda = -1$ is $\begin{pmatrix} -1 \\ 1 \end{pmatrix}$

2. On page 319, the characteristic polynomial $|A - \lambda I|$ was
$$(1 - \lambda)(2 - \lambda) - (3)(2)$$
which we simplified to     $(\lambda - 4)(\lambda + 1)$.

The algebraic multiplicity of $\lambda = -1$ is one.

An alternate definition of the algebraic multiplicity of, say, $\lambda = 8$ is the highest power of $(\lambda - 8)$ that divides evenly into the characteristic polynomial.


3. Pep talk before we begin: Remember that we are doing the Nine Steps to Compute $A^{100}$ so that we won't have to do the 100 matrix multiplications. The Nine Steps are labor-saving.

$$|A - \lambda I| = \begin{vmatrix} -2 - \lambda & 2 & -3 \\ 2 & 1 - \lambda & -6 \\ -1 & -2 & -\lambda \end{vmatrix} = -\lambda^3 - \lambda^2 + 21\lambda + 45 \text{ which we set}$$
equal to zero.

---

✶ This is an understatement which is formed by the negation ("not") of its opposite. It is an example of litotes [LIGHT-eh-tease]. If we weren't using litotes, "not as pleasant as you might wish," might be more directly expressed as "pure hell."

    Litotes is the flip side of hyperbole [high-PURR-bow-lee], which is exaggeration: "I tried a million times."

We need to solve $\lambda^3 + \lambda^2 - 21\lambda - 45 = 0$. (I multiplied through by $-1$.) Equivalently, we need to factor $\lambda^3 + \lambda^2 - 21\lambda - 45$.

In more general terms, if we are able to factor some fourth degree polynomial like          $24x^4 + 18x^3 + \ldots + 15$
into                    $(r_1x + s_1)(r_2x + s_2)(r_3x + s_3)(r_4x + s_4)$
then                        $r_1r_2r_3r_4$ would have to equal 24
and            $s_1s_2s_3s_4$ would have to equal 15.

So we can expect that the factoring of $\lambda^3 + \lambda^2 - 21\lambda - 45$ will look like $(\lambda - s_1)(\lambda - s_2)(\lambda - s_3)$ where $s_1s_2s_3 = -45$. That tells us we might first look at $s_1 = \pm 1, \pm 3, \pm 5, \pm 9, \pm 15$, and $\pm 45$ rather than just trying 7 or 98967204.

If $(\lambda - 3)$ is a factor of $\lambda^3 + \lambda^2 - 21\lambda - 45$, then $\lambda = 3$ should be a root of $\lambda^3 + \lambda^2 - 21\lambda - 45 = 0$. We try it: $3^3 + 3^2 - 21\cdot 3 - 45 \neq 0$. No luck.

Trying $\lambda = 5$, $5^3 + 5^2 - 21\cdot 5 - 45 = 0$, we have a winner. If this seems like feeling around in the dark to find the answer, you are right.

Since $\lambda = 5$ is a root of the equation of $\lambda^3 + \lambda^2 - 21\lambda - 45 = 0$, then $\lambda - 5$ divides evenly into $\lambda^3 + \lambda^2 - 21\lambda - 45$.

In advanced algebra we did this long division of polynomials. And you wondered what it would ever be used for. Now you know.

$$
\begin{array}{r}
\lambda^2 + 6\lambda \quad + 9 \\
\lambda - 5 \,\overline{)\lambda^3 + \lambda^2 - 21\lambda - 45} \\
\underline{\lambda^3 - 5\lambda^2} \\
6\lambda^2 - 21\lambda \\
\underline{6\lambda^2 - 30\lambda} \\
9\lambda - 45 \\
\underline{9\lambda - 45} \\
0
\end{array}
$$

So $\lambda^3 + \lambda^2 - 21\lambda - 45 = (\lambda - 5)(\lambda^2 + 6\lambda + 9)$
which then factors by high school algebra into $(\lambda - 5)(\lambda + 3)^2$.
So the roots of $\lambda^3 + \lambda^2 - 21\lambda - 45 = 0$ are $\lambda_1 = 5$, $\lambda_2 = -3$, and $\lambda_3 = -3$.

5 is an eigenvalue of algebraic multiplicity 1.

$-3$ is an eigenvalue of algebraic multiplicity 2.

Now to find the corresponding eigenvectors.

For $\lambda = 5$

$(A - \lambda I)x = 0$, which is $\begin{pmatrix} -2 - \lambda & 2 & -3 \\ 2 & 1 - \lambda & -6 \\ -1 & -2 & -\lambda \end{pmatrix} \begin{pmatrix} x_1 \\ x_2 \\ x_3 \end{pmatrix} = \begin{pmatrix} 0 \\ 0 \\ 0 \end{pmatrix}$

becomes a Gaussian elimination for $\begin{pmatrix} -7 & 2 & -3 & 0 \\ 2 & -4 & -6 & 0 \\ -1 & -2 & -5 & 0 \end{pmatrix}$

Interchanging rows 1 and 3 and creating some zeros: $\begin{pmatrix} 1 & 2 & 5 & 0 \\ 0 & 1 & 2 & 0 \\ 0 & 0 & 0 & 0 \end{pmatrix}$

which corresponds to the system: $\begin{cases} x_1 + 2x_2 + 5x_3 = 0 \\ x_2 + 2x_3 = 0 \end{cases}$

$x_3$ is a free variable. All we need is one particular eigenvector.
Letting $x_3$ equal $-1$ looks like it will make things pretty simple.
Then backsubstituting, we have $x_2 = 2$ and $x_1 = 1$.

$\begin{pmatrix} 1 \\ 2 \\ -1 \end{pmatrix}$ is an eigenvector corresponding to $\lambda = 5$.

For $\lambda = -3$

Things are nicer here. Gaussian elimination for $\begin{pmatrix} 1 & 2 & -3 & 0 \\ 2 & 4 & -6 & 0 \\ -1 & -2 & 3 & 0 \end{pmatrix}$

In one step we have $\begin{pmatrix} 1 & 2 & -3 & 0 \\ 0 & 0 & 0 & 0 \\ 0 & 0 & 0 & 0 \end{pmatrix}$ $\boxed{\text{One step: } -r_1 + r_2 \quad r_1 + r_3}$

The free variables are $x_2$ and $x_3$.
Letting $x_3 = 1$ and $x_2 = 0$, we have $x_1 = 3$.
Letting $x_3 = 0$ and $x_2 = 1$, we have $x_1 = -2$.

So the eigenvectors corresponding to $\lambda = -3$ are $\begin{pmatrix} 3 \\ 0 \\ 1 \end{pmatrix}$ and $\begin{pmatrix} -2 \\ 1 \\ 0 \end{pmatrix}$

Since this eigenvalue ($\lambda = -3$) has two (linearly independent) eigenvectors associated with it, we say that it has a **geometric multiplicity** of 2. (Note: The eigenvalues you found might be different than mine. It depends on which values of $x_3$ you chose.)

If an eigenvalue has an algebraic multiplicity of 7, then we are hoping that we will be able to find 7 linearly independent eigenvectors associated with it. In other words, we are hoping that it will have a geometric multiplicity that is also equal to 7.

That will allow us to do the next step in the Nine Steps.

**I, your reader, have a small question. What if my eigenvalue that has an algebraic multiplicity of 7 only has 6 linearly independent eigenvectors associated with it?**

Tut-tut. Perish the thought. With a big matrix with randomly chosen entries, the chances of having the geometric multiplicity be less than the algebraic multiplicity is about the same as the chances of your uncle mailing you a bowl of soup.

**Even if it were a bowl of soup made with fish feathers and my uncle has been dead for five years, you are telling me that it can happen. What can I do then?**

Easy. You compute $A^{100}$ the old-fashioned way: you multiply it out.

***Step Five***: You have the n eigenvectors for the matrix $A$. You hang them vertically (in columns) to create a matrix $P$.

The eigenvectors in the previous *Your Turn to Play*

were $\begin{pmatrix} 1 \\ 2 \\ -1 \end{pmatrix}$, $\begin{pmatrix} 3 \\ 0 \\ 1 \end{pmatrix}$ and $\begin{pmatrix} -2 \\ 1 \\ 0 \end{pmatrix}$

The matrix $P$ will be $\begin{pmatrix} 1 & 3 & -2 \\ 2 & 0 & 1 \\ -1 & 1 & 0 \end{pmatrix}$    I call it "hanging the vines."

**I call it smooshing the eigenvectors together. An easy step.**

***Step Six***: Compute $P^{-1}$.

This is the reason we wanted the eigenvectors (the columns of P) to be linearly independent. If they weren't linearly independent, then $P^{-1}$ wouldn't exist.

How to find $P^{-1}$: Create the augmented matrix $\begin{pmatrix} P & I \end{pmatrix}$ and use elementary row operations until the left half of the matrix becomes I. Then the right half will be $P^{-1}$. (We proved that this process will give you $P^{-1}$ back on page 70.)

---

*Your Turn to Play*

1. The matrix $P = \begin{pmatrix} 1 & 3 & -2 \\ 2 & 0 & 1 \\ -1 & 1 & 0 \end{pmatrix}$    Find $P^{-1}$.

. . . . . . . **COMPLETE SOLUTION** . . . . . . .

1.
$$\begin{pmatrix} 1 & 3 & -2 & 1 & 0 & 0 \\ 2 & 0 & 1 & 0 & 1 & 0 \\ -1 & 1 & 0 & 0 & 0 & 1 \end{pmatrix}$$

$-2r_1 + r_2$
$r_1 + r_3$
$$\begin{pmatrix} 1 & 3 & -2 & 1 & 0 & 0 \\ 0 & -6 & 5 & -2 & 1 & 0 \\ 0 & 4 & -2 & 1 & 0 & 1 \end{pmatrix}$$

Some people do $(3/2)r_2 + r_3$ at this point. I like to avoid fractions and decimals as long as possible.

$4r_1$
$2r_2$
$3r_3$
$$\begin{pmatrix} 4 & 12 & -8 & 4 & 0 & 0 \\ 0 & -12 & 10 & -4 & 2 & 0 \\ 0 & 12 & -6 & 3 & 0 & 3 \end{pmatrix}$$

$r_2 + r_1$
$r_2 + r_3$
$$\begin{pmatrix} 4 & 0 & 2 & 0 & 2 & 0 \\ 0 & -12 & 10 & -4 & 2 & 0 \\ 0 & 0 & 4 & -1 & 2 & 3 \end{pmatrix}$$

---

$2r_1$ $\qquad\qquad\qquad$ $\begin{pmatrix} 8 & 0 & 4 & 0 & 4 & 0 \\ 0 & -24 & 20 & -8 & 4 & 0 \\ 0 & 0 & 4 & -1 & 2 & 3 \end{pmatrix}$
$2r_2$

$-r_3 + r_1$ $\qquad\qquad$ $\begin{pmatrix} 8 & 0 & 0 & 1 & 2 & -3 \\ 0 & -24 & 0 & -3 & -6 & -15 \\ 0 & 0 & 4 & -1 & 2 & 3 \end{pmatrix}$
$-5r_3 + r_2$

and now I have to go to fractions

$r_1/8$ $\qquad\qquad\qquad$ $\begin{pmatrix} 1 & 0 & 0 & 1/8 & 1/4 & -3/8 \\ 0 & 1 & 0 & 1/8 & 1/4 & 5/8 \\ 0 & 0 & 1 & -1/4 & 1/2 & 3/4 \end{pmatrix}$
$r_2/(-24)$
$r_3/4$

$$\mathsf{P}^{-1} = \begin{pmatrix} 0.125 & 0.25 & -0.375 \\ 0.125 & 0.25 & 0.625 \\ -0.25 & 0.5 & 0.75 \end{pmatrix}$$

I changed to decimals, just to show you that I can do both fractions and decimals.

Want an easy step?

**Does the sun rise at dawn?**

*Step Seven*: Create the matrix $\mathsf{D}$.

$\mathsf{D}$ is the diagonal matrix whose entries on the diagonal are the eigenvalues computed in Step Three.

On page 325, the eigenvalues were $\lambda_1 = 5$, $\lambda_2 = -3$, and $\lambda_3 = -3$.

$$\mathsf{D} = \begin{pmatrix} 5 & 0 & 0 \\ 0 & -3 & 0 \\ 0 & 0 & -3 \end{pmatrix}$$

*Step Eight*: If you want $\mathsf{A}^{100}$, compute $\mathsf{D}^{100}$.

Raising a diagonal matrix to a power is also easy.

$$\mathsf{D}^{100} = \begin{pmatrix} 5^{100} & 0 & 0 \\ 0 & (-3)^{100} & 0 \\ 0 & 0 & (-3)^{100} \end{pmatrix}$$

Use your $\mathsf{y}^x$ button on your calculator if you want $5^{100}$ all worked out.

Here is where we put it all together.

We found P in Step Five by hanging the ~~vines~~ eigenvectors.

We found D in Step Six. It was the eigenvalues on the diagonal of a matrix that had zeros everywhere else.

We found $D^{100}$ by raising each entry in D to the $100^{th}$ power.

We found $P^{-1}$ by starting with ( P I ) and going to ( I $P^{-1}$ ).

**Step Nine**: You want $A^{100}$ without having to multiply A by itself a hundred times? Just multiply three matrices together . . .

$$A^{100} = P\,D^{100}\,P^{-1}$$

Now the one-step procedure I gave you back on page 310 may make a little more sense. I'll repeat it here so you won't have to turn back to see it.

Step 1: To find $A^{100}$, simply compute $PD^{100}P^{-1}$
where D is the diagonal matrix consisting of
the eigenvalues $\lambda_1, \lambda_2, \ldots, \lambda_n$ of A, and
where P is the matrix whose columns are the
eigenvectors $v_1, v_2, \ldots, v_n$ which correspond
to the eigenvalues $\lambda_1, \lambda_2, \ldots, \lambda_n$.

Some notes, now that we have done all Nine Steps.

♪#1: P and $P^{-1}$ can be used to **diagonalize** A.

We know $A^{100} = P\,D^{100}\,P^{-1}$ and that is true for any power of A. In particular, $A = P\,D\,P^{-1}$. Multiply this equation on the left by $P^{-1}$ and on the right by P and we get $P^{-1}AP = D$.

Of course, if you wanted to diagonalize A, it would be silly to compute P and $P^{-1}$ and then do $P^{-1}AP$.

*Your Turn to Play*

1. Why? (Please don't just look at the answer. By figuring it out on your own, you will get a better understanding of P and D.)

## ........COMPLETE SOLUTIONS........

1. In order to find P, you have to find all the eigenvectors (the "vines.")
In order to find all the eigenvectors, you have to first find all the
eigenvalues, $\lambda_1, \lambda_2, \ldots, \lambda_n$.

But the minute you have $\lambda_1, \lambda_2, \ldots, \lambda_n$, you immediately know D.

$$D = \begin{pmatrix} \lambda_1 & 0 & 0 & \ldots & 0 \\ 0 & \lambda_2 & 0 & \ldots & 0 \\ & & \ldots & & \\ 0 & 0 & 0 & \ldots & \lambda_n \end{pmatrix}$$

♪#2: Back on page 323, you took a big breath and found the eigenvalues
and corresponding eigenvectors for a 3×3 matrix. If you didn't make any
arithmetic errors, your arithmetic skills aren't that bad.*

There is a fairly easy way to check your final answers without
having to go over all your work line by line. On page 322, we noted that if
$\lambda$ and x are an eigenvalue and its corresponding eigenvector for some
matrix A, then $Ax = \lambda x$.

In the "big breath" computation, we found $\lambda_1 = 5$ and x $= \begin{pmatrix} 1 \\ 2 \\ -1 \end{pmatrix}$

and we can check those answers by seeing whether $Ax = \lambda x$.

$$Ax = \begin{pmatrix} -2 & 2 & -3 \\ 2 & 1 & -6 \\ -1 & -2 & 0 \end{pmatrix} \begin{pmatrix} 1 \\ 2 \\ -1 \end{pmatrix} = \begin{pmatrix} -2+4+3 \\ 2+2+6 \\ -1-4+0 \end{pmatrix} = \begin{pmatrix} 5 \\ 10 \\ -5 \end{pmatrix}$$

and    $\lambda x = 5\begin{pmatrix} 1 \\ 2 \\ -1 \end{pmatrix} = \begin{pmatrix} 5 \\ 10 \\ -5 \end{pmatrix}$    It checks.

---

* More litotes. The plain truth is that if you didn't make any arithmetic errors in that
page of computations, you are right up there somewhere between amazing and
incredible.

♪#3: Step Nine asserts that $A^{100} = P D^{100} P^{-1}$. For some people that seems a bit like magic. Why does it work? If you are curious, this note will explain the magic. If you aren't curious, just skip on down to ♪#4.

The easiest thing to show is how to go from $A = P D P^{-1}$ to $A^{100} = P D^{100} P^{-1}$.

Start with $A = P D P^{-1}$. Multiply it by itself a hundred times and you get $A^{100} = (P\ D\ P^{-1})(P\ D\ P^{-1})(P\ D\ P^{-1})(P\ D\ P^{-1})(P\ D\ P^{-1})(P\ D\ P^{-1})$ $(P\ D\ P^{-1})(P\ D\ P^{-1})(P\ D\ P^{-1})(P\ D\ P^{-1})(P\ D\ P^{-1})(P\ D\ P^{-1})(P\ D\ P^{-1})$ $(P\ D\ P^{-1})(P\ D\ P^{-1}) \ldots (P\ D\ P^{-1})$.

Then using the associative law for matrix multiplication,
$A^{100} = P\ D(P^{-1}P)D(P^{-1}P)D(P^{-1}P)D(P^{-1}P)D(P^{-1}P)D(P^{-1}P)D(P^{-1}P)$
$D(P^{-1}P)D(P^{-1}P)D(P^{-1}P)D(P^{-1}P)D(P^{-1}P)D(P^{-1}P)D(P^{-1}P)D(P^{-1}P)$
$D(P^{-1}P)D(P^{-1}P)D(P^{-1}P)D(P^{-1}P)D \ldots (P^{-1}P)DP^{-1}$.

Since $P^{-1}P = I$, we have
$A^{100} = P\ D\ I\ D\ I\ D\ I\ D\ I\ D\ I\ D\ I\ D\ I\ D\ I\ D\ I\ D\ I\ D\ I\ D\ I\ D\ I\ D\ I\ D$
$I\ D\ I\ D\ I\ D\ I\ D\ I\ D\ I\ D\ I\ D\ I\ D\ I\ D\ I\ D\ I\ D \ldots I\ D\ P^{-1}$.

And since $D\ I = D$,    $A^{100} = P D^{100} P^{-1}$.


The tougher thing to show is that $A = P D P^{-1}$ given our construction of $P$ as the hanging eigenvectors and $D$ as the diagonal matrix with diagonal entries equal to the eigenvalues, and $Ax = \lambda x$.

We will show that $A P = P D$, which is equivalent to $A = P D P^{-1}$ (by multiplying both sides of $A = P D P^{-1}$ on the right by $P$).

Our notation for this proof: $(P)_{ij} = p_{ij}$ (which means that the i-j$^{th}$ entry of $P$ is the number $p_{ij}$). Let $\lambda_1, \lambda_2, \ldots, \lambda_n$ be the eigenvalues of $A$ with corresponding eigenvectors $v_1, v_2, \ldots, v_n$. By the way we constructed $P$, the $v_1, v_2, \ldots, v_n$ are the columns of $P$.


From our definition of the multiplication of two matrices $A$ and $B$, we know that:

❋ the i-j$^{th}$ entry of $AB$ is the i$^{th}$ row of $A$ times the j$^{th}$ column of $B$.

Playing with this definition and looking at examples you make up for about five minutes will establish:

❋❋ the j$^{th}$ column of $AB$ is $A$ times the j$^{th}$ column of $B$.

and    ❋❋❋ the i$^{th}$ row of $AB$ is the i$^{th}$ row of $A$ times $B$.

So what is the $j^{th}$ column of AP? Using ✿✿ from the previous page, it is A times the $j^{th}$ column of P

which is equal to $A\mathsf{v}_j$         (since the $j^{th}$ column of P is $\mathsf{v}_j$)

which is equal to $\lambda_j \mathsf{v}_j$         (since $A\mathsf{x} = \lambda \mathsf{x}$)

which is equal to $\lambda_j$ times the $j^{th}$ column of P.

So $(AP)_{ij} = \lambda_j p_{ij}$                                   (line 1)

On the other hand, $(PD)_{ij}$ equals by ✿, the $i^{th}$ row of P times the $j^{th}$ column of D, which is

$$(p_{i1},\, p_{i2},\, \ldots,\, p_{in}) \begin{pmatrix} 0 \\ 0 \\ \cdots \\ 0 \\ \lambda_j \\ \cdots \\ 0 \end{pmatrix}$$

Thus, $(PD)_{ij} = p_{ij}\lambda_j$.                                   (line 2)

By lines 1 and 2, we have that corresponding entries of AP and PD are equal. This is the definition of the equality of matrices. So AP = PD.

♪#4: When I was learning about eigenvalues as a student in a classroom, the teacher wrote *eigenvalue* on the board. In my notes I wrote *ev*. It seemed that all the words in linear algebra have too many syllables. I thought *ev* looked cute.

When the teacher wrote *eigenvector* on the board, I was in deep trouble. I changed *ev* to *eval* and used *evec* for eigenvector. I'm all for making English as simple as possible.

The teacher headed in the opposite direction away from simplicity. Sometimes, instead of writing *eigenvalue*, he wrote *proper value* explaining that some people don't like the half-German and half-English word *eigenvalue*. And sometimes he wrote *characteristic value*. And sometimes he wrote *latent root* explaining that is what the social scientists call an eigenvalue.

And instead of *eigenvector*, he would spice things up by substituting *proper vector* or *characteristic vector* or *latent vector*.

In my notes, I kept writing *eval* and *evec*. On the day I was born, I remember my mom looking at me, shaking her head, and saying, "Simple." Was she using the imperative mood (a command), meaning that I should make things simple, or was she using the indicative mood (making a statement) and using *simple* as an adjective?

Fred looked at his new friends. The lion was starting to look a little hungry so Fred pulled a tennis ball out of his pocket and suggested, "Let's play a game of catch."

After Fred explained the rules,* he tossed the ball to someone who then threw it to someone else, or who threw it up in the air and caught it himself. . . .

Sometimes a player liked to throw the ball straight up and catch it himself. For example, the lion would throw the ball upwards and catch it himself 80% of the time. He would throw it to the monkey 10% of the time and to Fred 10% of the time.

$$\text{LION} \begin{pmatrix} 0.8 \\ 0.1 \\ 0.1 \end{pmatrix}$$

For the monkey, his throwing pattern was $\text{MONKEY} \begin{pmatrix} 0.2 \\ 0.7 \\ 0.1 \end{pmatrix}$ For Fred, $\text{FRED} \begin{pmatrix} 0.1 \\ 0.3 \\ 0.6 \end{pmatrix}$

The transition matrix combines these three probability vectors:

$$\begin{array}{ccc} \text{LION} & \text{MONKEY} & \text{FRED} \end{array}$$
$$A = \begin{pmatrix} 0.8 & 0.2 & 0.1 \\ 0.1 & 0.7 & 0.3 \\ 0.1 & 0.1 & 0.6 \end{pmatrix}$$

In the beginning, Fred had the ball. Initial vector $x = \begin{pmatrix} 0 \\ 0 \\ 1 \end{pmatrix}$

Then $Ax$ represents where the ball will probably be after one toss:

$$\begin{pmatrix} 0.8 & 0.2 & 0.1 \\ 0.1 & 0.7 & 0.3 \\ 0.1 & 0.1 & 0.6 \end{pmatrix} \begin{pmatrix} 0 \\ 0 \\ 1 \end{pmatrix} = \begin{pmatrix} 0.1 \\ 0.3 \\ 0.6 \end{pmatrix}$$

---

✶ Rules: You catch the ball and then throw it to someone.

After two tosses, $A^2x = \begin{pmatrix} 0.8 & 0.2 & 0.1 \\ 0.1 & 0.7 & 0.3 \\ 0.1 & 0.1 & 0.6 \end{pmatrix} \begin{pmatrix} 0.1 \\ 0.3 \\ 0.6 \end{pmatrix} = \begin{pmatrix} 0.2 \\ 0.4 \\ 0.4 \end{pmatrix}$

And $A^3x = \begin{pmatrix} 0.28 \\ 0.42 \\ 0.30 \end{pmatrix}$

Translation: After three tosses of the ball, we will expect that the lion will have the ball 28% of the time, the monkey 42% of the time, and Fred 30% of the time.

The question is: If they play catch all afternoon, what is the probability each of the players will have possession of the ball when they quit because it is too dark to play anymore? $A^\infty x = ?$

---

*Intermission*

And, maybe, a better question is, "Who cares?" After all, in the town you live in, how often will you find lions, monkeys, and Fred playing catch at your local park?

The answer is that a lot of people care, because, in a slightly disguised form, this "playing catch" happens all the time.

Example #1: There is a finite, fixed amount of gold on planet Earth. The various countries pass their gold back-and-forth each month.

Example #2: Each year people who live near KITTENS University pass between five different states: freshman, sophomore, junior, senior, and not-in-school. For example, sophomores this year have a 85% chance that they will be juniors next year (and a 10% chance they will still be sophomores next year, and a 5% they will not be in school).

Example #3: Suppose we classify each day at KITTENS University as either sunny, cloudy, rainy, or snowy. Then, if it is rainy today, the transition matrix A will give the probabilities of what the weather will be like tomorrow.

---

These playing-a-game-of-catch transition matrices A deal with *probabilities*. In contrast, the transition matrix at the beginning of this chapter that dealt with the changes in the populations of the lions and monkeys told you exactly what population numbers to expect as each year goes by.

These playing-a-game-of-catch transition matrices A are easy to spot: All the entries are nonnegative and the sum of each column is equal to one. Such matrices are called **stochastic matrices**. [steh-CASS-tick]

And when a stochastic matrix operates on some initial vector x and we compute x, Ax , $A^2x$, $A^3x$, $A^4x$, $A^5x$, $A^6x$, $A^7x$, . . . this is called a **Markov chain**.

---

*Your Turn to Play*

1. You own an ice cream store. (A fantasy of many people.) Your only competition is Carol's Creamery. Every day each person in town either visits your store, or Carol's store, or goes without ice cream.

People who visit your store today have an 80% chance of visiting your store tomorrow, 10% of switching to Carol's, and a 10% chance of not having ice cream tomorrow.

People who visit Carol's today have a 20% chance of switching to your store tomorrow, and a 70% chance of visiting Carol's tomorrow.

People who didn't have ice cream today, have a 10% chance of going to your store tomorrow, and a 30% chance of going to Carol's tomorrow.

Write the transition matrix A.

2. Is A a stochastic matrix?

3. After you and Carol have been in business for a long time, the fraction of customers at each store will begin to stabilize. At some point and for some v,  v, Av, $A^2v$, $A^3v$, $A^4v$, . . ., $A^{543}v$, $A^{544}v$, $A^{545}v$, $A^{546}v$. . . .

these will all start to become equal.

If, for example $A^{545}v = A^{544}v$, then     $A^{545}v - A^{544}v = 0$
                                                $A^{544}(Av - v) = 0$
          which will be true when     $Av - v = 0$

We want to find the **steady state vector** v such that $Av - v$ is equal to 0. In other words, a steady state vector v such that $Av = v$.

v is an eigenvector. What is the eigenvalue associated with it?

4. (The previous problem took 0.00382 seconds to solve. This one will take 7–13 minutes.) Find the steady state vector v such that $Av = v$. This

---

will tell us both: (1) what percentage of the time customers will choose your store or Carol's store or no ice cream on any particular day after your businesses have been established for a long time, and (2) what percentage of the time we might expect the ball to be in the "hands" of the lion, the monkey, and Fred after they have been playing catch all afternoon.

## .......COMPLETE SOLUTIONS.......

1.

YOUR STORE   CAROL'S   ☹

$$A = \begin{pmatrix} 0.8 & 0.2 & 0.1 \\ 0.1 & 0.7 & 0.3 \\ 0.1 & 0.1 & 0.6 \end{pmatrix}$$

If this matrix seems slightly familiar, it should be. It's also the playing catch matrix of three pages ago.

2. ✓ Are all the entries in A nonnegative? Yes.

   ✓ Is the sum of each column equal to one? Yes. It's stochastic.

3. Given square matrix A, if we can find a number $\lambda$ and a nonzero vector x so that multiplying x by A can be done more simply by multiplying x by a single number $\lambda$, then $\lambda$ is an eigenvalue of A and x is the eigenvector associated with $\lambda$. Ax $= \lambda$x (We did this on page 322.)

   In this problem we are given Av $=$ v. So $\lambda = 1$.

4. We are essentially doing Step 4 of the Nine-Step Procedure. We have $(A - \lambda I)v = 0$, where $\lambda = 1$, and want to find the eigenvector v.

   Solving $(A - I)v = 0$,
$$\begin{pmatrix} -0.2 & 0.2 & 0.1 & 0 \\ 0.1 & -0.3 & 0.3 & 0 \\ 0.1 & 0.1 & -0.4 & 0 \end{pmatrix}$$

$10r_1$
$10r_2$
$10r_3$
$$\begin{pmatrix} -2 & 2 & 1 & 0 \\ 1 & -3 & 3 & 0 \\ 1 & 1 & -4 & 0 \end{pmatrix}$$

$Ir_1r_3$
$$\begin{pmatrix} 1 & 1 & -4 & 0 \\ 1 & -3 & 3 & 0 \\ -2 & 2 & 1 & 0 \end{pmatrix}$$

$$-r_1 + r_2$$
$$2r_1 + r_3$$

$$\begin{pmatrix} 1 & 1 & -4 & 0 \\ 0 & -4 & 7 & 0 \\ 0 & 4 & -7 & 0 \end{pmatrix}$$

$$r_2 + r_3$$

$$\begin{pmatrix} 1 & 1 & -4 & 0 \\ 0 & -4 & 7 & 0 \\ 0 & 0 & 0 & 0 \end{pmatrix}$$

This is equivalent to the system $\begin{cases} v_1 + v_2 - 4v_3 = 0 \\ -4v_2 + 7v_3 = 0 \\ 0 = 0 \end{cases}$

$v_3$ is a free variable. We may set it to any nonzero value. (If we set it equal to zero, then both $v_2$ and $v_1$ will be zero. This is not permitted since eigenvectors are not supposed to equal the zero vector **0**.)

If I set $v_3$ equal to 1, then I get fractions for $v_2$ and $v_1$. Too much work.

Looking at $-4v_2 + 7v_3 = 0$, I let $v_3$ equal 4. Then $v_2$ equals 7, and $v_1$ equals 9 (from the first equation).

So $\begin{pmatrix} 9 \\ 7 \\ 4 \end{pmatrix}$ solves the system, but there is one last step needed. The vector **v** must express the percentage visiting your store, or Carol's store, or neither store. We know the customers are in the continued ratio 9:7:4, but for a probability vector, the sum of the entries must equal one.

What do I multiply **v** by in order to have $v_1 + v_2 + v_3 = 1$?
How about $\dfrac{1}{20}$?    $(20 = 9 + 7 + 4)$  (This trick always works.)

Thus the percentage of the time the customers visit your store or Carol's or stay home after the businesses have been established for a long time is

given by the steady state vector $\begin{pmatrix} 9/20 \\ 7/20 \\ 4/20 \end{pmatrix}$  or  $\begin{pmatrix} 0.45 \\ 0.35 \\ 0.20 \end{pmatrix}$  or  $\begin{pmatrix} 45\% \\ 35\% \\ 20\% \end{pmatrix}$

In the extreme case, $A^{n+1}v = A^n v$ will only occur in the limiting case when $n \to \infty$.  $Av = v$ will still make $A^{\infty}Av = A^{\infty}v$ true.

SMALL PROOF:

$$Av = v$$
$$Av - v = 0$$
$$A^{\infty}(Av - v) = 0$$
$$A^{\infty}Av - A^{\infty}v = 0$$
$$A^{\infty}Av = A^{\infty}v$$

You can't always find a $v$ so that the Markov chain $v$, $Av$, $A^2v$, $A^3v$, $A^4v$, . . . settles down to some single steady state vector.  Just like some eight-year-old boys trying to take a nap in the afternoon, they just keep wiggling around.

But there is one very common case in which the stochastic matrix $A$ will have a steady state vector.  It is when $A$ is **regular**.  And before you, my reader, can interrupt me with the obvious question, let me define *regular*.

𝔇𝔢𝔣𝔦𝔫𝔦𝔱𝔦𝔬𝔫: A stochastic matrix is regular if every entry of $A$ (or of some power of $A$) has only positive entries.

So, regular matrices always have an eigenvalue equal to one.

And now more Anguish.  Oops, I mean English.  (They sound so much alike.)  For the eigenvector which corresponds to an eigenvalue of 1, which we call the steady state vector, my teacher would sometimes write on the board *stationary vector*.  Or sometimes *fixed probability vector*.

For regular matrices, finding the steady state vector $v$ doesn't depend on the initial vector $x$.  When Fred was playing catch with the lion and the monkey, in the beginning Fred had the ball.

$$x = \begin{pmatrix} 0 \\ 0 \\ 1 \end{pmatrix}$$

We computed in the *Your Turn to Play* on the previous page that after they had played for a long while, we might expect that the lion would have possession of the ball 45% of the time, the monkey 35% of the time, and Fred 20% of the time.  It really didn't matter who had the ball when they started the game.

In contrast, suppose the lion had a nasty habit of keeping the ball whenever he received it.  Instead of the lion's throwing pattern being

LION
$$\begin{pmatrix} 0.8 \\ 0.1 \\ 0.1 \end{pmatrix}$$

it would be 
LION
$$\begin{pmatrix} 1 \\ 0 \\ 0 \end{pmatrix}$$ 
which is very unpleasant for the other players.

Sometimes, lions are not very much fun to play with.

This situation in which the lion just keeps the ball is called an **absorbing state**.  Once you enter that state, you never leave it.  Atheists think that death is an absorbing state.

Now suppose that the monkey didn't understand the idea of "playing nice" and would also keep the ball when he received it.  Since he was afraid of what the lion would do if he kept the ball, he would scamper up into the trees with the ball.

The monkey's throwing pattern would be 
$$\begin{pmatrix} 0 \\ 1 \\ 0 \end{pmatrix}$$

—another absorbing state.

Everything depended on whom Fred first threw the ball to.  It was going to be Fred➙Lion➙Lion➙Lion➙Lion➙Lion➙Lion➙Lion➙Lion or it was going to be Fred➙Monkey➙Monkey➙Monkey➙Monkey.

Of course, Fred could have tossed the ball to himself for a while, but once he threw it to someone else, the game was over.

LION  MONKEY  FRED

The transition matrix is 
$$\begin{pmatrix} 1 & 0 & 0.1 \\ 0 & 1 & 0.3 \\ 0 & 0 & 0.6 \end{pmatrix}$$

Note that this matrix is not regular—its entries are not all positive, and no power of it has all-positive entries.  Also note that the implication is that regular implies the existence of a steady state vector: Regular → Steady.  We do not have Steady → Regular.  An implication does not imply its converse. —another sampler from Carrie's store wall.

Fred thought this game of catch would be a lot of fun, but the game was soon over.  As you can see . . .

When Betty held out her hand to Fred, he knew it was time to get going.  He took her hand, and the three of them headed back to KITTENS University.

---

**Emmet**

---

1. Confirm that $\begin{pmatrix} 1 \\ -1 \\ 0 \end{pmatrix}$ is an eigenvector of $\begin{pmatrix} 3 & 1 & 1 \\ 2 & 4 & 1 \\ 1 & 1 & 3 \end{pmatrix}$

and find the eigenvalue corresponding to it.

2. What Handy Fact allows you to say that the determinant of the identity matrix is always equal to one? (The Handy Facts are listed starting on page 314. The identity matrix was defined on page 62.)

3. Two square matrices A and B are **similar** if you can find a matrix C such that $A = C B C^{-1}$. In this chapter, two matrices were shown to be similar. Which are they?

4. 𝔗𝔥𝔢𝔬𝔯𝔢𝔪: If two matrices A and B are similar, they have the same characteristic polynomial.

We will need a lemma* to prove this theorem. The lemma is proved by just looking at the i-j$^{th}$ entries of each side of the equation.

𝔏𝔢𝔪𝔪𝔞: For any scalar $\lambda$ and matrices A, B, and C

$$A\lambda B = \lambda AB$$
$$A(B + C) = AB + AC$$
$$(B + C)A = BA + CA \quad \text{We skip the proof of the Lemma.}$$

Here is the proof of the theorem. It's what I call a meat-grinder proof—five thousand semi-obvious, but boring, steps. *Supply the reasons for each equality.* Each equal sign is numbered. Again, the Handy Facts are listed starting on page 314. (The reason we are looking at this is that it provides plenty of practice with the properties of determinants.) $|A|$ means the determinant of A.

$$|A - \lambda I| \overset{①}{=} |CBC^{-1} - \lambda I| \overset{②}{=} |CBC^{-1} - \lambda CC^{-1}| \overset{③}{=} |CBC^{-1} - C\lambda C^{-1}|$$

$$\overset{④}{=} |CBC^{-1} - C\lambda IC^{-1}| \overset{⑤}{=} |C(B - \lambda I)C^{-1}| \overset{⑥}{=} |C||B - \lambda I||C^{-1}|$$

$$\overset{⑦}{=} |C||C^{-1}||B - \lambda I| \overset{⑧}{=} |CC^{-1}||B - \lambda I| \overset{⑨}{=} |I||B - \lambda I| \overset{⑩}{=} |B - \lambda I| \quad ■$$

---

✶ A lemma is a small theorem that is used to help prove a bigger theorem.

*answers*

1.  If x is an eigenvector of A, then for some number $\lambda$, Ax = $\lambda$x.

We compute Ax:    $\begin{pmatrix} 3 & 1 & 1 \\ 2 & 4 & 1 \\ 1 & 1 & 3 \end{pmatrix} \begin{pmatrix} 1 \\ -1 \\ 0 \end{pmatrix} = \begin{pmatrix} 2 \\ -2 \\ 0 \end{pmatrix}$    which is 2x. So $\lambda = 2$.

2.  H.F.#2 states that if A is upper triangular, then |A| is equal to the product of the entries on the diagonal. The identity matrix I is certainly upper triangular. The product of its diagonal entries is the product of ones.

3.  We showed that A = P D P$^{-1}$ on page 330. P is the matrix whose columns are eigenvectors. D is the matrix whose diagonal entries are the eigenvalues corresponding to the eigenvectors of P.

4.      ① definition of similar matrices

   ② I = CC$^{-1}$ by definition of C$^{-1}$

   ③ by the Lemma

   ④ property of I: for any A, IA = AI = A. In this case, C$^{-1}$ = IC$^{-1}$

   ⑤ by the Lemma

   ⑥ Handy Fact #8   | AB | = | A || B |

   ⑦ Determinants are numbers. The commutative law holds for multiplying numbers. $921 \times 5 = 5 \times 921$

   ⑧ Handy Fact #8

   ⑨ definition of C$^{-1}$

   ⑩ problem 2 of this City

---

**Teman**

---

DIFFERENTIAL EQUATIONS

1.  In calculus, when we had the differential equation $\frac{dy}{dx}$ = 5y, we solved it by separating the variables      $\frac{dy}{y} = 5dx$

and then integrated both sides      $\ln y = 5x + C$

and then solved for y      $y = e^{5x + C}$

$y = e^{5x}e^{C}$

and letting $e^{C} = c$      $y = ce^{5x}$.

The original differential equation could have been written as $y' = 5y$ where y is a function of x.

Now, since this is linear algebra, the logical extension would be to consider systems of equations. For example, $\begin{cases} y_1' = y_1 + 4y_2 \\ y_2' = 2y_1 + 3y_2 \end{cases}$

If we define the vector $y = (y_1, y_2)^T$, then the system could be written as $y' = Ay$ where $A = \begin{pmatrix} 1 & 4 \\ 2 & 3 \end{pmatrix}$ or $Ay = y'$.

We are so used to seeing $Ax = b$ that we are going to change variables. In calculus, the most common situation is when y is a function of x. The second most common situation is when x is a function of t. Instead of $\frac{dy}{dx}$, which was abbreviated as $y'$, we abbreviated $\frac{dx}{dt}$ as $\dot{x}$.

So $\begin{cases} y_1' = y_1 + 4y_2 \\ y_2' = 2y_1 + 3y_2 \end{cases}$ will become $\begin{cases} \dot{x}_1 = x_1 + 4x_2 \\ \dot{x}_2 = 2x_1 + 3x_2 \end{cases}$

and we will get to write $Ax = \dot{x}$.

---

### How to Solve Systems of Linear, First-Order Differential Equations

(Second-order differential equations involve y".)

1. To solve $Ax = \dot{x}$, first find the eigenvalues of A: $\lambda_1$ and $\lambda_2$.
2. Then find the eigenvectors $v_1$ and $v_2$ associated with $\lambda_1$ and $\lambda_2$.
3. Create matrix P in which the columns of P are the eigenvectors. This is what I called "hanging the vines" in **Step Five**, page 327.
4. Create vector $u = ( c_1 e^{\lambda_1 t}, c_2 e^{\lambda_2 t})^T$.
5. Then the solution to $Ax = \dot{x}$ is $x = Pu$.

---

The above box is for systems of two equations in which A is 2×2. If A were 3×3, you would first find eigenvalues $\lambda_1$, $\lambda_2$, and $\lambda_3$, etc.

You are a door-to-door salesman selling **Fred's Fountain Pens**. As people in the neighborhood buy Freds, the more delighted they are with fountain-pen writing, and the more they tell their neighbors, who then, in turn, want to buy Fred's Fountain Pens. Your job gets easier and easier.

*Nothing Flows Like a Fred*

Let $x_1$ = the number of fine point pens that you have sold up to the moment t in time.

Let $x_2$ = the number of broad point pens that you have sold up to the moment t in time.

Fountain pen owners tend to want to buy more pens. Especially people who own the broad point pens tend to also want the fine point pens. Your rate of sales of the fine point pens $\dot{x}_1$ equals $x_1 + 4x_2$.

People who own either the fine point or the broad point pens tend to want to buy more broad point pens: $\dot{x}_2 = 2x_1 + 3x_2$.

$$\text{Solve the system} \quad \begin{cases} \dot{x}_1 = x_1 + 4x_2 \\ \dot{x}_2 = 2x_1 + 3x_2 \end{cases}$$

2. In the box on the previous page, we said that the solution to $Ax = \dot{x}$ is $x = Pu$. The three-part proof of this is pretty straightforward. I've separated the three parts by dotted lines. *Supply the reasons for each equality.* Each equal sign is numbered.

$$\dot{x} \overset{①}{=} \frac{dx}{dt} \overset{②}{=} \frac{d\,Pu}{dt} \overset{③}{=} P\frac{d\,u}{dt} \overset{④}{=} P \text{ times the matrix whose } i^{th} \text{ row}$$
$$\text{is } \lambda_i c_i e^{\lambda_i t}.$$

................................................................

$$Ax \overset{⑤}{=} APu \overset{⑥}{=} PDu \overset{⑦}{=} P\begin{pmatrix} \lambda_1 & 0 & 0 & \cdots & 0 \\ 0 & \lambda_2 & 0 & \cdots & 0 \\ & & \cdots & & \\ 0 & 0 & 0 \cdots 0 & \lambda_n \end{pmatrix}\begin{pmatrix} c_1 e^{\lambda_1 t} \\ c_2 e^{\lambda_2 t} \\ \cdots \\ c_n e^{\lambda_n t} \end{pmatrix} \overset{⑧}{=} P \text{ times the matrix whose } i^{th} \text{ row is } \lambda_i c_i e^{\lambda_i t}.$$

................................................................

So $\dot{x} \overset{⑨}{=} Ax$.

................................................................

*answers*

1. $A = \begin{pmatrix} 1 & 4 \\ 2 & 3 \end{pmatrix}$     $\begin{vmatrix} 1-\lambda & 4 \\ 2 & 3-\lambda \end{vmatrix} = \lambda^2 - 4\lambda - 5$ which we set equal to zero and solve.

We find that $\lambda_1 = 5$ and $\lambda_2 = -1$.

<hr>

**For $\lambda = 5$**   $(A - \lambda I)x = 0$ becomes $(A - 5I)x = 0$

which is solved by Gaussian elimination $\begin{pmatrix} -4 & 4 & 0 \\ 2 & -2 & 0 \end{pmatrix}$

$x_2$ becomes a free variable. We let it equal 1. Then $x_1 = 1$.
The eigenvector associated with $\lambda = 5$ is $v_1 = (1, 1)^T$.

<hr>

**For $\lambda = -1$**

$(A - (-1)x) = 0$    $\begin{pmatrix} 2 & 4 & 0 \\ 2 & 4 & 0 \end{pmatrix}$ where $x_2$ becomes a free variable.

Let $x_2 = 1$, and $x_1$ will equal $-2$.    $v_2 = (-2, 1)^T$.

"Hanging the vines," $P = \begin{pmatrix} 1 & -2 \\ 1 & 1 \end{pmatrix}$      $u = \begin{pmatrix} c_1 e^{5t} \\ c_2 e^{-t} \end{pmatrix}$

$x = Pu = \begin{pmatrix} c_1 e^{5t} - 2c_2 e^{-t} \\ c_1 e^{5t} + c_2 e^{-t} \end{pmatrix}$   or   $\begin{cases} x_1 = c_1 e^{5t} - 2c_2 e^{-t} \\ x_2 = c_1 e^{5t} + c_2 e^{-t} \end{cases}$

SMALL REVIEW OF DIFFERENTIAL EQUATIONS TERMINOLOGY . . .

This is called the **general solution** to the system of differential equations. To find a **particular solution** (i.e. to find values for $c_1$ and $c_2$) we need to be given additional information.

For example, if we know that at $t = 57$, we sold $x_1 = 889$ fine point pens and $x_2 = 3987$ broad point pens, we could substitute these into the general solution and solve for $c_1$ and $c_2$ by the methods of Chapter 1. Since the independent variable $t$ is the same for both of the dependent variables, $x_1$ and $x_2$, these are called **initial conditions**.

If, on the other hand, we know that at $t = 33$, $x_1 = 608$ fine point pens, and that at $t = 2$, $x_2 = 44$ broad point pens, we have been given **boundary point** conditions—the two subsidiary conditions were given at different values of the independent variable $t$.

2.   ① definition of $\dot{x}$

    ② $x = Pu$ is given.

    ③ $P$ consists of just scalars (numbers).

       Recall from calculus: $\dfrac{d\, 9872x^6}{dx} = 9872 \dfrac{d\, x^6}{dx}$

    ④ the $i^{th}$ row of $u$ is $c_i e^{\lambda_i t}$ and the derivative of $c_i e^{\lambda_i t}$ is $\lambda_i c_i e^{\lambda_i t}$.

    ⑤ $x = Pu$ is given.

    ⑥ We showed that $AP = PD$ two-thirds of the way down page 332.

⑦ definition of D (See the *Your Turn to Play* on page 331.)

⑧ definition of the multiplication of matrices

⑨ the transitive property of equality (Since $\dot{x}$ and A$x$ both equal "P times the matrix whose i$^{th}$ row is $\lambda_i c_i e^{\lambda_i t}$," they are equal to each other.)

---

## Wamsutter

1. If A is a 3×3 matrix, find the value of r in $|\,5A\,| = r|\,A\,|$.

2. The determinant
$$\begin{vmatrix} 1 & 2 & 3 & 4 \\ 5 & 6 & 7 & 8 \\ 9 & 10 & 11 & 12 \\ 13 & 14 & 15 & 16 \end{vmatrix}$$
looks very neat and orderly, but I would hate to expand by minors in order to compute its value. Handy Fact #7 (on page 316) suggested that you could add a multiple of any row to any other row, and the value of the determinant would remain the same. (That is also true for columns.)

If I subtract column one from column two*, and subtract column three from column 4, I get
$$\begin{vmatrix} 1 & 1 & 3 & 1 \\ 5 & 1 & 7 & 1 \\ 9 & 1 & 11 & 1 \\ 13 & 1 & 15 & 1 \end{vmatrix}$$
and then if I subtract column two from column 4, I get a column of zeros. Thus, the original determinant is equal to zero by H.F.#3.

I did it with columns. You do it with rows.

3. Looking at problem 1, complete the statement: If A is an n×n matrix, then $|\,rA\,| = ?$

4. Why can't the set of all eigenvectors associated with a particular eigenvalue $\lambda$ be a vector space?

---

\* Officially, I am adding minus one times column one to column two.

*answers*

1. 5A is a scalar times a matrix. That multiplies each entry in A by 5. By **Handy Fact #4**, multiplying any row of A by r multiplies the determinant by r. We have multiplied all three rows of A by 5. This will multiply the determinant by 125.

2. The obvious thing to do is subtract row one from row two and subtract row three from row four. Then row two and row four will be identical. Subtract row two from row four, and row four will be all zeros.

But who wants to do things the obvious way? If, instead, you subtract row one from row three and subtract row two from row four, rows three and four will be identical. Then subtract row three from row four, and row four will be all zeros.

3. $|rA| = r^n |A|$.

4. It is *almost* a vector space. Suppose $v_1$ and $v_2$ were eigenvectors associated with $\lambda = 47$ and some matrix A. Then $v_1 + v_2$ would also have the property:

$$A(v_1 + v_2) = Av_1 + Av_2 = 47v_1 + 47v_2 = 47(v_1 + v_2) \quad \text{(line 1)}$$

and also for any number, say 552, we would have the property:

$$A(552v_1) = 552Av_1 = 552(47)v_1 = 47(552v_1).$$

So you ask, "Where's the problem? It sure looks like a vector space to me."

The answer is that the set of all eigenvectors associated with $\lambda = 47$ is not closed under addition. Back on line 1, I showed that $v_1 + v_2$ seems to act like an eigenvector, but in the case where $v_1$ and $v_2$ add to 0, we run into difficulty. In **Step Four** we gave the procedure for finding an eigenvector: *Take the eigenvalue, put it into* $(A - \lambda I)x = 0$ *and solve. There will always be an infinite number of answers. Pick any* <u>nonzero</u> *answer.*

If we look at the set of all eigenvectors associated with a particular eigenvalue $\lambda$ plus the zero vector, then we have a vector space.

---

**Zuni**

---

1. Show that $\begin{pmatrix} 2 & 0 \\ 2 & 2 \end{pmatrix}$ is not diagonalizable.

2. Ever since Chapter 1, Robin had been looking for work for a long time.* Today, she headed off to **eager eddie's employment egency** to see if she could find work. The place reminded her of the poet Edward Estlin Cummings who always used to write his name as e. e. cummings, except that Cummings knew how to spell *agency*.

When Robin arrived at Eddie's, she found out that he only dealt with two job categories: unemployed actresses and unemployed rock stars.

He knew how to find jobs for these people. In the first week that Eddie opened his "egency," he had one actress and two rock stars come in, and he found jobs for both of them. $\begin{pmatrix} 1 \\ 2 \end{pmatrix}$ The next week, the word got out, and he found jobs for five actresses and seven rock stars. $\begin{pmatrix} 5 \\ 7 \end{pmatrix}$ The next week it was 22 actresses and 26 rock stars. $\begin{pmatrix} 22 \\ 26 \end{pmatrix}$

Each actress who found employment through Eddie told three actress friends and one of her rock star friends, and they all came to see Eddie. Each rock star told one actress and three rock stars, and they all came to see Eddie. The transition matrix is $\begin{pmatrix} 3 & 1 \\ 1 & 3 \end{pmatrix}$ and week number one was $\begin{pmatrix} 1 \\ 2 \end{pmatrix}$. Find how many people Eddie was working with in his eleventh week of being in business. (Eddie's eleventh week is $A^{10}x$.)

3. Robin knew exactly what job to apply for at Eddie's. She would work in crowd control since in Eddie's eleventh week there would be 1,572,352 actresses and 1,573,376 rock stars there at Eddie's looking for work.

Robin was hired at Eddie's. During the first several weeks she knew that the crowd control work would be light, and she would have plenty of time for coffee breaks.

This is the story of her three coffee breaks (before she was fired).

---

\* Things hadn't worked out at **Wally's Window Washing**, or at **Willy's Walnuts**, or at **Wanda's Wallpapering**, or at **William's Welding**, or at **Harry's Hamburgers**, or at **Frank's Franks**.

Coffee Break #1: Robin always liked $\begin{pmatrix} 1 \\ 1 \end{pmatrix}$ as an initial vector. She would keep that initial vector during all her coffee breaks. While she had her doughnuts and coffee she was going to try out different transition matrices.

During her first break she started with the identity matrix $\begin{pmatrix} 1 & 0 \\ 0 & 1 \end{pmatrix}$ and wrote out the sequence x, Ax, A$^2$x, A$^3$x, . . .

$\begin{pmatrix} 1 \\ 1 \end{pmatrix}$, $\begin{pmatrix} 1 \\ 1 \end{pmatrix}$, $\begin{pmatrix} 1 \\ 1 \end{pmatrix}$, $\begin{pmatrix} 1 \\ 1 \end{pmatrix}$, $\begin{pmatrix} 1 \\ 1 \end{pmatrix}$, $\begin{pmatrix} 1 \\ 1 \end{pmatrix}$, $\begin{pmatrix} 1 \\ 1 \end{pmatrix}$, $\begin{pmatrix} 1 \\ 1 \end{pmatrix}$, but this got boring.

Coffee Break #2: For variety, she tried $\begin{pmatrix} 1 & 1 \\ 0 & 1 \end{pmatrix}$ and wrote out x, Ax, A$^2$x, A$^3$x, . . .

$\begin{pmatrix} 1 \\ 1 \end{pmatrix}$, $\begin{pmatrix} 2 \\ 1 \end{pmatrix}$, $\begin{pmatrix} 3 \\ 1 \end{pmatrix}$, $\begin{pmatrix} 4 \\ 1 \end{pmatrix}$, $\begin{pmatrix} 5 \\ 1 \end{pmatrix}$, $\begin{pmatrix} 6 \\ 1 \end{pmatrix}$, $\begin{pmatrix} 7 \\ 1 \end{pmatrix}$, $\begin{pmatrix} 8 \\ 1 \end{pmatrix}$, which brought a little life to her break. She liked the first row entries: 1, 2, 3, 4, 5. . . . They formed a nice pattern.

Coffee Break #3: Things got much more intriguing when she tried $\begin{pmatrix} 1 & 1 \\ 1 & 0 \end{pmatrix}$ and got $\begin{pmatrix} 1 \\ 1 \end{pmatrix}$, $\begin{pmatrix} 2 \\ 1 \end{pmatrix}$, $\begin{pmatrix} 3 \\ 2 \end{pmatrix}$, $\begin{pmatrix} 5 \\ 3 \end{pmatrix}$, $\begin{pmatrix} 8 \\ 5 \end{pmatrix}$, $\begin{pmatrix} 13 \\ 8 \end{pmatrix}$, where the first row entries

are 1, 1, 2, 3, 5, 8, 13, 21, 34, 55, 89, 144, 233, 377, 610, 987, 1597, 2584, 4181, 6765, 10946, 17711, 28657, 46368, 75025, 121393, 196418, 317811, 514229, 832040, 1346269, 2178309, 3524578, 5702887, 9227465, 14930352, 24157817, 39088169, 63245986, 102334155, 165580141, 267914296, 433494437, 701408733, 1134903170, 1836311903, 2971215073, 4807526976, 7778742049, 12586269025, 20365011074, 32951280099, 53316291173, 86267571272, 139583862445, 225851433717, 365435296162, 591286729879, 956722026041, 1548008755920, 2504730781961, 4052739537881, 6557470319842, 10610209857723, 17167680177565, 27777890035288, 44945570212853, 72723460248141, 117669030460994, 190392490709135, 308061521170129, 498454011879264, 806515533049393, 1304969544928657, 2111485077978050, 3416454622906707, 5527939700884757, 8944394323791464, 14472334024676221, 23416728348467685, 37889062373143906, 61305790721611591, 99194853094755497, 160500643816367088, 259695496911122585, 420196140727489673, 679891637638612258, 1100087778366101931, 1779979416004714189, 2880067194370816120, 4660046610375530309, 7540113804746346429, 12200160415121876738, 19740274219868223167, 31940434634990099905, 51680708854858323072, 83621143489848422977, 135301852344706746049, 218922995834555169026, 354224848179261915075, 573147844013817084101, 927372692193078999176, 1500520536206896083277, 2427893228399975082453, 3928413764606871165730, 6356306993006846248183, 10284720757613717143913, 16641027750620563662096, 26925748508234281076009, 43566776258854844738105, 70492524767089125814114, 114059301025943910552219, 184551825793033836366333, 298611126818977066918552, 483162952612010163284885, 781774079430987230203437, 1264937030204297393488322, 2046711111473984623691759, 3311648141585890011180081, 5358359254990966640871840, 8670007398507948658051921, 14028366653498915298923761, 22698374052006863956975682, 36726740705505779255889443, 59425114757512643212875125, 96151855463018422468774568, 155576970220531065681649693, 251728825683549488150424261, 407305795904080553832073954, 659034621587630041982498215, 1066340417491710595814572169, 1725375039079340637797070384, 2791715456571051233611642553, 4517090495650391871408712937, 7308805052221443105020355490, 11825896447817834976429068427, 19134702400093278081449423917, 30960598847965113057878492344, 50095301248058391139327916261, 81055900096023504197206408605, 131151201344081895336534324866, 212207101440105399533174073471, 343358302784187294870257058337, 555565404224292694404015791808, 898923707008479989274290850145, 1454489111232772683678306641953, 2353412818241252672952597492098, 3807901929474025356630904134051, 6161314747715278092953501626149, 9969216677189303538626146592...

As you can tell, Robin took a long coffee break. Robin had stumbled upon the **Fibonacci numbers**. Each term in the sequence is the sum of the previous two terms. It would be easy to write a *Life of Fred: Fibonacci Numbers* book—there are so many things to say about those numbers.

     Here's just one little item. You may have noticed that as you go out in the sequence of Fibonacci numbers, dividing one term by its predecessor starts to approach a limit. 1/1 = 1, 2/1 = 2, 3/2 = 1.5, 5/3 = 1.666, 8/5 = 1.6, 13/8 = 1.625, 21/13 = 1.61538. . . . And if you whip out your calculator and divide the last two numbers listed above, in the fine print,

99692166771893033862144405760200 ÷

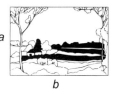

61613147477152780295835016 26149 the answer starts to
look very close to the **golden ratio** ϕ which creates the
most artistically pleasing rectangle to humans. *b/a* is
equal to the golden ratio.

In geometry we do a ruler-and-compass construction of the **golden
rectangle** in four easy steps.

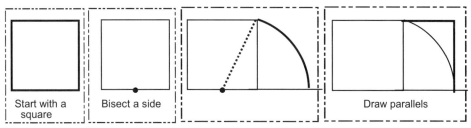

| Start with a square | Bisect a side | | Draw parallels |

So $\phi = \lim\limits_{i \to \infty} \dfrac{F_{i+1}}{F_i}$ where $F_i$ is the i[th] term in the Fibonacci sequence.

In this chapter we have only been considering transitions from one
state to the next state: $x, Ax, A^2x, A^3x, A^4x, A^5x, A^6x, A^7x. \dots$ Each state
only depended on its immediate predecessor. In her *Coffee Break* #3,
Robin found a way of making each state depend on the previous *two*
*states*.

Your question: How might this be significant if you someday apply
linear algebra to the real world?

***answers***

1. The characteristic equation is $\lambda^2 - 4\lambda + 4 = 0$ which gives $\lambda = 2$ (with an
algebraic multiplicity of 2) as the eigenvalue. Using *Step Four* to find
the corresponding eigenvectors, we arrive at $\begin{pmatrix} 0 & 0 & 0 \\ 2 & 0 & 0 \end{pmatrix}$. Let $x_2 = 1$.
Then $x_1 = 0$. One eigenvector is $(0 \quad 1)^T$. But there is no second
eigenvector. The geometric multiplicity of $\lambda = 2$ is one. Your uncle
mailed you a bowl of soup. (See page 327.) We said on page 327 that
*with a big matrix with randomly chosen entries, the chances of having the*
*geometric multiplicity be less than the algebraic multiplicity is* very small.
In order to get this non-diagonalizable matrix, I picked a small matrix with
carefully chosen entries.

2. For A, the eigenvalues are $\lambda = 2$ and any nonzero multiple of $\begin{pmatrix} -1 \\ 1 \end{pmatrix}$ and $\lambda = 4$ and any nonzero multiple of $\begin{pmatrix} 1 \\ 1 \end{pmatrix}$

P is $\begin{pmatrix} -1 & 1 \\ 1 & 1 \end{pmatrix}$ and to find $P^{-1}$ we do elementary row operations on $\begin{pmatrix} -1 & 1 & 1 & 0 \\ 1 & 1 & 0 & 1 \end{pmatrix}$ until we arrive at $\begin{pmatrix} 1 & 0 & -\frac{1}{2} & \frac{1}{2} \\ 0 & 1 & \frac{1}{2} & \frac{1}{2} \end{pmatrix}$

$A^{10}x = PD^{10}P^{-1}x = \begin{pmatrix} -1 & 1 \\ 1 & 1 \end{pmatrix}\begin{pmatrix} 2^{10} & 0 \\ 0 & 4^{10} \end{pmatrix}\begin{pmatrix} -\frac{1}{2} & \frac{1}{2} \\ \frac{1}{2} & \frac{1}{2} \end{pmatrix}\begin{pmatrix} 1 \\ 2 \end{pmatrix} = \begin{pmatrix} -2^9 + \frac{3}{2}4^{10} \\ 2^9 + \frac{3}{2}4^{10} \end{pmatrix}$

$= \begin{pmatrix} 1,572,352 \\ 1,573,376 \end{pmatrix}$

3. There are some people (engineers, physicists, economists, biologists) who take the pure mathematics of linear algebra and actually apply it to real-life situations. Their procedure is invariably the same: ① Look at the complicated world; ② Make a model; ③ Keep simplifying it until you can apply the mathematics; ④ Apply the math.

For example, suppose ① we want to know how many shirts there are in the Great Woods.

We make the model ②: There is Betty's blue shirt with pearl buttons, Alexander's brown shirt with nutmeg buttons, and Fred's shirt with ducks printed on it with plastic buttons.

Simplify ③: Betty' blue shirt. Alexander's brown shirt. Fred's duck shirt.

Keep simplifying ③: Betty's shirt. Alexander's shirt. Fred's shirt.

Keep simplifying ③: shirt + shirt + shirt

Keep simplifying ③: 1 + 1 + 1.

Apply the math ④: 1 + 1 + 1 = 3.

The difficulty is in step ③. As you simplify you lose information.

If our model $x$, $Ax$, $A^2x$, $A^3x$, $A^4x$, $A^5x$, $A^6x$, $A^7x$ . . . only depends on the previous state, it may not predict the future states as accurately as a model that *looks back to more than just one previous state.*

Blanche's Better Blankets and Carrie's Comfy Covers both made $120 yesterday. But Blanche has lost money annually for the last ten years, whereas Carrie has made money. Which store would you rather invest in?

## Darlington

1.  Being a den mother for the Cub Scouts isn't always easy.  When Madison was looking through the Cub pack archives, she found *The Complete History of Den Mothers*, which cataloged the story of each den mother in her pack for the last hundred years or so.

   She found out that two-thirds of the first-year den mothers signed up to work for a second year.

   Two-thirds of the second-year mothers signed up for a third year.

   None of the mothers stayed beyond the third year, since by that time their boys were out of Cub Scouts.

   Write the transition matrix.

2.  Is the transition matrix regular?

3.  Show that the probability that a den mother chosen at random is in her first year is 9/19, in her second year, 6/19, and in her third year, 4/19.

4.  When Madison's son turned 11, he joined the Boy Scouts, and Madison volunteered to be the treasurer for the troop.  She (correctly) guessed that her apartment wouldn't suffer the damage that it had when she was a den mother.  She learned from *The Complete History of Boy Scout Troop Treasurers* that 90% of all first-year treasurers went on to serve a second year.

   80% of all second-year treasurers went on to serve a third year.

   70% of all third-year treasurers went on to serve a fourth year.

   And 60% of all treasurers who had served four or more years continued for another year.

The transition matrix is  $A = \begin{pmatrix} 0.1 & 0.2 & 0.3 & 0.4 \\ 0.9 & 0 & 0 & 0 \\ 0 & 0.8 & 0 & 0 \\ 0 & 0 & 0.7 & 0.6 \end{pmatrix}$

   $A$ is obviously stochastic since all of the entries are nonnegative and the sum of each column is equal to one.  If you were to investigate whether $A$ is regular, you would first compute $A^2$, which is

$$\begin{pmatrix} 0.19 & 0.26 & 0.31 & 0.28 \\ 0.09 & 0.18 & 0.27 & 0.36 \\ 0.72 & 0 & 0 & 0 \\ 0 & 0.56 & 0.42 & 0.36 \end{pmatrix}$$

and then, since there are still four entries that are zero, you would compute $A^3$, which is

$$\begin{pmatrix} 0.253 & 0.286 & 0.253 & 0.244 \\ 0.171 & 0.234 & 0.279 & 0.252 \\ 0.072 & 0.144 & 0.216 & 0.288 \\ 0.504 & 0.336 & 0.252 & 0.216 \end{pmatrix}$$

and you might be thinking to yourself . . .

**Wait a minute! I, your reader, will tell you what I'm thinking. You don't have to guess. This part of linear algebra is yucky. I hate all the arithmetic. I notice that throughout the book, you have been pretty merciful when it comes to excluding problems that involved a lot of arithmetic. Some books have a ton of the stuff.**

**And I must say that I appreciate your finding** $A^2$ **and** $A^3$ **and not making me do that.**

You are welcome. But I have to confess—I also hate arithmetic. I didn't compute $A^2$ and $A^3$.

**You mean you just made up those numbers!?**

No. Those are the actual numbers for $A^2$ and $A^3$. It's time that I reveal a dirty little secret. It might make your life out there in the real world a lot easier when it comes time to doing big problems in linear algebra.

**Tell me. I'm all ears.**

Okay.


A Dirty Little Secret: There are a million computer programs out there that will do matrix multiplication for you. All you do is type in the two matrices and the product will pop out. Some of the programs are plain and some are fancy. I just use the spreadsheet program that came with my word processing program (WordPerfect). I didn't have to spend a nickel more. And it will also find the determinant of a square matrix. And the inverse of a matrix.

There are big, hairy programs that will do just about any linear algebra computation that you can think of. Some are quite expen$ive. Some you can find on the Internet that are free.

But what those programs don't give you is the understanding of what is going on. You can ask some of them to find the eigenvectors of a matrix. What this book tells you is what an eigenvector is.

Now, where were we? You were given A, which is the transition matrix for Boy Scout treasurers. We know A is regular, since $A^3$ has no zero entries. We are going to find out is the probability that today, the treasurer is a first-year treasurer, or a second-year treasurer, etc. We follow the usual procedure and form $(A - I)v = 0$ and get ready to

solve $\begin{pmatrix} -0.9 & 0.2 & 0.3 & 0.4 & 0 \\ 0.9 & -1 & 0 & 0 & 0 \\ 0 & 0.8 & -1 & 0 & 0 \\ 0 & 0 & 0.7 & -0.4 & 0 \end{pmatrix}$ by Gaussian elimination.

But wait. Let's save some work. First, multiply each row by 10 to eliminate the decimals.

Second, we know that A is regular. If we went through all the arithmetic of Gaussian elimination, one of the rows will have all zeros in it.* By Chapter 2, we know one of the variables will be a free variable and we will have tons of possible answers. The last line of that augmented matrix translates into $7v_3 - 4v_4 = 0$. Consider $v_4$ as the free variable. Let $v_4 = 7$ to make the arithmetic easier. Please compute $v_3$, $v_2$, and $v_1$, and then divide through by the appropriate number so that the sum of the entries will equal one.

### answers

1.
$A = \begin{pmatrix} 1/3 & 1/3 & 1 \\ 2/3 & 0 & 0 \\ 0 & 2/3 & 0 \end{pmatrix}$

---

* Here's the logic. Normally, a homogeneous system (that's one where all the equations equal zero, $Ax = 0$) has a single solution, namely, $x = 0$. But since A is regular, we know there is a nonzero solution in addition to $x = 0$. But systems of equations either have one solution, an infinite number of solutions, or no solution. (That's what Chapters 1, 2, and 3 were all about.) Since it has more than one solution, it will have an infinite number of solutions. We will have a free variable, and one of the rows must turn into an all-zero row. In algebra, we called this having fewer equations than unknowns.

2. A has four zero entries. $A^2$ has two zero entries. But $A^3$ has all positive entries. Therefore, A is regular.

3. Since A is regular we know that $\lambda = 1$ is an eigenvalue of A. We want to find the steady state vector v, such that $Av = v$.

$(A - I)v = 0$.

Using Gaussian elimination, $\begin{pmatrix} -2/3 & 1/3 & 1 & 0 \\ 2/3 & -1 & 0 & 0 \\ 0 & 2/3 & -1 & 0 \end{pmatrix} \rightarrow \begin{pmatrix} 2 & -1 & -3 & 0 \\ 2 & -3 & 0 & 0 \\ 0 & 2 & -3 & 0 \end{pmatrix}$

$\rightarrow \ldots \rightarrow \begin{pmatrix} 2 & -1 & 3 & 0 \\ 0 & -2 & 3 & 0 \\ 0 & 0 & 0 & 0 \end{pmatrix}$

which corresponds to the system $\begin{cases} 2v_1 - v_2 + 3v_3 = 0 \\ -2v_2 + 3v_3 = 0 \end{cases}$

$v_3$ is a free variable, and we can let it equal any nonzero number. Originally, I tried $v_3 = 2$, but I encountered fractions when I got to $v_1$. If I let $v_3 = 4$, then $v_2$ will be 6, and $v_1 = 9$.

The solution vector is $\begin{pmatrix} 9 \\ 6 \\ 4 \end{pmatrix}$ which I multiply by 1/19 (19 = 9 + 6 + 4)

to obtain $\begin{pmatrix} 9/19 \\ 6/19 \\ 4/19 \end{pmatrix}$

4. The augmented matrix is equivalent

to the system $\begin{cases} -9v_1 + 2v_2 + 3v_3 + 4v_4 = 0 \\ 9v_1 - 10v_2 = 0 \\ 8v_2 - 10v_3 = 0 \\ 7v_3 - 4v_4 = 0 \end{cases}$

We start with the fourth equation and let $v_4 = 7$, since that will make $v_3$ an integer rather than a fraction. $v_3 = 4$. Substituting $v_3 = 4$ into the third equation, $8v_2 - 10v_3 = 0$, we find that $v_2 = 5$. Putting $v_2 = 5$ into the second equation, we find that $v_1 = 50/9$. So $v = (50/9, 5, 4, 7)^T$. Multiply by 9 to eliminate the fractions: $(50, 45, 36, 63)^T$.

Note that $(50, 45, 36, 63)^T$ checks in the first equation: $-(9)(50) + 2(45) + 3(36) + 4(63)$ does equal zero.

As the last step, we need to turn $(50, 45, 36, 63)^T$ into a probability vector. We multiply it by 1/194 (where 194 = 50 + 45 + 36 + 63). Final answer: $(50/194, 45/194, 36/194, 63/194)^T$ which is approximately equal to $(26\%, 23\%, 19\%, 32\%)^T$. So, for example, we would expect that there is a 23% chance that a Boy Scout treasurer would be in a second year of service.

**Joliet**

1. If **A** and **B** are both n×n matrices, show that **AB** and **BA** have the same eigenvalues.

2. Jackie was new on the police force, a rookie. Each year Jackie would either move up one rank or be demoted one rank. To keep things simple, let's suppose there are four ranks:

Civilian  ↔  Rookie  ↔  Police Chief  ↔  Retired

Jackie

     Each year there was two-thirds of a chance of being promoted, i.e. of moving to the right on the Civilian ↔ Rookie ↔ Police Chief ↔ Retired chart.

     Each year there was one-third of a chance being demoted, i.e. of moving to the left.

     If you became a civilian, you stayed a civilian. If you became retired, you stayed retired. These are absorbing states.

     So each year Jackie would get bounced around, eventually to become either a civilian or be retired.

     Understandably, Jackie asked, "Hey, what are my chances of being retired?" In this and the next several questions, we are going to slowly work toward the answer of 4/7. Except for Fred, four-sevenths is not what someone might call the obvious answer to Jackie's question. Your question: Assuming Fred's answer is correct, what is the chance that Jackie will eventually be demoted to civilian?

Retired?

It's 4/7

3. Write the transition matrix **A**.
Hint: The third column will be $(0 \quad \frac{1}{3} \quad 0 \quad \frac{2}{3})^{\mathrm{T}}$.

4. Is **A** regular?

5. Find the eigenvalues of **A**.

6. Find an eigenvector associated with $\lambda = \frac{-\sqrt{2}}{3}$

Hint: The approximate answer will be $(1 \quad -4.41 \quad 6.24 \quad -2.83)^{\mathrm{T}}$.

*Intermission*

When I was working out this problem, I got as far as finding $P$. I used many pages and made several arithmetic errors. As you know by now, one arithmetic error is fatal.

The next step was to find $P^{-1}$. I augmented $P$ with the identity matrix, ( $P\ I$ ) and began using elementary row operations in order to obtain ( $I\ P^{-1}$ ).

When some of the entries became $\dfrac{-3-\sqrt{2}}{-3+\sqrt{2}}$

I knew that it was time to use `A Dirty Little Secret` and find $P^{-1}$ using a little computer help.

I typed in $P$ and requested $P^{-1}$.

I wanted $A^{\infty}$ which is equal to $PD^{\infty}P^{-1}$.

$$\text{Since } D = \begin{pmatrix} 1 & 0 & 0 & 0 \\ 0 & 1 & 0 & 0 \\ 0 & 0 & \frac{\sqrt{2}}{3} & 0 \\ 0 & 0 & 0 & \frac{-\sqrt{2}}{3} \end{pmatrix} \quad \text{we have } D^{\infty} = \begin{pmatrix} 1 & 0 & 0 & 0 \\ 0 & 1 & 0 & 0 \\ 0 & 0 & 0 & 0 \\ 0 & 0 & 0 & 0 \end{pmatrix}$$

I asked the computer to multiply out $PD^{\infty}P^{-1}$ (which is $A^{\infty}$) and it told me

$$\text{that } A^{\infty} = \begin{pmatrix} 1 & 3/7 & 1/7 & 0 \\ 0 & 0 & 0 & 0 \\ 0 & 0 & 0 & 0 \\ 0 & 4/7 & 6/7 & 1 \end{pmatrix}$$

7. Jackie's initial state as a rookie was $(0\quad 1\quad 0\quad 0)^{\mathrm{T}}$. Multiply $A^{\infty}$ times Jackie's initial state. This will give you the probability of Jackie being in each of the four ranks after many years have passed.

8. Suppose, at some point in time, Jackie was promoted to police chief. At that point, what would be Jackie's chances of eventually being retired?

***answers***

1. Suppose $\lambda$ is an eigenvalue of $\mathbf{AB}$. Then there exists a vector $\mathbf{v}$ such that $\mathbf{ABv} = \lambda\mathbf{v}$. Multiply on the left by $\mathbf{B}$: $\mathbf{BABv} = \mathbf{B}\lambda\mathbf{v}$. By the lemma of the first City (Emmet), $\mathbf{BABv} = \lambda\mathbf{Bv}$. Let $\mathbf{Bv}$ equal $\mathbf{w}$. $\mathbf{BAw} = \lambda\mathbf{w}$. This establishes that $\lambda$ is an eigenvalue of $\mathbf{BA}$.

2. Jackie will eventually either be retired or a civilian. If there is 4/7 of a chance of retirement, then there is 3/7 of a chance of being a civilian.

3.
$$\mathbf{A} = \begin{pmatrix} 1 & 1/3 & 0 & 0 \\ 0 & 0 & 1/3 & 0 \\ 0 & 2/3 & 0 & 0 \\ 0 & 0 & 2/3 & 1 \end{pmatrix}$$

4. For any power of $\mathbf{A}$, the first column of $\mathbf{A}^n$ will always be $(1 \ \ 0 \ \ 0 \ \ 0)^{\mathrm{T}}$ so $\mathbf{A}$ is not regular.

5. $\lambda = 1, \dfrac{\sqrt{2}}{3}, \dfrac{-\sqrt{2}}{3}$      $\lambda = 1$ has an algebraic multiplicity of two.

6. Any nonzero multiple of $(1 \quad -3 - \sqrt{2} \quad 6/\sqrt{2} + 2 \quad -2\sqrt{2})^{\mathrm{T}}$.

7.
$$\begin{pmatrix} 1 & 3/7 & 1/7 & 0 \\ 0 & 0 & 0 & 0 \\ 0 & 0 & 0 & 0 \\ 0 & 4/7 & 6/7 & 1 \end{pmatrix} \begin{pmatrix} 0 \\ 1 \\ 0 \\ 0 \end{pmatrix} = \begin{pmatrix} 3/7 \\ 0 \\ 0 \\ 4/7 \end{pmatrix}$$

8. The initial state would be $(0 \ \ 0 \ \ 1 \ \ 0)^{\mathrm{T}}$.

$\mathbf{A}^{\infty}$ times that initial state will equal $\begin{pmatrix} 1/7 \\ 0 \\ 0 \\ 6/7 \end{pmatrix}$

so there is a six-sevenths of a chance Jackie will eventually be retired.

# Last Night

Last night I got a call from my son-in-law, Dan. He began, "Suppose you have a bag with five balls in it—all different colors. You draw one out, look at it, and put it back in the bag. After 20 draws, what is the probability that you will see each color at least once?"

Dan knew that I would enjoy such a question.* Dan and my daughter Jill have several copies of *Life of Fred: Statistics,* and Chapters 2 and 3 which deal with probability.

I took a wild stab at it and said, "$1 - (\frac{4}{5})^{20}$."

But that wasn't right. That was just the probability that the red ball would show up at least once.**

The probability that the red ball would show up at least once is not independent of the probability that the green ball would show up at least once. (Translation: If the green ball never shows up, that affects the chances of the red ball showing up at least once.)

I told Dan that I needed to think about it and said I would call him back. I then headed off to my copy of *LOF: Statistics* to look up conditional probability in Chapter 3. In that chapter, Fred had just been arrested for vagrancy, and the policeman had asserted that 99% of all vagrants are homeless.***

That could be written as $\mathcal{P}(H \mid V) = 0.99$—the probability of being homeless given that he is a vagrant. Fred had to argue in

the handcuffs didn't fit Fred's small wrists

---

* Randy Pausch wrote in his book, *The Last Lecture*, that there are "two types of families:
    (1)  Those that need a dictionary to get through dinner.
    (2)  Those who don't."

** The probability that the red ball would not show up on the first draw is 0.8. Each draw is independent. The probability that the red ball would not show up on any of the first 20 draws is $(0.8)^{20}$. So the probability that the red ball would show up at least once is $1 - (0.8)^{20}$.

*** Fred was homeless because the math office that he had lived in for years had burned down in Chapter 1.

court that what was important wasn't $P(H \mid V)$, but $P(V \mid H)$, which is the probability that he was a vagrant given the fact that he was homeless. He needed to use Bayes' Theorem to switch from $P(H \mid V)$ to $P(V \mid H)$:

$$P(V \mid H) = \frac{P(V)\, P(H \mid V)}{P(V)\, P(H \mid V) + P(\bar{V})\, P(H \mid \bar{V})}$$

(You can see why I needed to head back to the book. I didn't have that formula memorized.)

Yesterday, I had written up one of the six Cities for this final chapter of *LOF: Linear Algebra*. Then it hit me. I thought to myself: Oh no! This is a linear algebra problem! In fact, it's a linear algebra problem from this Chapter 4.

Instead of the lion, the monkey, and Fred tossing a ball around and computing the probability of where the ball would be after each toss, I needed to compute the probability of the total number of different colors I would see after each draw.

I defined $x_n = \begin{pmatrix} \text{the probability exactly one color has appeared after the n}^{th}\text{ draw} \\ \text{the probability exactly two colors have appeared after the n}^{th}\text{ draw} \\ \text{the probability exactly three colors have appeared after the n}^{th}\text{ draw} \\ \text{the probability exactly four colors have appeared after the n}^{th}\text{ draw} \\ \text{the probability exactly five colors have appeared after the n}^{th}\text{ draw} \end{pmatrix}$

So after the first draw we have $\begin{pmatrix} 1 \\ 0 \\ 0 \\ 0 \\ 0 \end{pmatrix}$ since there is 100% probability that after one draw we will have seen exactly one color.

This vector, which is the probabilities after one draw, is $x_1$.

All I needed now was the transition matrix $A$. Recall, that when the lion had the ball in their game of catch, there was an 80% chance he would keep the ball by throwing it up in the air and catching it, a 10% chance he would throw it to the monkey, and a 10% chance he would throw it to Fred.

So the lion's part of the transition matrix was $\begin{pmatrix} \text{LION} \\ 0.8 \\ 0.1 \\ 0.1 \end{pmatrix}$

The first column of the transition matrix $\mathbf{A}$ for Dan's problem will correspond to the first state of having exactly one color appear. We need to figure out after another draw, what are the chances that we will still have just one color appear, the chances that a second color will appear, that a third color will appear, a fourth color will appear, and a fifth color will appear.

If only one color has appeared, and we do another draw, there is no chance that we will have three, or four, or five colors appear. There is a 20% chance that after that draw we will still only have one color and an 80% chance that a second color will appear.

In short, the first column of $\mathbf{A}$ will be $\begin{pmatrix} 0.2 \\ 0.8 \\ 0 \\ 0 \\ 0 \end{pmatrix}$

All of $\mathbf{A}$ is

| | ONE COLOR HAS APPEARED | TWO COLORS HAVE APPEARED | THREE COLORS HAVE APPEARED | FOUR COLORS HAVE APPEARED | FIVE COLORS HAVE APPEARED | |
|---|---|---|---|---|---|---|
| | 0.2 | 0 | 0 | 0 | 0 | → CHANCES FOR A TOTAL OF ONE COLOR AFTER THE DRAW |
| | 0.8 | 0.4 | 0 | 0 | 0 | → CHANCES FOR A TOTAL OF TWO COLORS AFTER THE DRAW |
| | 0 | 0.6 | 0.6 | 0 | 0 | → CHANCES FOR A TOTAL OF THREE COLORS AFTER THE DRAW |
| | 0 | 0 | 0.4 | 0.8 | 0 | → CHANCES FOR A TOTAL OF FOUR COLORS AFTER THE DRAW |
| | 0 | 0 | 0 | 0.2 | 1 | → CHANCES FOR A TOTAL OF FIVE COLORS AFTER THE DRAW |

Note that the fifth column is an absorbing state. Once you have five colors appear, nothing new will happen.

I called Dan back and told him that $\mathbf{x}_{20}$ was $\mathbf{A}^{19}\mathbf{x}_1$.

That was a sweet moment in linear algebra.

a happy Stan

# *Index*

If you'd like to visit Fred at his official Web site

and see what other books have been written about him

FredGauss.com